INTELLIGENT TRANSPORT SYSTEMS

PRADIP KUMAR SARKAR
Professor and Former Head
Department of Transport Planning
School of Planning and Architecture, New Delhi

AMIT KUMAR JAIN
Officer on Special Duty in Railway Board
Ministry of Railways, New Delhi

PHI Learning Private Limited
Delhi-110092
2018

₹ 550.00

INTELLIGENT TRANSPORT SYSTEMS
Pradip Kumar Sarkar and Amit Kumar Jain

ISBN-978-93-87472-06-8 (Text)
978-81-935938-4-4 (eBook)

Published by Asoke K. Ghosh, PHI Learning Private Limited, Rimjhim House, 111, Patparganj Industrial Estate, Delhi-110092 and Printed by Mudrak, 30-A, Patparganj, Delhi-110091.

Contents

Foreword 1

Today many of the Indian cities are going for metro rail systems. Metro railways have already started operations in 9 cities—Kolkata, Delhi, Bengaluru, Jaipur, Gurugram, Mumbai, Chennai, Kochi and Lucknow. It is under different stages of development in many other cities. Government of India has decided to develop metro railways in all cities having a population more than 2 million. Such a large scale development in the field of metro rail systems requires trained and educated manpower to effectively manage these projects not only during construction phase but also for operations and maintenance. Presently, the knowledge on metro rail system is mostly imparted by training schools of metro railways. The intelligent transport system (ITS) is fundamental to present day metro rail systems, including train control system, train management system, fare collection system, track and traction management system, etc.

I am pleased to find this book on 'Intelligent Transport Systems' which comprehensively coveres ITS for both railways and roadways. Along with holistic coverage of road-based transportation system, it encompasses all the relevant topics concerning metro railway systems. A mix of academic and practical approach has been adopted to present different topics in the book. Almost every concept has been explained using theory as well as real-life examples. The book has included even the upcoming technologies such as unmanned train operations (UTO), autonomous cars, biometric fare collection systems, etc. I am quite confident that this book will be highly useful for professionals, planners, operators, trainers, faculty members and the students.

Dr. Mangu Singh
Managing Director
Delhi Metro Rail Corporation Limited

Foreword 2

An efficient and effective transportation system has emerged as a vital component for any thriving and progressive city. Efficient urban mobility and public transport are the key components of the Smart City Mission of Government of India. Conventionally, the need of urban transport has been met with more or wider roads, more flyovers, more buses, fixed signals, traffic police, etc. However, all these measures, although useful, could not fulfil the expectation of the present day users who demand an intelligent system which can smartly manage traffic congestion, smartly monitor the functioning of traffic systems and intervene intelligently wherever require, provide information on real-time basis or in advance wherever possible and so on. The law enforcement functions are also being taken over by intelligent systems such as red light violation detection system, speed limit detection system, automatic number plate recognition system, etc. The technology has changed the way the transportation management in a city was performed. The introduction of intelligent transport system (ITS) has emerged as a game changer for traffic and transportation systems in an urban agglomerate.

In this context, the study of ITS has become fundamental for any professional or student closely or remotely connected with urban transportation systems, town planning, traffic engineering, civil engineering, etc. Although a sea of material on ITS is available on the Internet, there has been a need of structured and organised material covering fundamentals of ITS which can be handy for professionals, students, trainers and teachers.

I am pleased to write this foreword for this excellent book on ITS that covers every relevant topic on ITS for road as well as rail systems. The unique feature of the book is a perfect blend of theory and practical in the form of real-life case studies. I find this book most up to date, covering even emerging technologies such as electric vehicles, autonomous cars, automated highway systems, 5G mobile technology, biometric ticketing systems and intelligent trains.

I am confident that the readers will find this book extremely useful. I highly recommend this book for students of urban transportation, traffic engineering, civil engineering, town planning, architecture and professionals dealing with these subjects, law enforcing agencies, metro train companies, railways, etc.

Prof. Satish Chandra
Director
CSIR—Central Road Research Institute
New Delhi

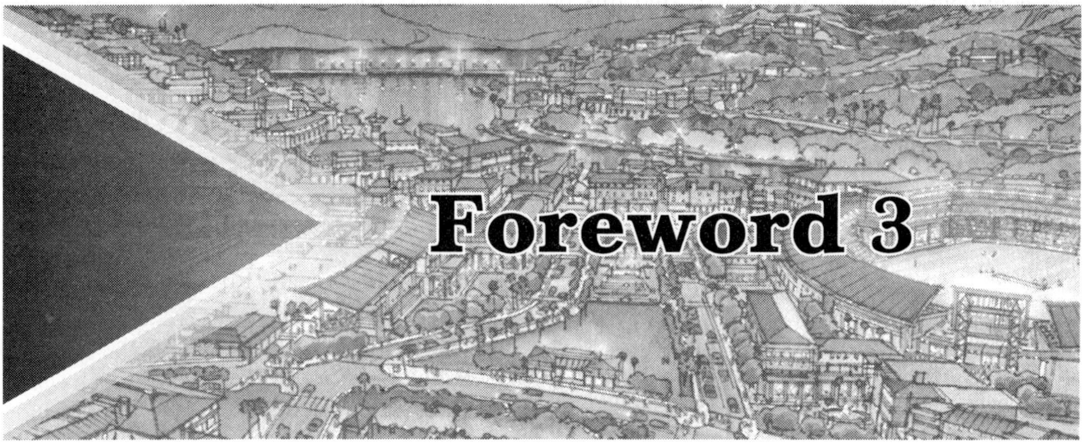

Foreword 3

India is going through a phase of fast pace urbanisation. While around 350 million people live in cities today, this number is projected to double in the next 25 years. This will make it more than the combined population of US, UK, Gemany, France and Italy. This rapid growth in our urban population will create a very huge demand for transportation and will severely aggravate the problems of traffic congestion, environmental degradation, traffic accidents, pollution, and our energy security. Remedial measures will become even more important.

The last decade has witnessed a revolution in the development of information and communication technologies (ICTs). This has made a significant difference to our daily lives. Adoption of ICT based intelligent transportation system (ITS) has changed the way transportation systems function and brought about disruptive changes in how we travel and move from place to place.

Perthaps ITS is the only answer to burgeoning transportation porblems being faced by our cities. Not surprisingly, the adopation of ITS has become a focus area for the Government of India also. However, the adoption of ITS is still at a nascent stage in the country, despite notable changes in some progressive cities. There is a need for more widespread use of ITS and the first step in this direction would be to provide for high quality ITS education in our learning programs. Only adequately informed and trained professionals can facillitate the adoption of ITS at the required scale.

I am delighted to find that this book on 'Intelligent Transport Systems' in the present context has been authored by Prof. Sarkar and Mr. Jain. I find this book comprehensive encompassing all relevant topics on ITS such as ITS architecture, building blocks, traffic management, advanced vehicle control systems, ITS for vulnerable road users, funding options, standards and specifications for ITS tools, ITS for rail-road systems and related best practices.

The practical approach adopted in the book makes it equally useful both for students and professionals. The authors have made concerted efforts to collate all relevant information on ITS in this book in a systematic manner. I am sure that this book will prove highly beneficial for urban transport professionals, traffic law enforcing agencies, policy makers, academicians and students.

O.P. Agarwal
Chief Executive Officer
World Resources Institute India
Ex-Urabn Transport Adviser to the World Bank, and
OSD (UT), Ministry of Urban Development

Preface

The change is a continuous process. However, an unprecedented pace of change is being witnessed in recent times and the credit goes to the developments taking place in information technology (IT). The transport sector is not untouched of the transformation wave generated by IT. The wide range of IT applications in the field of transport sector has manifested in the form of intelligent transport system (ITS). The ITS is an advanced transport application aiming to provide innovative services related to different modes of transport to improve safety and operation of traffic condition. It enables users to be better informed about the traffic and travel condition in order to make the travellers' journey safer and smoother with the help of smarter transport networks.

Applying information technology to the country's transportation network delivers five key classes of benefits viz., increasing driver and pedestrian safety; improving the operational performance of the transportation network, particularly by reducing congestion; enhancing personal mobility and convenience; delivering environmental benefits; and boosting productivity, expanding economic and employment growth.

The importance of ITS for an effective and efficient transport system in a country needs no emphasis. However, it is easy to prepare plans but difficult to implement it in practice. Some of the main issues in the deployment of ITS in India are: an underdeveloped road network, mixed traffic conditions, severe budget restrictions, rapid urbanisation and growth, lack of resources for maintenance and operation, less demand for automation, lack of interest among government decision makers, and lack of user awareness coupled with lack of political will and initiatives.

Further, there is a lack of understanding and education on ITS among the decision makers, transport planners, town planners and sector practitioners. Hence, continuing education and training for new engineers as well as practicing engineers is required to keep pace with development in ITS technologies. There is a general agreement among all the people involved in transportation in public and private sectors as well as the academicians that it is essential to introduce ITS education in curriculum, thus various academic institutions and many academic programs have been devised to address these needs.

ITS is an interdisciplinary field as it involves understanding of various subjects like computers, electronics, IT, civil engineering, transportation and so on. For this reason, ITS groups consist of professionals from diverse backgrounds and academic levels. It has been realised that the traditional syllabus in civil engineering is not able to cater the need of training engineers adequately in ITS, and so there is the need of cross-disciplinary education programs. The interdisciplinary factor should encourage teamwork in ITS groups of public and private organisations which consist of engineers, non-engineers and technicians from different disciplines. The students of ITS should be educated not only in traditional civil engineering, but also in the fields of communication, traffic surveillance, system analysis as well as the social and institutional matters. Besides formal educational programs and degrees, there is a need of continuing education through seminars, meetings, activities, in-service training and in-house training. The chief areas of interest in continuing education may be system architecture, funding opportunities, partnerships of public and private sectors and the national ITS needs.

Combining theory and practice, the book is broadly divided into three parts. The **road-based** (first) part of the book covers the challenges and opportunities of ITS, the architecture of ITS, building blocks of ITS, various detection, identification and collection methods of ITS, traffic management systems, transport demand management, public transport management, advanced vehicle control systems, safety of vulnerable road users, and ITS for law enforcement. The **rail-based** (second) part of the book describes intelligent train control system, smart trains, intelligent train maintenance system and automatic fare collection (AFC) which is equally applicable for roadways also. Whereas, the last part of the book deals with funding options for implementation of ITS, performance and evaluation criteria for ITS tools, and ITS technologies for smart cities some success stories of ITS application followed by standards and specifications for ITS tools. Further each part has been subdivided into a number of subparts dealing exclusively ITS related applications for road and rail.

At the end of each chapter, there are questions for practice and bibliography for further study on the subject. The readers are encouraged to read the chapters thoroughly in the sequence as presented in the book and attempt to answer the questions for better understanding of the issue. The applications of ITS are included in all the chapters followed by the theory of the technology in use. For better understanding of the technological aspects of ITS, a chapter on Building Blocks of ITS has been included which describes the fundamentals of the various technologies used in ITS such as theory of electromagnetic waves, telecommunication technologies, data and information processing.

The subject matter of ITS is too vast to be covered in a book. Everyday new developments are taking place in ITS. However, an attempt has been made to equip the readers with most relevant and recent knowledge available in the field. The upcoming applications of ITS such as autonomous cars, cooperative ITS, 5G mobile technology, etc. have also been introduced to keep pace with the subject.

Knowledge of standards and specifications for ITS tools is of utmost importance for selecting a suitable option meeting the requirement of a city from the wide array of ITS tools available in the market. The functional requirements in the form of standards and specifications for different ITS tools have been included in the last chapter of the book for the guidance of professionals as well as academicians.

The goal of writing the book is to create a comprehensive work on fundamentals of ITS for readers. The authors have made every effort to meet this objective and produce a work worthy of meeting the needs of the students, academicians and transport professionals to better understand the ITS as a subject and to enable them to craft a strategy for effective implementation of ITS in a city. The authors would like to acknowledge the help rendered by Amal Jose, Rohit Kasyap of Transport Planning students of SPA. Any constructive comments for improving the contents will be warmly appreciated.

Pradip Kumar Sarkar
Amit Kumar Jain

Abbreviations

AC	Alternate Current
ACC	Adaptive Cruise Control
ACM	Asset Condition Monitoring
ADAS	Advanced Driver Assistance System
ADCR	All Door Close Relay
AEC	Automation Enforcement Centre
AERARS	Accident Emergency Response and Routing Software
AFC	Automatic Fare Collection
ALPR	Automatic Licence Plate Recognition
APC	Automatic Passenger Counter
API	Advance Passenger Information
ASS	Advanced Signal System
ATIS	Advanced Traveller Information System
ATM	Active Traffic Management
ATMS	Advanced Traffic Management System
ATO	Automatic Train Operation
ATP	Automatic Train Protection
ATS	Automatic Train Supervision
AVAU	Automatic Voice Announcement Unit
AVCC	Automatic Vehicle Classification and Counting
AVCS	Advanced Vehicle Control and Safety
AVCSS	Advanced Vehicle Control and Safety System
AVI	Automatic Vehicle Identifiers

AVL	Automatic Vehicle Location
AVLC	Automatic Vehicle Location and Control
BAP	Battery-Assisted Passive
BCU	Brake Control Unit
BECU	Brake Electronic Control Unit
BOOT	Build, Own, Operate, and Transfer
BOT	Build, Operate and Transfer
BQS	Bus Queue Shelter
CAL	Crossings Adaptive Lighting
CALM	Continuous Air-interface Long and Medium
CATC	Continuous Automatic Train Control
CATT	Centre for Advanced Transportation Technology
CBA	Cost-Benefit Analysis
CBD	Central Business District
CBI	Computer Based Interlocking
CBM	Condition Based Maintenance
CBTC	Communication Based Train Control
CC	Central Computer
CCS	Centralised Computer System
CEPT	Centre for Environmental Planning and Technology
CF&F	Cooperative Freight and Fleet
CI	Converter-Inverter
CONCERT	Country-wide Network of Computerised Enhanced Reservation and Ticketing
CRF	Central Road Fund
CRIS	Centre for Railway Information System
CSC	Contactless Smart Card
CST	Contactless Smart Token
CTC	Central Traffic Control
CVIS	Cooperative Vehicle Infrastructure System
CVO	Commercial Vehicle Operation
DAS	Driver Assistance System
DCC	Dublin City Council
DCU	Door Control Unit
DDM	Driver Drowsiness Monitoring
DFD	Data Flow Diagrams
DLPCOS	Door Proving Loop Cut Out Switch
DMIC	Delhi Mumbai Integrated Corridor
DMU	Diesel Multiple Unit

DOT	Department of Transport
DPR	Door Proving Relay
DSRC	Dedicated Short Range Communications
DSS	Decision Support System
DTC	Delhi Transport Corporation
DTO	Driverless Train Operation
EADS	Electric Arcs Detection System
ECP	Electronically Controlled Pneumatic
EMCIS	European and Mediterranean Conference on Information System
EMU	Electric Multiple Unit
EOC	Emergency Operation Centre
EOD	Equipment Operating Data
ERP	Electronic Road Pricing
ERR	Economic Rate Return
ETA	Estimated Time of Arrival
EV	Electric Vehicle
EVI	Electronic Vehicle Identification
EVM	Emergency Vehicle Management
FOIS	Freight Operations Information System
GIFT	Gujarat International Finance Tec-City
GIS	Graphic Information System
GLIDE	Green Link Determination
GNCTD	Government of National Capital Territory of Delhi
GoA	Grades of Automation
GoI	Government of India
HAZMAT	Hazardous Material
HCC	Hindustan Construction Company
HF	High Frequency
HVAC	Heating, Ventilating, and Air Conditioning
ICATET	International Conference on Advanced Trends in Engineering and Technology
ICE	Internal Combustion Engine
ICMS	Integrated Coach Management System
ICT	Information and Communication Technology
ICTSL	Indore City Transport Services Ltd.
IMP	Incident Management Plan
IMS	Intelligent Maintenance System
IPT	Intermediate Public Transport
IRCTC	Indian Railway Catering and Tourism Corporation

IRR	Internal Rate of Return
ISA	Intelligent Speed Adaptation
ISM	Industrial, Scientific and Medical
ITACA	Intelligent Traffic Adaptive Control Area
ITCS	Intelligent Train Control System
ITS	Intelligent Transportation Systems
IUT	Institute of Urban Transport
IVU	In-Vehicle Unit
JNNURM	Jawaharlal Nehru National Urban Renewal Mission
KMC	Knowledge Management Centre
KSRTC	Karnataka State Road Transport Corporation
LCS	Lane Control Signal
LF	Low Frequency
LKS	Lane Keeping System
LTA	Land Transport Authority
MCG	Mobile Communication Gateway
MCGM	Municipal Corporation of Greater Mumbai
MMI	Man Machine Interface
MMITSS	Multi-Modal Intelligent Traffic Signal System
MMRDA	Mumbai Metropolitan Region Development Authority
MoRTH	Ministry of Road Transport and Highways
MoUAE	Ministry of Urban Affairs and Employment
MU	Multiple Unit
MVS	Mobile Video Surveillance
NFC	Near Field Communication
NTES	National Train Enquiry System
NUTP	National Urban Transport Policy
NVW	Night Vision and Warning
OBU	On-Board Unit
OCC	Operation Control Centre
OD	Origin Destination
ODM	Original Design Manufacturer
OFC	Optical Fibre Cable
OLAP	Online Analytical Processing
P&E	Performance & Evaluation
PA	Public Announcement
PATH	Partners for Advanced Transit and Highways
PCU	Passenger Car Unit

PDS	Pedestrian Detection System
PEA	Passenger Emergency Alarms
PIDS	Passenger Information Display System
PMS	Parking Management System
POTS	Plain Old Telephone System
PPP	Public Private Partnership
PRG	Priority Request Generator
PRS	Passenger Reservation System
PSAP	Public Service Answering Point
PSO	Public Service Obligation
PSU	Public Sector Undertaking
PT	Public Transport
PTW	Powered Two Wheelers
PWM	Pulse Width Modulation
RACV	Royal Automobile Club of Victoria
RCM	Reliability-Centred Maintenance
RCS	Rail Corrugation System
RFID	Radio Frequency Identification
RITIS	Regional Integrated Transportation Information System
RLVDS	Red Light Violation Detection System
RPS	Rail Profile System
RSU	Road Side Unit
RTLS	Real-Time Locating System
RTPI	Real-Time Passenger Information
S&D	Scheduling and Dispatching
SATSaNG	Software Aided Train Scheduling and Network Governance
SCADA	Supervisory Control and Data Acquisition
SCK	Smart City Kochi
SCOT	Shared Computer Operated Transport
SCOOT	Split Cycle Offset Optimisation Technique
SD	Standard Deviation
SEZ	Special Economic Zone
SIR	Special Investment Region
SMS	Short Messaging Service
SOV	Single-Occupant Vehicle
SP	Signal Priority
STO	Semi-automatic Train Operation
SUM	Smart Urban Management

SUTP	Sustainable Urban Transport Project
TAS	Travel Advisory System
TBC	Traction Brake Controller
TCMS	Train Control Management System
TDM	Transport Demand Management
TDS	Train Diagnostic System
TfL	Transport for London
TIMS	Train Integrated Management System
TMC	Traffic Management Centre
TMCC	Travel Management Coordination Center
TOT	Ticket Operating Terminal
TRC	Track Recording Car
TSP	Traffic Signal Priority
UHF	Ultra High Frequency
UTO	Unmanned Train Operation
UTS	Unreserved Ticket System
VCB	Vacuum Circuit Breaker
VES	Violation Enforcement System
VMT	Vehicle Miles Traveled
VMU	Vehicle Mounted Unit
VRU	Vulnerable Road User
VTL	Virtual Trip Line
VVD	Video Vehicle Detection
VVVF	Variable Voltage and Variable Frequency
WAP	Wireless Application Protocol
WiM	Weigh-in-Motion
WPC	Wireless Planning and Coordination
WWVDS	Wrong Way Vehicle Detection System

1 Introduction to Intelligent Transport System

1.1 HISTORICAL BACKGROUND

In developing countries like India, as the traffic on the existing road system in a city increases, congestion causes a severe crisis on road. Most of the cities in India have been witnessing a steep rise in vehicular growth due to accelerated economic development in the country resulting in severe traffic congestion and haphazard movement on the road network and the growth has been quite rapid in the past decade (Figure 1.1). This results in increase in traffic congestion on the network.

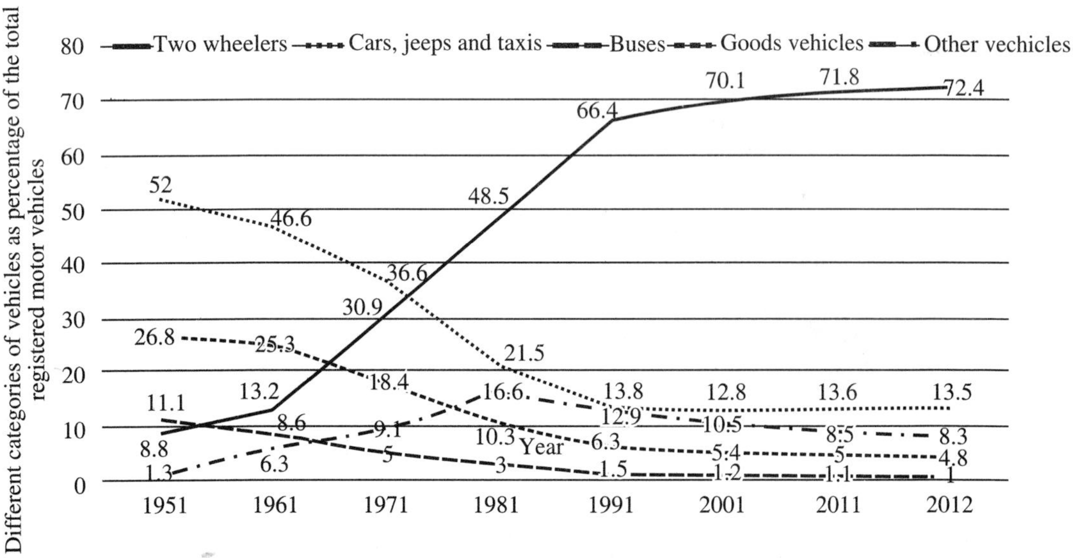

FIGURE 1.1 Composition of registered vehicles as compared to total registered vehicles (*Source:* Road Transport Year Book, India, 2012 and Statistical Year Book, India, 2015).

This ultimately gives rise to the traffic congestion on the network. As per the Ministry of Road Transport and Highways (MoRTH), the number of motorised vehicles has increased by 21 million in 1991 to 142 million in 2011 in India registering nearly 10% increases in a very year. The over-concentration of vehicles in a very few selected cities, notably in metropolitan cities is one of the major problems. Between the year 2000 and 2009, the compound annual growth rate of urbanisation is 3.16% and automobile is 9.6%. According to the Basic Road Statistics 2011, the urban street length went up from 123120 km in 1981 to 411840 km in 2011, i.e., increases by 3.35 times. Vehicles per million population have expanded by 219% while urban streets for every million of vehicles expanded by just 124%, in the recent decade. The street space for vehicles has diminished (from 0.18 km for each vehicle to 0.01 for every vehicle) in the recent couple of decades, resulting in high level of congestion in all urban areas. Figure 1.2 shows the vicious circle of urban transport due to which congestion occurs on road network.

Further when it is difficult to augment the supply facilities to improve traffic conditions, smart IT based traffic management is the only option available to engineers and planners. Medium- and long-term solutions like widening roads, providing flyovers, and construction of by-passes and urban expressways are costly and time consuming. Traffic management solutions are simple and inexpensive and can help to tide over the crisis for sometime. It can be said that most of the central cities of the world like London, Paris, New York, Hong Kong, Singapore, etc. do not have any opportunities to augment the supply of road network in terms of widening the road network or construction of new road. They are managing the cities with sustainable development by virtue of bringing information technology and ITS solutions to enhance the safety and mobility of the cities.

These solutions can be adopted quickly and efficiently as a part of short-term measures to minimise traffic congestion. It is a set of relatively low-cost strategies and methods used to make better use of the existing transportation system.

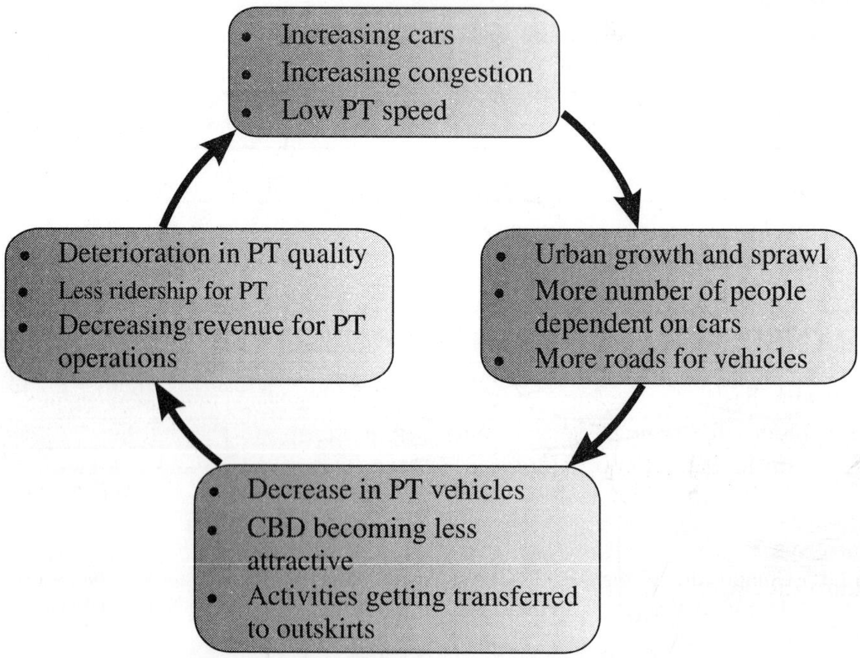

FIGURE 1.2 Vicious circle of urban transport.

1.2 URBANISATION

Urbanisation may be a product of demographic explosion and poverty induced rural-urban migration. It is happening due to combined effect of urban pull and rural push. Globalisation, liberalisation, and privatisation are instrumental to negative process for urbanisation in India.

The number of urban agglomerations/towns has increased from 1827 in the year 1901 to 7935 in the year 2011 (Census of India, 2011). The population residing in urban areas has increased from 258 million in 1901 to 377 million in 2011 (Census of India, 2011).

The population of India is 1.21 billion as per the Census of India, 2011 and it is anticipated to grow at the rate of 1.5–1.8% each year by 2030. Urban population in India was 17% in 1951 which has almost doubled to 32% in 2011 and is likely to extend up to around 40% by the year 2040.

However, the number of metropolitan cities, those with a million plus population, has increased sharply over this era. From thirty five in 2001, the number of metropolitan cities rose to fifty as per the Census of India, 2011. Out of those fifty, eight cities namely Mumbai, Delhi, Kolkata, Chennai, Hyderabad, Bangalore, Ahmedabad, and Pune experience a population more than five million.

1.3 MOTORISATION

According to Motor Transport Statistics in 1987, there have been 12.6 million vehicles and ten years later in 1997 this range had increased three-fold to 37.2 million that shows 20% annual growth for that decade. Vehicle population is calculable to be over fifty million by 2001 additionally showing identical rate continuing for the any years.

A majority of cars in India are focused in urban centres and it is noteworthy to mention that 32% of those vehicles are plying in metropolitan cities alone that represent simply around only 15% of the whole population. It is fascinating to notice that Delhi contains around 1.4% of the Indian population accounts for over 7% of all motor vehicles within the country. There are already over 9 million registered motor vehicles in Delhi and concerning 1200 vehicles are being registered daily. Two wheelers and cars account for over 85% of the vehicle population in most of the metropolitan cities. They account for a minimum of 90% of total vehicles in Ahmedabad, Bhopal, Coimbatore, Delhi, Kanpur, Lucknow, Nagpur, Vadodara, Varanasi, and Vishakhapatnam. Two-wheeler motor vehicles alone account for over 80% of the whole vehicles in range of metropolitan cities.

1.4 TRANSPORT SYSTEM CHARACTERISTICS

Transport system has a wide range of transportation facilities that represent a variety of transportation modes for both passengers and freight transport. These following components make up the multi-modal transportation system:

(a) Roadways
(b) Railways
(c) Metro/LRT/Mono Rail/PRT

(d) Personalised Vehicles
(e) Bus Transit Systems
(f) Bike Lanes and Trails
(g) Aviation System
(h) Inland Water Transport System

For an efficient transportation system to exist, each mode of the transportation system must be connected to and mutually supportive of each other. Each part of the system is important and provides accessibility and mobility to meet the travel requirements of residents and other travellers, or to transport various types of freight.

Understanding the various types of transportation infrastructure as well as understanding how it is utilised, will allow making decisions about investing in transportation infrastructure of all types. Following are the elements of transportation system in urban areas:

(a) Major Transportation Corridors
(b) Functional Classification System
(c) Collector and Local Streets
(d) Bridges and Interchanges
(e) Freight Transport
(f) Aviation Facilities
(g) Public Transportation
(h) Bicycle and Pedestrian Systems

There are three elements which contributes in the level of urban travel demand in India that is expanding significantly throughout the years. The main is the increase in the population, as the urbanisation procedure has shown that the population size in urban territory doubled in around two decades. The second variable is the mobility rate that characterises as the normal average number of trips per person per day. The mobility rate in urban India is expanding throughout over the years. For example, in Delhi, the normal number of trips by individual every day has expanded from 0.49 in 1969 to 1.10 in 2001. According to the report of Ministry of Urban Affairs and Employment (MoUAE) the trip rates for Mumbai, Kolkata, Chennai, Hyderabad, Bangalore, Ahmedabad, and Pune are 1.26, 1.26, 1.22, 1.05, 1.20, 1.57, and 1.48 individually. The third component adding to travel interest is the increment in trip length because of an expansion in the physical development of the city. For instance, the normal trip length in Delhi has expanded from 5.4 km in 1969 to around 13.5 km in 2001. The normal trip length on Delhi Transport Corporation (DTC) transports has expanded from 6.4 km in 1972 to about 18.0 km at present. Currently, it is estimated that the average trip length of four mega cities varies from 12.7 to 13.5 km. There is also a change in the pattern of trip distribution; more and more trips are being made in urban areas for work, followed by education. For example, more than 60% of the total trips in Mumbai are meant for work and around 31% for education.

1.5 TRANSPORT PROBLEMS AND ISSUES

(a) ***Traffic congestion and parking difficulties:*** Congestion and shortages in parking supply are the most common transport issues in huge urban agglomerations, especially connected with motorisation and the dispersion of cars, which has resulted in the increase in demand of transport infrastructure. While the supply of transport infrastructure has not helped much to augment the capacity significantly, motorisation has given rise not only increased traffic congestion but also to the demand for additional parking space.

(b) ***Longer commuting:*** Apart from the traffic congestion, people are spending an increasing amount of time commuting between their residence and workplace. A critical element behind this pattern is identified with the trend related to residential affordability with regard to housing located further away from central areas (where most of the employment remains). In most metropolitan areas, workers prefer to stay away from the places of work which are generally characterised by high property value and high rent resulting in longer commuting time.

(c) ***Public transport inadequacy:*** Many public transit systems or parts of them are either over or under used. Short or poor supply of public transport system results in crowding in peak hours. Poor operation of public transport system results in low ridership which makes many services financially unsustainable. Critical endowments and cross-financing are the possible results when public transport systems are not well managed and operated. Therefore, public transit systems cannot generate adequate revenue from the revenue box to pay to cover its operational and capital expenses.

(d) ***Difficulties for non-motorised transport (NMT):*** Because of the intense traffic, the enjoyment of walking by the pedestrian, bikes and vehicles is impaired if the facilities for NMT are not properly addressed in the road network design. It is therefore imperative to improve the efficiency of NMT operation by proper planning and design the road network so as to create the level of conduce environment friendly design for cyclists and pedestrian.

(e) ***High maintenance costs:*** Financial deficiencies have been facing for aging transport infrastructure, for maintenance costs as well as pressures to upgrade to more modern infrastructure. Therefore effort is needed to timely address the maintenance problems of the transport infrastructure. There is a need for special maintenance fund to be created either through allocated budget or through innovative measures.

(f) ***Environmental impacts and energy consumption:*** Pollution, including noise, generated by the motorised gasoline-driven vehicles has become a serious impediment to the quality of life and even to the health of urban populations. Further, energy consumption by urban transportation has dramatically increased which is significantly dependent on petroleum and diesel-based fuel.

(g) ***Accidents and safety:*** Growing traffic in urban areas is linked with a growing number of accidents and fatalities, especially in developing countries. As traffic increases, people feel less safe to use the streets. Every year more than 5,00,000 people are victims of road accidents including around 1,40,000 fatalities in India alone.

(h) ***Lack of traffic enforcement:*** Major reasons for road accidents is primarily lack of awareness of road safety and traffic rules and regulations among the users. Frequent traffic violations with respect to red light jumping, driving beyond speed limit, indiscipline behaviours of pedestrian traffic at intersection or mid-block, rash driving and road rage on urban as well as rural areas are the major concerns for road safety hazards.

1.5.1 What is Needed?

In India, a quickly developing economy, the issue of road traffic congestion is intensely felt in every major city. This is fundamentally on the grounds that infrastructure growth is not enough as compared with expansion in number of vehicles, due to a number of factors. Secondly, Indian traffic being non-lane based and chaotic, is largely different from the western traffic. Thus, intelligent transport systems (ITS), used for efficient traffic management in developed countries, cannot be used. ITS techniques have to undergo adaptation and innovation to suit the contrasting traffic characteristics of Indian roads.

One of the major reasons that ITS service has garnered public and professional interest is the concern created by the continuing disparity between the growth in surface transportation travel demand and the relatively insignificant addition of travel capacity. This combination has resulted in increased regional roadway congestion, greater uncertainty in travel time estimates, and higher perceived costs in safety and productivity. Increasing transportation capacity by building roads and other related infrastructure is not a feasible solution in many urban areas due to the high costs as well as environmental and associated societal concerns. While information technologies enable elements within the transportation system—vehicles, roads, traffic lights, message signs, etc.—to become intelligent by embedding them with microchips and sensors and empowering them to communicate with each other through wireless technologies. With the help of information and communication technologies supported with computers, Internet, electronics and electrical operational system, it has been possible to integrate vehicles and vehicle infrastructure to send and receive and respond information between each other to enable traffic operate smoothly and safely. This is the fundamental concept of intelligent transport system aiming at providing necessary traffic-related information to all road users including passengers and freight so that traffic moves on the network by using best possible routes which are safe and efficient and cost effective.

1.6 INTELLIGENT TRANSPORT SYSTEM (ITS)

Intelligent transport system (ITS) is an advanced transport application aiming at providing innovative services relating to different modes of transport to improve safety and operation of traffic condition and enables various types of users to be better informed about the traffic and travel condition in order to make the travellers' journey safer and efficient with the help of 'smarter' use of transport networks. Figure 1.3 shows the Passenger Information Display System (PIDS) board giving information to commuters. ITS is a system in which information and communication technologies are applied in the field of transport, including infrastructure, vehicles and users allowing the provision of online information in all areas of public and private administration.

FIGURE 1.3 PIDS board giving information to commuters, DMRC (reproduced with permission from source).

These systems involve vehicles, drivers, passengers, road operators, and managers including all types of stakeholders through their interaction with each other. Intelligent traffic management techniques also enable various types of users to be better informed and guide and help their journey make safer, more coordinated to achieve comprehensive mobility for all. Therefore, it is necessary to develop system ITS architecture. As the definition of National ITS Architecture, USA, the National ITS Architecture provides a common framework for planning, defining, and integrating ITS. It defines the functions that must be performed by subsystems, where these functions reside (e.g., field, traffic management centre, in vehicle), the interfaces and architecture flows to/from the subsystems, and the communications requirements for the architecture flows. It is a mature product that reflects the contributions of a broad cross-section of the ITS community (e.g., transportation practitioners, systems engineers, system developers, technology specialists).

1.6.1 ITS is Important for Solving Problems in Urban Areas

Applying information technology to the country's transportation network delivers five key classes of benefits by:

(a) Increasing driver and pedestrian safety
(b) Improving the operational performance of the transportation network, particularly by reducing congestion
(c) Enhancing personal mobility and convenience
(d) Delivering environmental benefits
(e) Boosting productivity and expanding economic and employment growth

1.6.2 Function of ITS Components

Intelligent transportation systems include a wider and growing range of technologies and applications. ITS America has classified the intelligent transport system into eight functional areas as mentioned in Table 1.1.

TABLE 1.1 Functional areas of ITS

S.No.	ITS category	ITS applications
1	Advanced Traveller Information Systems (ATIS)	Real-time Traffic Information Provision
		Route Guidance/Navigation Systems
		Parking Information
		Roadside Weather Information Systems
2	Advanced Transportation Management Systems (ATMS)	Traffic Operations Centres (TOCs)
		Adaptive Traffic Signal Control
		Dynamic Message Signs (or "Variable" Message Signs)
		Ramp Metering
3	ITS-Enabled Transportation Pricing Systems	Electronic Toll Collection (ETC)
		Congestion Pricing/Electronic Road Pricing (ERP)
		Fee-Based Express (HOT) Lanes
		Vehicle Miles Travelled (VMT) Usage Fees
		Variable Parking Fees
4	Advanced Public Transportation Systems (APTS)	Real-time Status Information for Public Transit System (e.g., Bus, Subway, Rail)
		Automatic Vehicle Location (AVL)
		Electronic Fare Payment (e.g. Smart Cards)
5	Vehicle-to-Infrastructure Integration (VII) and Vehicle-to-Vehicle Integration (V2V)	Cooperative Intersection Collision Avoidance System (CICAS)
		Intelligent Speed Adaptation (ISA)
6	Commercial Vehicles Operation (CVO)	Commercial Vehicle Preclearance
		Commercial Vehicle Administrative Processes
		Automated Roadside Safety Inspection
		Commercial Vehicle On-board Safety Monitoring
		Commercial Vehicle Fleet Management
7	Advanced Rural Highway Management Systems (ARHMS)	Advanced Freeway Management Systems
		Advanced Rural Transportation System
8	Emergency Management Systems (EMS) and Safety	Emergency Notification and Personal Security
		Emergency Vehicle Management
		Hazardous Materials and Incident Notification
		Public Travel Security

1.6.3 Importance of ITS Architecture

The traditional meaning of the term "architecture" is related to the design of buildings and other physical structures. In modern times, however, the term has a broader meaning and is used to mean the planning and design of any kind of system. This is especially true in the case of complex systems that have many interrelated parts, which put together to work in an integrated manner to fulfil a common objective. ITS is a complex system; so it is common to see the term "architecture" used with ITS. ITS architecture defines a framework within which a system can be planned, designed and built. Intelligent transportation system (ITS) is interrelated systems that work together to deliver transportation services. Integration of these systems requires architecture to illustrate to arrive at a consensus on the approach to be taken by a group of stakeholders regarding their particular systems. An ITS architecture defines the systems that allows to interface and interconnect with various subsystems relating to the flow of information exchanges between these subsystems.

1.6.4 Need for Developing ITS to Regulate Traffic in Urban Area

In view of the above, it is extremely imperative to develop traffic management plan using the emerging technology, i.e., ITS. In this book, an attempt has been made to appreciate the functioning of ITS architecture with respect to physical, functional and organisational architecture which is the central structure on which the traffic management plan is to be evolved for addressing comprehensive mobility and safety for all types of users including passenger and freight traffic. Various technologies and detection/identification system are also put in place for comprehensive understanding. Along with this, the conventional traffic management coupled with ITS-driven traffic management has been presented in details along with examples and applications. Finally the evaluation of performance of ITS demonstrates the various parameters to be examined for decision making.

1.7 STRUCTURE OF THE BOOK

This book is broadly divided into two parts namely on road base ITS system and rail-based ITS system followed by intelligent ticketing system, ITS funding options, ITS for smart cities and some case studies on ITS implementation. Further each part has been subdivided in a number of subparts dealing exclusively ITS-related applications for road and rail. As far as the first part of ITS is concerned, the first chapter addresses the overall concept of ITS while the second chapter deals with challenges and opportunities in ITS to promote in India. The third and fourth chapter explain the architecture of ITS and technology building blocks of ITS respectively. The fifth chapter deals with various detection, identification and collection methods of ITS which are primarily based on various sensor technologies used for ITS applications. The sixth chapter of the book is devoted to the applications of traffic management system components using intelligent transport system where various phases of traffic management system using ITS through ITS architecture have been demonstrated while the seventh chapter deals with the concept and applications of travel demand management. As public transport is the backbone and lifeline for sustainable transport system, it is therefore felt to cover the public transport system using ITS in eighth chapter, while the ninth chapter deals with advanced vehicle control and safety service and emergency management systems. As safety of the vulnerable road user is increasingly becoming important and integral part of the overall road safety programme, it is therefore included in the tenth chapter. The eleventh chapter explains how ITS tools can be used for law enforcement. Chapters 12 to 15 are devoted to ITS for railways starting from introduction to railways, intelligent train control system, intelligent trains, intelligent train maintenance system. Chapter 16 discusses various aspects of intelligent ticketing system and automatic fare collection (AFC) which are equally applicable to roadways and railways. The various funding options for implementation of ITS have been discussed in Chapter 17. Chapter 18 is focused on performance and evaluation criteria for ITS tools which are very important for developing specification and evaluation of ITS tools. The ITS technologies required for smart cities have been presented in Chapter 19 followed by success stories of ITS application in Chapter 20.

The Chapter 21 which is the last chapter presents the standards and specification for mostly used ITS tools such as Intelligent Traffic management system, red light violation detection, speed detection, CCTV, APNR etc.

Questions

1.1 What do you understand by urbanisation? Please describe trends of urbanisation in India.

1.2 What are different components of a multi-modal transportation system?

1.3 Please explain problems associated with the transportation system in a city.

1.4 What do you understand by ITS?

1.5 What are different functional areas of ITS?

1.6 What do you understand by ITS architecture? Please explain the importance of ITS architecture.

1.7 Explain the need of ITS to regulate traffic in urban areas.

Bibliography

Balaji, P.G. and D. Srinivasan, "Type-2 Fuzzy Logic Based Urban Traffic Management", Engineering Applications of Artificial Intelligence, 2011.

Boomington/Monroe County Metropolitan Planning Organisation, "Regional Intellingent Transportation System Architecture", September 2008.

Ezell, Stephen, *Intelligent Transportation Systems,* The Information Technology and Innovation Foundation (ITIF), January 2010.

Institute of Urban Transport and Centre for Study of Science, Technology and Policy, Review of Urban Transport in India.

Intelligent Transport Society of Korea, 2005.

Intelligent Transportation Society of America, "Primer on Vehicle-Infrastructure Integration", VII White Paper Series, 2005.

Intelligent Transportation Society of America, North American Intelligent Transportation Systems: ITS Industry Sectors and State Program–Market Data Analysis, Phase 1, White Paper, December 2009.

Japan Highway Industry Development Organisation, *ITS Handbook Japan 2007–2008,* 2007.

Kim, Jeffrey, "Strategies for Developing an Intelligent Transportation Systems Industrial Base in South Korea", Articles, Cambridge 2003.

Sadek, A. and M.A. Chowdhary, *Fundamentals of Intelligent Transportation Systems Planning,* Artech House Inc., US, 2003.

Singh, Padam, "Urbanisation and urban transport in India: The sketch for a policy", published in Workshop on Integrated Urban Transport Policy: Transportation, Land Use and the Environment in Indian Cities, Hyderabad, 2004.

US Department of Transportation, "National ITS Architecture Implementation Strategy", September 1998.

US Department of Transportation, Research and Innovative Technology Administration, ITS Strategic Research 114 Plan 2010–2014, Transforming Transportation Through Connectivity, January 2010.

Vimala, M. and Unnikrishnan Namoodiri, "Fast Growing Cities and Emergence of Urban Slums in Kerala", *Asia Pacific Journal of Research*, pp. 108–119, Vol. I, Issue XXXII, October 2015.

2 Challenges and Opportunities in Intelligent Transport System

2.1 INTRODUCTION

Urbanisation in India has followed a slow pace than in other developing countries. The proportion of the population in urban areas in India has been only 31% based on the 2011 census. However, rate of urbanisation has been accelerated as the country sets to a more rapid growth thanks to economic reforms, improving prosperity and significant increase in middle class.

As per an article 'India's urbanisation: A closer look' published in McKinsey Quarterly, July 2010—"Urban expansion in India will happen at a speed quite unlike anything the country or the world has seen before. It took nearly 40 years (from 1971 to 2008) for India's urban population to rise by nearly 230 million; it will take only half that time to add the next 250 million. This expansion will affect almost every state. For the first time in India's history, five of its largest states will have more of their population living in cities than in villages."

This large scale urbanisation has given rise to the problem of traffic congestion not only in metro cities but also in smaller cities. The conventional solution to increase road space to address traffic congestion has further exacerbated the situation by attracting more private vehicles. The technological solutions are required to mitigate the problems of traffic congestion. Intelligent transportation system (ITS) is a tested option to mitigate traffic congestion problems. ITS can be broadly defined as the use of technology for improving transportation systems. The major objective of ITS is to evaluate, develop, analyse and integrate new technologies and concepts to achieve traffic efficiency, improve environmental quality, save energy, conserve time, and enhance safety and comfort for drivers, pedestrians, and other traffic groups.

2.2 CHALLENGES AND OPPORTUNITIES IN ITS

Some of the main issues facing the deployment of ITS in India are: an underdeveloped road network, severe budget restrictions, rapid urbanisation and growth, lack of resources for maintenance and operation, less demand for automation, lack of interest among government decision makers, and lack of user awareness coupled with lack of political will and initiatives.

The main reasons for the challenges are how to start the implementation of ITS with political support, whether to start piecemeal way or by developing ITS architecture at national/state or city level, what kind of technologies to be adopted whether old or advanced technologies. Apart from these major reasons, the following aspects need to be considered as a part of addressing the problems for implementing ITS programme.

(a) ***Mixed traffic:*** The diverse range of vehicular speed (pedestrian, bicycle, LMV's, HMV's, animal drawn carts), wide variety of vehicles (including pedestrian traffic) are plying on the Indian roads. Automated vehicle identification (AVI) and automated vehicle location (AVL) are expensive and therefore it needs public participation. Moreover, Area Traffic Control (ATC) is also very simple to implement for heterogeneous traffic.

(b) ***Poor lane discipline:*** The lane discipline is not followed by the most of the Indian cities. As a result of it, most of the ITS applications which are successfully implemented in developed countries cannot be carried out by reproducing it in Indian situation. For example detectors, which are lane-based, are inapplicable due to the above reasons. Also more systems are required for automatic number plate recognition system and speed detection system.

(c) ***Lack of pedestrian facilities:*** The most of the cities in India are not equipped with the proper network of footpath. The pedestrians use the same space which are allocated for vehicles. This needs to be looked seriously to instal ITS.

(d) ***Lack of service lanes and on street parking*:** In many urban areas, the major road does not have a service lane which results into the parking of vehicle on the road. It is the source of traffic congestion. As a result, the ITS application is unable to perform the tasks up to their desired level.

(e) ***Non-standards accessories:*** The commercial vehicles are plying with different fittings even with fixing proper registration plate number, it means, the vehicles with different look and fittings create an enforcement issue while being detected through RFID. For instance, the number on the registration plates is often written in local language and also with some different font size which are very difficult to capture the correct data through automatic number plate recognition system.

(f) ***Insufficient electric power supply:*** The various part of the country is not equipped with the proper continuous electric power supply which will hinder the performance of ITS working. Even non-continuous supply of the electricity leads to failure in the operation of ITS system.

(g) ***Inefficient technical support:*** The equipment and software required for the efficient ITS system are not updated by the supplier as well as the users are also not aware of the updated equipment or latest versions in case of software. It may be procured due to limited fund where the cost of system is low but its maintenance cost or operating cost is likely to be more at many situations. Therefore, this is another challenge posing in selecting an appropriate technology/ system ITS operation.

(h) ***Communication network:*** In case, the ITS application uses the mobile service-based communication system for transferring of data, then traffic on communication network itself delay or chokes the data transfer from one place to other. The full potential of the ITS can not be exploited without a fast and effective communication system.

(i) ***Poor maintenance:*** The system required the proper and time-bound maintenance for efficient working of the system. But if the appropriate care has not been taken due to lack of audit and poor administration, the system suffers a setback as it happened for Delhi's operation of Split Cycle Offset Optimisation Technique (SCOOT) system sometime back as some part of the road network in Delhi in the SCOOT operated area experienced overlaying of roads. SCOOT is a tool for managing and controlling traffic signals in urban areas. It is an adaptive system that responds automatically to fluctuations in traffic flow through the use of on-street detectors embedded in the road. With the continuous overlaying of the road pavement, SCOOT system became obsolete without addressing this problem for upgradation of the system.

(j) ***Constraints on site and engineering issues:*** Sometime system is selected without analysis of the site constraints that result in the rise of frequent problems while maintaining the other utility services. Even, fixing a camera on the elevated structure, which is passing all along, the residential areas may leads to social issue.

(k) ***No equipment standardisation:*** There are not such standard followed for the ITS equipment.

(l) ***Deficiency in ITS architecture:*** In India, most of the cities are implementing the ITS solution to fulfil the need of local areas but they are not designed to address any expansion of that system to fulfil the need of region and also to integrate with region as well as national level. There is a need of setting up fully functional traffic management centres for coordinating the urban and regional ITS activities to get the benefit in complete manner.

2.3 ITS—TODAY AND TOMORROW

Urban transportation is certainly to be a key feature of the twenty-first century with its high-urbanised life. Transportation will gain more significance, as individual lifestyles will change with more flexible work culture and greater leisurely activities. Moreover, transportation will be required for the cargo to be brought in the cities. Therefore, in future, the transportation will have to depend on super intelligence to be built as an integrated component of the vehicles, control systems and planning tools, etc. Super intelligence stands for the smart and intelligence of the system as a whole reaching higher levels for disseminating real-life information to online information continuously. This is primarily due to the integration of intelligence in all the components of the system like intelligent vehicles, intelligent infrastructure, intelligent control system, intelligent commuters, intelligent cargo and intelligent devices information.

As regards, super intelligent vehicles, they need no drivers. The intelligence makes the system autonomous and automatic. The intelligence, along with driving the vehicle responds to the signals from other vehicles or the infrastructure.

The Singapore-MIT Alliance for Research and Technology (SMART) launched Singapore's very first locally-developed driverless car that is designed for operations on public roads in August 2016 (Figure 2.1). Adapted from its prototype driverless golf cart, this driverless car dubbed SCOT (shared computer operated transport) is operationally-ready for the public roads.

FIGURE 2.1 Singapore's first self-driving car (*Source:* https://cdn.macrumors.com/article-new/2016/08/r.jpeg).

Unlike other driverless cars which are retrofitted with expensive 3-D laser sensors, SCOT relies on low-cost off-the-shelf LIDAR sensors which enable the car to drive autonomously, independent of the global positioning system (GPS). This unique feature allows it to drive even in tunnels and places where GPS signals would be hindered.

An intelligent infrastructure can communicate with other aspects of the transport system. Such infrastructure may include LEDs on the road surface for providing warnings about incidents, self-healing concrete which makes the roads and bridges more durable and the bike lanes, which can produce electricity out of sunlight.

In intelligent control systems, traffic streams are guided or directed by real-time information of the condition of traffic ahead. These systems can also find solutions for certain traffic situations by responding to vehicles and associated transport infrastructures. There are also smart grids, which are able to communicate with vehicles and bring electricity to them.

With the intelligent vehicles and infrastructure, commuters also become smart, intelligently using the devices like the smart glasses, and smart navigation tools which are able to suggest alternative routes in case of the possibility of traffic jams. These smart applications may be customised or may adapt as per the preferences of commuters and respond to their daily habits. In addition of being able to take a photo or video, smart glasses are capable to integrate augmented reality with your own, feeding you live information during your activities, projecting images at a high resolution, and even letting you manipulate 3D objects with ease; it is only a matter of time before devices like smart glasses become a part of our daily lives.

The cargo may function intelligently and efficiently by communicating to its destination. In such a smart cargo system, the packages are enabled to interact with their surroundings and take decisions according to the context using RFID technology.

However, for the efficient operation of the intelligent transport system, huge amount of data needs to be processed, analysed and acted upon. With processing capabilities of the computers increasing day by day, there will be great opportunities for artificial intelligence dealing Fuzzy logic, artificial neural network and genetic algorithm to be deployed in the field of ITS. The concepts of augmented reality may also put into practice when the devices become faster and smarter. We may imagine the future smart cities to be driven by new technologies leading to smooth and automatic transportation.

2.4 ITS TRAINING AND EDUCATION NEEDS

ITS holds the key of a very bright future, but at the same time, it is mired with many challenges for the transportation community. There is an urgent need for continuing education and training for new engineers as well as practicing engineers keeping in mind the fast pace of changes in ITS technologies. Some of the facts with regard to the need for upgradation with skill of ITS and imparting education on ITS in the curriculum are as follows:

(a) There is a general agreement among all the people involved in transportation in public and private sectors as well as the academicians that it is essential to provide ITS education on both graduate and undergraduate levels.

(b) The awareness regarding the needs of education in private and public sectors in ITS is percolating to the academic institution and many academic programmes are devised to address these needs. A number of prestigious academic institutions are introducing changes in the existing curricula to introduce new courses or revising the already existing courses to include ITS issues and components.

(c) ITS is an inter-disciplinary field as it involves many disciplines like computers, electronics, IT, civil engineering, transportation, etc. That is the reason why ITS groups consist of professionals from diverse backgrounds and academic levels. It has been realised that the traditional syllabus in civil engineering is not able to cater the need of training engineers adequately in ITS and that there is the need of cross-disciplinary education programmes. The inter-disciplinary factor should encourage teamwork in ITS groups of public and private organisations which consist of engineers, non-engineers and technicians from different disciplines.

(d) The students of ITS should be educated not only in traditional civil engineering, but also in fields like communication, traffic surveillance, system analysis as well as the social and institutional matters.

(e) A dedicated degree in ITS may confine ITS education to some narrow specialised courses. On the other hand, there is need to develop a core ITS programme including various disciplines like civil systems, electrical, industrial, mechanical and computer engineering.

(f) The projects of ITS are usually conducted in teams for which group work becomes essential. Therefore, it is essential that during their academic experience in ITS, students from different disciplines should work in teams and get exposed to the actual ITS projects.

(g) Along with the formal educational programmes and degrees, there is a need of continuing education through seminars, meetings, activities, in-service training and in-house training. The chief areas of interest in continuing education may be system architecture, funding opportunities, partnerships of public and private sectors and the national ITS needs.

2.5 KNOWLEDGE MANAGEMENT CENTRE, INSTITUTE OF URBAN TRANSPORT (IUT) DELHI

Introducing intelligent transport systems for traffic management is one of the stated objective of National Urban Transport Policy (NUTP) 2006. As discussed above, there is a need to develop knowledge base on ITS applications to meet the objectives of NUTP. Under the Sustainable Urban Transport Project (SUTP) of Government of India, a knowledge management centre (KMC) as a learning repository including a database (collectively called KMC) has been set up, for use by all the stakeholders working in the domain of urban transport. The Website of the knowledge management and database centre of the IUT may be viewed at http://kmciutindia.org/.

The portal has been developed to facilitate systematic collection, assimilation, transformation, loading, interpretation and analysis of data in evolving future policies, programmes and strategies besides enhancing the level of awareness of citizens on urban transport matters. The KMC is expected to achieve certain specific objectives that are related to the different stakeholders as mentioned above that are as follows:

(a) ***Knowledge and learning platform:*** KMC acts as a national knowledge and learning platform for diverse set of users in transportation ecosystem with just-in-time knowledge sharing with these users.

(b) ***Decision support system for policy makers:*** KMC provides available information and acts as a decision support system for policy makers. KMC helps decision-making/benchmarking/cross-referencing/academic research/business/industry/consultants, who can prepare better plans, devise more efficient schemes, make more rational priorities for public expenditure and so on. The academic researchers can also be benefitted the KMC in producing high quality empirical research. Similarly, businesses/industry/consultancy can be benefitted from the knowledge base available with KMC.

(c) ***Aligning the policy, plans, and programmes with emerging situations:*** The KMC helps in achieving this objective includes updated and analysed transportation data of cities, best practices, information on technology and vendor options and existing rules, policies, Acts, bye-laws, standards and guidelines.

(d) ***Conducting and facilitating meaningful research:*** The KMC helps diverse stakeholders in efficient and effective performance delivery in the transportation ecosystem.

(e) ***Converting unstructured information into organised output:*** KMC offers support in organising and summarising the transportation related raw data into meaningful information for the stakeholders.

(f) ***User-friendly modes of knowledge sharing outputs:*** KMC is able to provide delivery of knowledge outputs in form of thematic maps, graphs, tables, etc.

(g) ***Interactive platform:*** KMC also provides interactive platform for users of various spectrums dealing with urban transport issues.

The KMC has been developed to help stakeholders in various stages of design, development, implementation or operations to make timely decisions based on documented best practices and vast correlative data available on KMC platform. The knowledge database available with KMC may be of immense help to develop an ITS system for a city based on experience of other cities if the case studies of ITS are duly documented and preserved. KMC can play a vital role in guiding the cities for implementation of ITS projects more efficiently and effectively.

2.6 ROLE AND IMPORTANCE OF ITS IN CONTEXT OF INDIAN TRANSPORT SYSTEM

With increase in prosperity and availability of low cost vehicles like Nano Car in India, the number of personal automobiles on Indian urban roads has increased significantly so much so that it is believed that India, in a few years will be ahead of China in the fastest growing car market. Moreover, there is a constant migration of rural population towards urban areas which results in a greater demand on transportation infrastructure in the existing situation. A number of cities in the country are already over populated and the government has imposed a physical limit on the additional infrastructure in a city. These factors make the management of transport infrastructure quite complicated.

The Centre for Excellence in Urban Transport, IIT Chennai in its study has flagged following issues which are frequently encountered by the developing countries like India in the application of ITS:

(a) Underdeveloped road network and poor maintenance of existing road network
(b) Financial constraints—strict budget restrictions
(c) Fast pace of urbanisation
(d) Lack of financial and physical resources for maintenance and operation
(e) Less demand for automation
(f) Lack of interest among the policy makers and the government
(g) Inadequate awareness among the users of transport facilities

Although, there are many small scale ITS projects which have been commissioned in cities like New Delhi, Bangalore, Pune, Indore and Chennai, the emphasis of these projects has been on standalone development like parking information, area wide signal control, electronic toll collection, Web-based traveller information, etc. Most of these projects are indeed kind of pilot studies being conducted on a small scale. Thus, the operation of a comprehensive ITS in India is still absent, and there is a need for a systematic approach to the implementation of comprehensive ITS. Further, more efforts are needed to focus on other ITS concepts like emergency management, congestion management, advanced traffic management systems, commercial vehicle operation, advanced vehicle control systems, advanced traveller information systems, etc.

The implementation of ITS needs to be done comprehensively at a network level rather than in small corridors. But there are a number of challenges that need to be addressed for the implementation of ITS in India. The following approach may be adopted to deal with these challenges:

(a) Development of a national ITS standard for the application of various ITS systems.
(b) Setting up a national ITS centre of excellence, which should document all ITS projects with detailed information about the design, implementation, lessons learnt, cost benefit details, etc.
(c) Setting up fully functional integrated traffic management centres for the coordination of the ITS activities on the urban and the regional levels.
(d) Developing and installing automated traffic data collection devices.
(e) Setting up a national archive for ITS data.
(f) Developing models and algorithms for implementation of the ITS.
(g) Boosting more interaction and coordination among the academia, industries and government agencies for generating greater interest and undertaking more projects in the area of the ITS.

2.7 OPPORTUNITY FOR SECTOR GROWTH OF ITS

The large scale urbanization in India offers huge opportunity for growth of ITS market in the country. The prominent drivers of growth for the ITS market in India are:

(a) Increase in number of vehicles has exacerbated existing traffic congestion, especially in metropolitan cities. Delhi, Mumbai, Kolkata and Bengaluru have only 5% of India's population, but 14% of the total registered vehicles which causes heavy traffic congestion, deterioration in air quality, and increase in noise levels. There is an urgent need to implement ITS tools to manage the traffic, ease congestion and control pollution in the cites.
(b) As a part of National Urban Transport Policy 2006, Government of India (GoI) has been providing financial support to cities for implementing: metro rail, BRT, monorail and other forms of public transport. Every big city in India is coming up with projects of metro rail, BRTs, monorail, etc. The modern public transport systems are using ITS tools for operation, maintenance, fare collection, integration, public information, etc. Thus, it opens up huge opportunities in many sectors like passenger information systems, smart cards and integrated ticketing, and parking management.
(c) Freight is mostly a disorganised sector in India and has huge potential for development and deployment of ITS tools to improve the operational efficiency of the freight transport.

2.8 SUMMARY

With the increasing growth of Indian economy and the population explosion, there is a lot of emphasis on applications of traffic management in the metropolitan cities and towns. Although

India has already initiated intelligent transport systems for organising traffic, there is the urgent need of more integration of advanced technology with traffic management. The application of ITS in vehicles, infrastructure, traffic management and traveller information has led to much safer mobility of people and freight in countries like the USA, UK, Japan, Middle East and Canada etc. But in India, it is still in the stage of infancy. India cannot model its ITS on the successful existing ITS of other countries because of basic cultural, geographical and practical differences among the cities. There is the need for modifying and adapting the existing concepts to make them suitable in Indian transportation.

For designing an intensive ITS programme, there should be further development and implementation of technologies like electronic equipment such as sensors, detectors, communication devices and the application of global navigation satellite system. There is the need of developing models of the ITS concepts to suit the traffic needs of our country. But this kind of development must co-exist with conservation of natural resources, energy efficiency and sustainability. There is the need of the development of human skills for developing, implementing and managing the emerging technologies for the ITS design and application. Besides the development of ITS technology to transport system sustainable and environment friendly, it has a huge potential not only for capacity building programme but also offers a mechanism for generating huge number of jobs in the area of ITS.

Questions

2.1 What are the challenges for implementation of ITS in India?

2.2 What are the opportunities for development of ITS in India?

2.3 Write a short note on ITS—today and tomorrow.

2.4 What is the need of training and education for implementation of ITS?

2.5 Explain role and importance of ITS in context of Indian transport system.

Bibliography

Atkinson, Robert D. and Daniel D. Castro, "Digital Quality of Life: Understanding the Personal and Social Benefits of the Information Technology Revolution," Information Technology and Innovation Foundation, October 1, 2008, 106. (http://www.itif.org/index.php?id=179)

Begg, D. (2014), "A 2050 vision for London: what are the implications of driverless transport". (Commissioned by Clear Channel)

Berthelsen, E., "Movement analytics delivers an untapped revenue opportunity for mobile operators", Machina Research White Paper for INRIX, February 2015. (https://machinaresearch.com, www.INRIX.com)

Dutt, Panchagnula Krishna and Alok Tripathi, "Integration of Intelligent Transport Systems in Indian Cities", *Journal of Civil Engineering and Environmental Technology*, Print ISSN: 2349-8404; Online ISSN: 2349-879X; Vol. 2, pp. 721–724, No. 8; April–June, 2015.

India's urbanisation: A closer look, an article published in McKinsey Quarterly, July 2010. (http://www.mckinsey.com/global-themes/urbanisation/indias-urbanisation-a-closer-look)

Intelligent Transport Systems in India, Snapshot a Report by European Business and Technology Centre (EBTC). (http://ebtc.eu/pdf/120913_SNA_Snapshot_Intelligent-transport-systems-in-India.pdf)

Intelligent Transport Systems: Synthesis report and issues and challenges under Indian conditions, Centre of Excellence in Urban Transport, IIT Chennai, December 2010. (http://www.itsinternational.com/event-news/intertraffic-2012/indian-minister-explains-future-transportation-opportunities-at-intertraffic/)

Intelligent Transportation Systems for Sustainable Development in Asia and the Pacific, Working Paper by the Information and Communications Technology and Disaster Risk Reduction Division. (http://www.unescap.org/sites/default/files/ITS.pdf)

National Road Transport Policy, Ministry of Road Transport and Highways, Government of India.

National Urban Transport Policy, Government of India.

Sakar, Pradip Kumar, et al., *Transport Planning: Principles, Practices and Policies*, PHI Learning, Delhi, 2015.

Sengupta, Dibyendu, Intelligent Transport Systems in India, 2014, EBTC, available at: http://ebtc.eu/index.php/sector/transport/transport-news/278-insight-intelligent-transport-systems-in-india.

Singapore MIT Alliance for Research & Technology. (https://smart.mit.edu/news-events/news/smart-nus-driverless-car-trials-launched-at-one-north-)

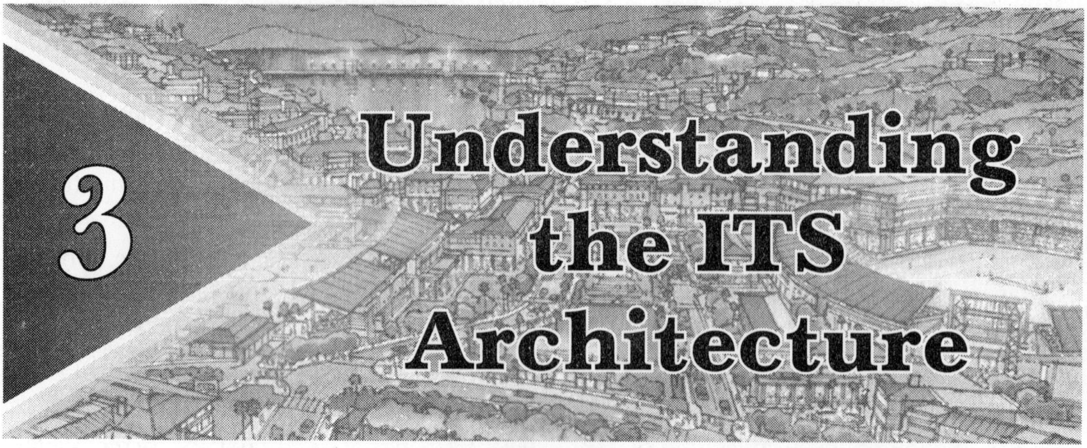

3 Understanding the ITS Architecture

3.1 INTRODUCTION

An ITS architecture is the framework within which a system of ITS projects is developed. It defines different components of ITS and the interconnections and information flow between the components. The primary components of an ITS architecture are subsystems and information flows.

Subsystems: Subsystems are individual pieces of the overall ITS which perform specific functions such as providing traveller information, managing traffic, or responding to emergencies. Subsystems can be associated with specific organisations/departments/agencies such as public safety agencies, transportation services, emergency management agencies, or transit providers. They are either sources or users of information or both provided by other subsystems within the boundary of ITS architecture. Subsystems may include centre systems, roadside equipment, vehicle equipment and traveller devices related to ITS.

Information flows: Information flows define the information that is transferred and exchanged between subsystems such as traffic information, or surveillance and sensor control data. They depict integration of ITS by illustrating the information links between subsystems. This integration in ITS, is not limited to technical but also include institutional dimension. The system interfaces require cooperation and shared responsibilities on the part of owners and operators of each participating system.

ITS architecture can be defined as comprehensive framework or picture of functions that must be performed (user services) through physical entities where these functions reside in the subsystems and interfaces between subsystems (information flows) for communications requirements for interfaces and stakeholder roles.

The ITS architecture offers a framework for conceptualising, planning, defining, developing and integrating various components of intelligent transportation systems. It describes the way different ITS components should interact with each other to address the transportation problems with a wide variety of options. It is similar to building architecture wherein an architect plan and design a building considering various functions of subsystems such as bed room, living room, bathroom and kitchen to be integrated in the building. ITS architecture describes various functions and detail responsibilities to each stakeholder of ITS. A common and standard ITS architecture is recommended throughout a region so that it can integrate various subsystems and also address various issues concerning different agencies. The following issues need to be kept in mind while developing an effective ITS Architecture:

(a) ***Interoperability:*** The ITS subsystems are required to be used by many departments like traffic police, transport department, municipality, public transport operators, etc. The ITS architecture should be so designed and developed that the information collected, function implemented or any equipment installed are interoperable by different departments/agencies in different states and regions.

(b) ***Information sharing capabilities:*** Huge amount of data is generated by ITS. This data can be analysed to improve and manage traffic system in the city. The ITS architecture shall be designed to facilitate seamless sharing of information among various different agencies.

(c) ***Sharing of infrastructure:*** The infrastructure like regional communication towers constructed by various private agencies may be used for ITS operations to optimise the cost of the ITS implementation. The provision of sharing of existing resources/infrastructure may be built in the ITS architecture.

3.2 FUNCTIONALITIES REQUIRED FOR USER SERVICE

The provision of a user service required a number of functions to be performed by the system. Each functionality of a user service needs to be defined so that the system can be designed accordingly. For illustration, the user service requirements for a traffic control user are presented in Box 3.1.

Box 3.1 User service requirements for traffic control user

Objective: To provides the capability to efficiently manage the movement of traffic on streets and highways.

Functions:

- Traffic flow optimisation
- Traffic surveillance
- Traffic control
- Information sharing

User service requirement:

- Traffic control shall include a traffic flow optimidation function to optimise traffic flow.
- Traffic flow optimisation shall employ control strategies that maximise traffic-movement efficiency.
- Traffic flow optimisation shall include a wide area optimisation capability, to cover several jurisdictions.
- Wide area optimisation shall combine the control of network signal systems with the control of express ways.
- Wide area optimisation shall facilitate preferential treatment for transit vehicles to promote public transport in a city.
- Traffic surveillance function should be part of Traffic control (TC).

3.3 LOGICAL ARCHITECTURE

A number of functions are needed to meet user service requirements. The logical architecture defines a set of functions and information flows that are needed to meet user service requirements. The interaction of different components of ITS are defined by it as a logical process. The processes and data flows involved in a particular function can be represented graphically by data flow diagrams (DFDs). Figure 3.1 describes the interaction of 'Traffic Management Process' with other processes. Each process is divided into more subprocesses. The subprocess is further divided into subprocesses which are termed *process specifications* (p-specs) at lowest level. These p-specs are required to be performed to meet user services requirements.

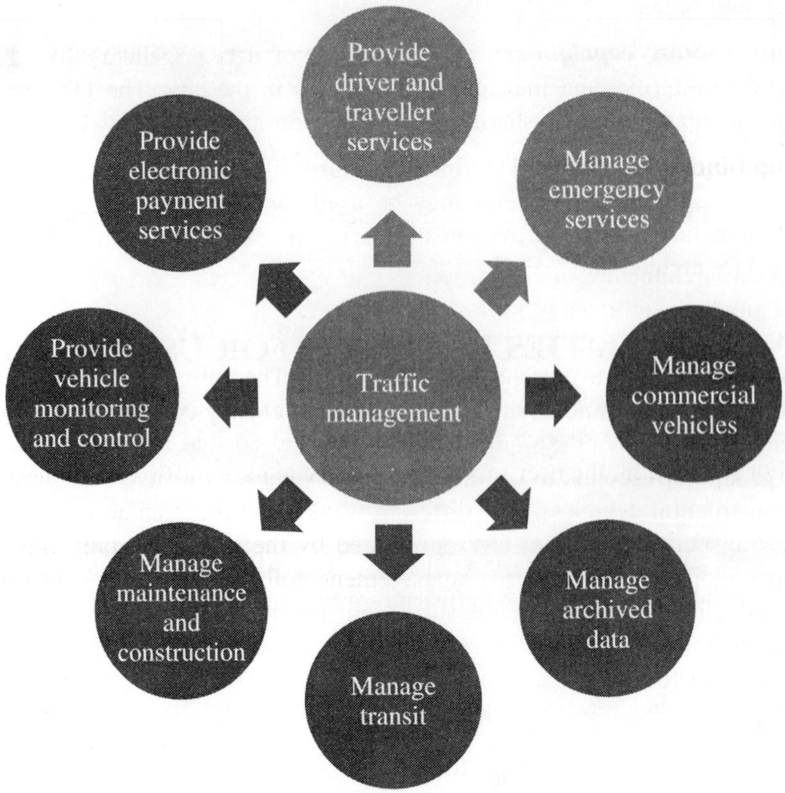

FIGURE 3.1 High level ITS logical architecture.

3.4 PHYSICAL ARCHITECTURE

The functions described in logical architecture which serve the same purpose are grouped into subsystems. These subsystems form a physical entity to deliver functions. The data flow of logical architecture is also integrated to define interfaces between subsystems. Figure 3.2 describes the functions *A* and *B* of logical architecture assigned to subsystem *A* in physical architecture. Both the architecture forms the core of ITS.

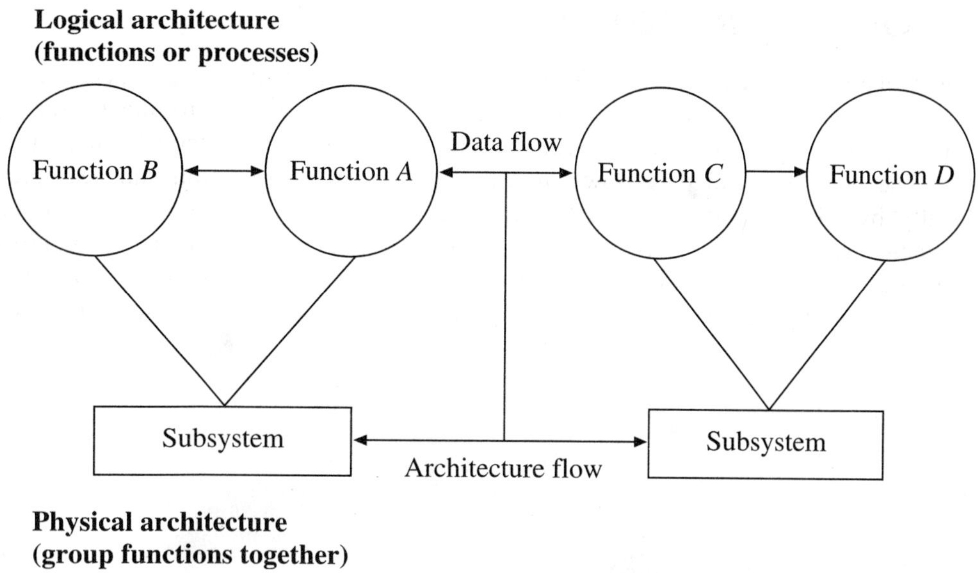

FIGURE 3.2 Function from logical to the physical architecture.

The physical architecture of ITS describes the physical subsystems and architectural flows based on the logical architecture. The subsystems can be broadly classified in four groups as centres, field, vehicle, and travellers. The subsystems and communications that comprise the physical architecture are presented in Figure 3.3. The subsystems represent collection of functions required for the same transportation need and closely correspond to physical elements of transportation management system.

Vehicle group corresponds to different types of vehicles (normal, emergency, commercial, transit and construction vehicles). The different ways a traveller can access information on the status of the transportation system are represented by the traveller group. Similarly, the units of ITS comprise of roadway, security management, toll collection, parking management and commercial vehicle check. The centre subsystem comprises traffic management, emergency management, toll administration, transit management, etc. as described in Figure 3.3. There are following four different types of communication systems used in ITS application:

(a) Fixed point to fixed point communications
(b) Wide area wireless communications
(c) Vehicle to vehicle communications
(d) Field to vehicle communications

Traffic management subsystem is connected with communications to get real-time information of the transportation system through roadway subsystem which comprises of signal control, detectors, camera, Variable Message Sign (VMS), etc.

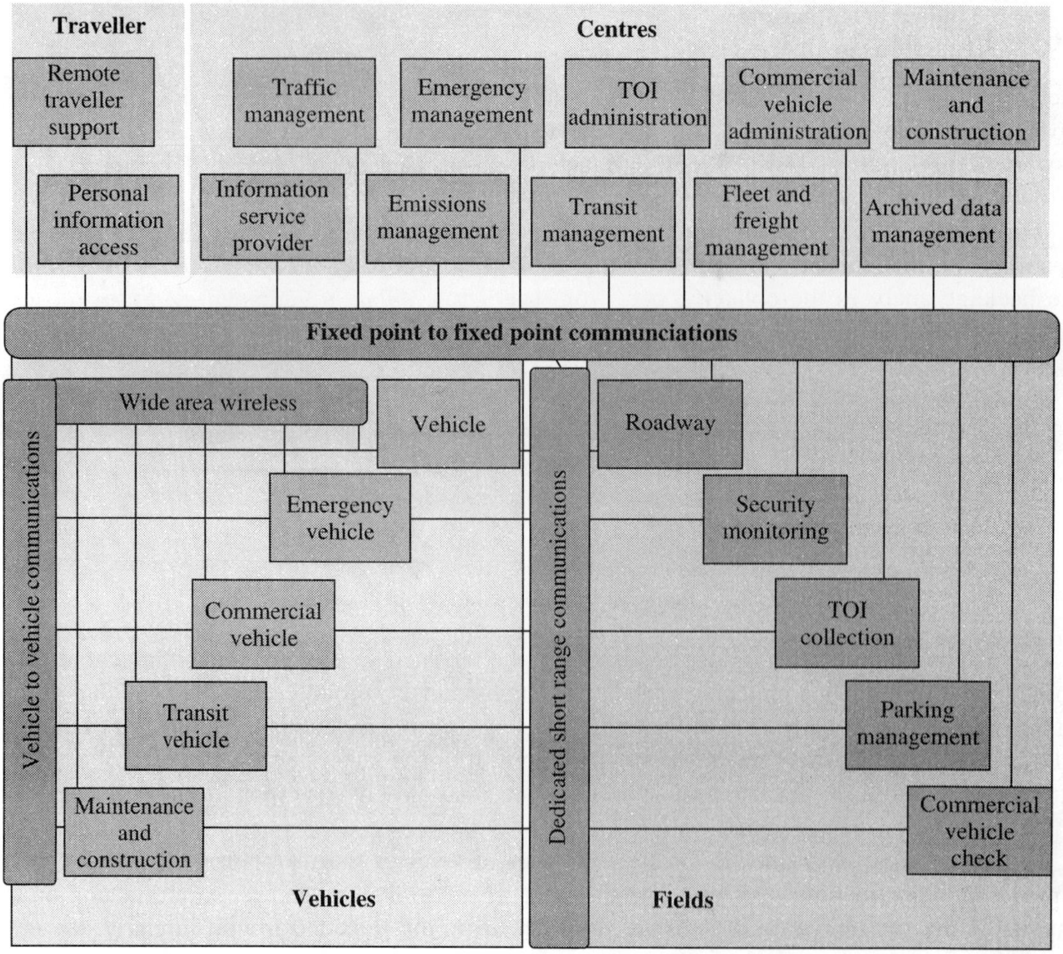

FIGURE 3.3 ITS physical architecture showing subsystems and communications.

3.5 ORGANISATIONAL ARCHITECTURE

A number of departments/organisations/stakeholders are engaged in a system like traffic management. Organisational architecture represents the involvement of various organisations/stakeholders associated to be a part of the traffic management centre (or central control centre) such as national highways authority of India (NHAI), traffic police, municipal corporation, public work department (PWD), mobile/Wi-Fi service providers, utility service provider companies, etc. for the control and management of traffic movement, incident management, maintenance of street lights, traveller information service, commercial vehicle management, electronic payment systems, safety services, emergency management services, etc. The organisational architecture is also used to define role, responsibilities and involvement of various stakeholders in a system. A clear distribution of work amongst all the agencies is always desirable to bridge any gap and to analyse the reasons of any failure of the system so that corrective action may be taken.

3.6 EQUIPMENT PACKAGES

The similar functions of a particular subsystem can be grouped together and implemented by a package of hardware and software facilities. The grouping helps in cost optimisation and easy implementation of ITS services. The equipment package is developed by grouping a set of functions of a particular subsystem.

For example, a traffic management centre (TMC) equipment package may provide the capability for traffic managers to monitor and manage the traffic flow at signalised intersections. The package analyses the collected data from traffic surveillance equipment and implements control plans for signalised intersections. TMC signal control equipment package contains five p-specs:

(a) Traffic operation personnel traffic interface
(b) Process traffic data
(c) Select strategy
(d) Determine indicator state for road management
(e) Output control data for roads

3.7 MARKET PACKAGE

A set of equipment packages that are required to work together to provide a given transportation service may be clubbed to form a market package. Most market packages are made up of equipment packages from two or more subsystems. The market package are designed to address specific transportation problems and needs.

The market package for TMC provides the central control and monitoring equipment, communication links and the signal control equipment that support local street control or arterial traffic management. The various signal control systems dynamically adjusted control plans and strategies based on current traffic conditions and priority requests.

3.8 NEED OF ITS ARCHITECTURE TO SOLVE PROBLEMS IN URBAN AREA

ITS architecture is a useful tool for integrating ITS technique into conceptualising and planning process. The ITS architecture describes the comprehensive set of data that is required to be shared by various agencies of transportation network. With the knowledge of what data must be shared, these agencies can develop a common interest in cooperating planning efforts between all transportation projects.

As vehicle miles travelled (VMT) and congestion has been growing rapidly in the cities with no end in sight, precision technology has become essential for transportation planning. As urban areas are expanding and more roadways are interlinked with one another, operations and maintenance of this technology is imperative. Therefore, ITS architecture has been developed as the best way to oversee the physical and virtual networks interacting with ITS. At the same time, as ITS technology grows, the ITS architecture also needs to adapt to new ITS technology.

3.9 EXAMPLES OF ITS ARCHITECTURE: ITS ARCHITECTURE IN DIFFERENT COUNTRIES

The major developed countries of the world have taken the lead in developing ITS system architectures customised to their own needs. Some notable examples are US, the European Union, and Japan. Many other countries, both developed and developing, have developed and defined their own national ITS architectures based on these architectures.

3.9.1 National ITS Architecture of US

The National ITS Architecture of US provides a common framework for planning, defining, developing, and integrating intelligent transportation systems. It is a comprehensive and rational that reflects the involvement of a broad cross-section of the ITS community (transportation practitioners, systems engineers, system developers, technology specialists, consultants, etc.). The architecture defines:

(a) The functions/utilities (e.g. gather traffic information or request a route) that are required for ITS.

(b) The physical entities or subsystems where these functions/utilities reside (e.g., the field or the vehicle).

(c) The information flows and data flows combine these functions and physical subsystems together into an integrated system.

Architecture layers

The National ITS Architecture offers a framework for planning, programming, and implementing intelligent transportation systems. The architecture framework is comprising of two technical layers, a transportation layer and a communication layer, which operate in the context of an institutional layer.

Institutional layer: It comprises of the institutions, policies, funding mechanisms, and processes that are required for effective implementation, operation, and maintenance of an intelligent transportation system. The institutional layer is considered as the base because strong institutional support and effective decisions are foundation to an effective ITS programme. The objectives and requirements for ITS are formalised in institutional layer.

Transportation layer: It is used where the transportation solutions are described in terms of the subsystems and interfaces along with underlying functionality and data definitions which are required for each transportation service. This layer is the heart of the National ITS Architecture.

Communications layer: It provides for the accurate and timely sharing of information between systems to support the transportation solutions.

User services bundles and user service

The user service bundle consists of following services:

(a) *Travel and traffic management user services:*
   (i) Pre-trip travel information
   (ii) En-route driver information

(iii) Route guidance
(iv) Ride matching and reservation
(v) Traveller services information
(vi) Traffic control
(vii) Incident management
(viii) Travel demand management
(ix) Emissions testing and mitigation
(x) Highway rail intersection

(b) *Public transportation management (user services bundle) user services:*
(i) Public transportation management
(ii) En-route transit information
(iii) Personalised public transit
(iv) Public travel security

(c) *Electronic payment (user services bundle) user services:*
(i) Electronic payment services

(d) *Commercial vehicle operations (user services bundle) user services:*
(i) Commercial vehicle electronic clearance
(ii) Automated roadside safety inspection
(iii) On-board safety and security monitoring
(iv) Commercial vehicle administrative processes
(v) Hazardous materials security and incident response
(vi) Freight mobility

(e) *Emergency management (user services bundle) user services:*
(i) Emergency notification and personal security
(ii) Emergency vehicle management
(iii) Disaster response and evacuation

(f) *Advanced vehicle safety systems (user services bundle) user services:*
(i) Longitudinal collision avoidance
(ii) Lateral collision avoidance
(iii) Intersection collision avoidance
(iv) Vision enhancement for crash avoidance
(v) Safety readiness
(vi) Pre-crash restraint deployment
(vii) Automated vehicle operation

(g) *Information management (user services bundle) user services:*
(i) Archived data

(h) *Maintenance and construction management (user services bundle) user services:*
(i) Maintenance and construction operations

Different users of the National Architecture use the logical architecture presented here in different ways:

(a) ***Public sector agencies:*** Most public sector employees are not be required to deal with the logical architecture directly, however, they are required to review the software operation to verify that it meets their requirements. They may use the logical architecture to write their own system and interface requirements specifications.

(b) ***Consultants:*** They often are required to create logical and physical architectures that are traceable to their user's requirements. Hence, integration and organisation of the physical and the logical architectures is of paramount interest to them. The additional detail provided by the logical architecture may also support developers as they begin to implement a project.

3.9.2 ITS Architecture for Canada

The ITS Architecture for Canada provides a comprehensive framework for integration to guide the coordinated deployment of ITS programmes within the public and private sectors. It offers a starting point from where stakeholders of ITS can work together to achieve compatibility among different ITS elements to ensure integrated ITS deployment for a region. The architecture defines interaction among physical components of the transportation system including travellers, vehicles, roadside devices, and control centres. It also specifies the information and communications system requirements, how data should be exchanged and used, and the standards required to facilitate information sharing. Overall, the ITS Architecture for Canada describes the functionality of different ITS components and the information flows required among ITS elements to achieve total system goals.

The various components of the logical architecture required for the architecture framework have also been defined, including components such as user service requirements, process specifications, data flows, and data flow diagrams. The logical architecture of the ITS architecture has been developed in parallel with the physical architecture. This is unlike the US work which developed the logical architecture first and then physical architecture.

Regional ITS architecture

The regional ITS architecture is an important tool used in transportation planning, programming, and project implementation. It can help in identifying opportunities for making ITS investments in a more cost-effective fashion. It defines how to develop a regional ITS architecture, which will be a cornerstone of planning for effective inter-agency coordination and for deployment and operation of technology-based projects.

User services bundles and user services of canadian architecture

The user services of the ITS Architecture for Canada have been organised into 8 user service bundles as against 7 bundles in the US National ITS Architecture. In case of Canada, the travel and traffic management user service bundle of the US Architecture has been separated into two separate bundles: (a) Travel information services and (b) Traffic management services.

The ITS Architecture for Canada includes 35 user services. Of these 35 user services, 6 user services are developed specifically for the ITS Architecture for Canada. The remaining 29 user services are based on the 31 user services from the US National Architecture. A number of the US user services are either combined into single user services, or divided into separate user services as shown below:

(a) *Traveller information services*

- (i) Traveller information
- (ii) Route guidance and navigation
- (iii) Ride matching and reservation
- (iv) Traveller services and reservations

(b) *Traffic management services*

- (i) Traffic control
- (ii) Incident management
- (iii) Travel demand management
- (iv) Environmental conditions management
- (v) Operations and maintenance
- (vi) Automated dynamic warning and enforcement
- (vii) Non-vehicular road user safety
- (viii) Multi-modal junction safety and control

(c) *Public transport services*

- (i) Public transport management
- (ii) En-route transit information
- (iii) Demand responsive transit
- (iv) Public travel security

(d) *Electronic payment services*

- (i) Electronic payment services

(e) *Commercial vehicle operations*

- (i) Commercial vehicle electronic clearance
- (ii) Automated roadside safety inspection
- (iii) On-board safety monitoring
- (iv) Commercial vehicle administrative processes
- (v) Intermodal freight management
- (vi) Commercial fleet management

(f) *Emergency management services*

- (i) Emergency notification and personal security
- (ii) Hazardous material planning and incident response
- (iii) Disaster response and management
- (iv) Emergency vehicle management

(g) *Vehicle safety and control systems*

- (i) Vehicle-based collision avoidance
- (ii) Infrastructure-based collision avoidance
- (iii) Sensor-based driving safety enhancement
- (iv) Safety readiness
- (v) Pre-collision restraint deployment
- (vi) Automated vehicle operation

(h) *Information warehousing services*
  (i) Weather and environmental data management
  (ii) Archived data management

3.9.3 ITS Architecture Framework for Europe

Intelligent Transport Systems (ITS) Architecture for Europe is also called 'The FRAME Architecture' (Figure 3.4). The FRAME Architecture describes the top-level requirements and functionalities, or the use cases, for almost all the ITS applications and services which have been considered for implementation somewhere in the European Union. It is at a "level" such that it can be used as a reference by all ITS architects in Europe, and is intended to be the foundation for building the other types of architecture which may be necessary. The FRAME Architecture enables them to guarantee compliance at the interfaces of other systems so that seamless services can be offered to cross-border travellers, and an open European market of compatible components can be developed.

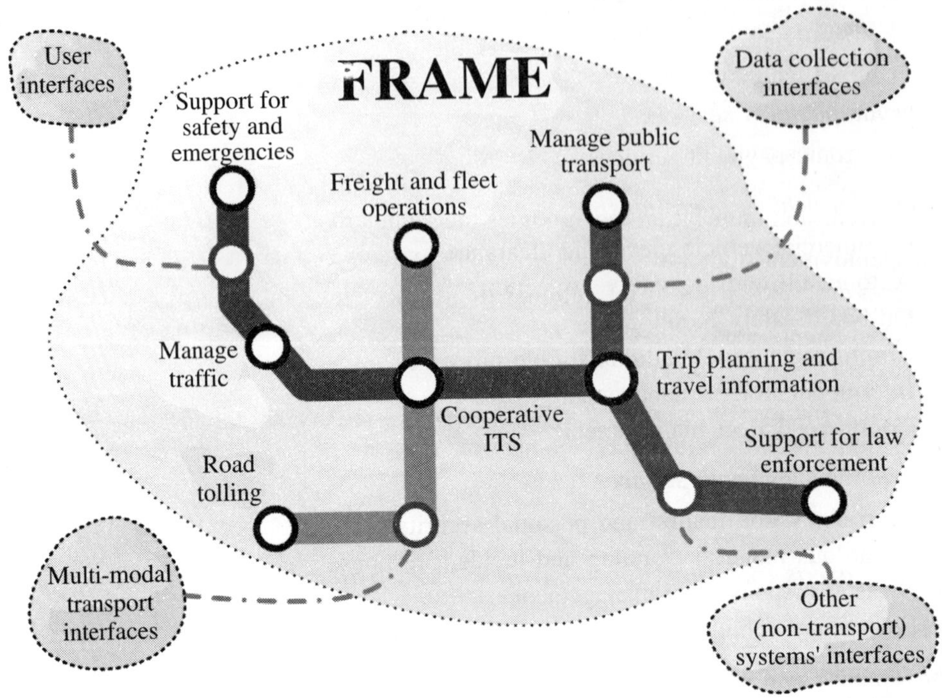

FIGURE 3.4 The Frame Architecture of ITS in Europe (*Source*: http://frame-online.eu/, reproduced with permission from the FRAME Forum.)

The FRAME Architecture was developed as per recommendations from the high level group on transport telematics duly supported by a resolution of the Council of Ministers. It was developed and first published by the EC funded project KAREN in October 2000. It has since been modified and updated to include user tools and expand the range of ITS services it can support, including cooperative systems (connected vehicles as it is known in the USA). The underlying objective of this initiative was to encourage the deployment of (mainly road-based) ITS in Europe by creating a framework which would provide a systematic basis for planning ITS implementations, facilitate their integration when multiple systems were to be deployed, and help to ensure inter-operability, including across European borders.

The FRAME expresses architecture from a number of viewpoints such as:

(a) User needs to define what the stakeholders want an ITS tool to provide in terms of services to be delivered.

(b) Functional architecture defines the functionalities needed by the ITS tools to meet the user needs. The functional architecture consists of functional areas, which are further divided into functions. Each area has data flow diagrams (DFDs) to describe how the functions relate to each other, to Data Stores, and to the outside world through the data flows.

(c) Physical architecture describes how functions can be combined into physical locations to form implementable systems, considering any user needs that have physical as well as functional requirements. It consists of a series of "Example Systems" and provides a broad methodology for deployment and implementation.

(d) Communications architecture describes the communications links which are required to support physical data flows. It provides an analysis of the communications requirements for several of the "Example Systems" from the physical architecture, and describes current communication technologies and standards.

(e) Deployment study have shown how to deploy the systems derived from the architecture and describes ways to migrate existing systems to conform to FRAME.

(f) Cost benefit study helps in predicting the likely costs and benefits that can be expected from deploying the architecture.

The functionality within the 'Functional View' as part of the FRAME Architecture covers many ITS domains, which enables it to support a wide range of ITS related services. For better appreciation and convenience, this functionality has been further divided into the functional areas shown in Figure 3.5.

Functional areas and functions—european ITS framework architecture

(a) *General*

(i) Architectural properties
(ii) Data exchange
(iii) Adaptability
(iv) Constraints
(v) Continuity
(vi) Cost/Benefit
(vii) Expandability
(viii) Maintainability
(ix) Quality of data content
(x) Robustness
(xi) Safety
(xii) Security
(xiii) User friendliness
(xiv) Special needs

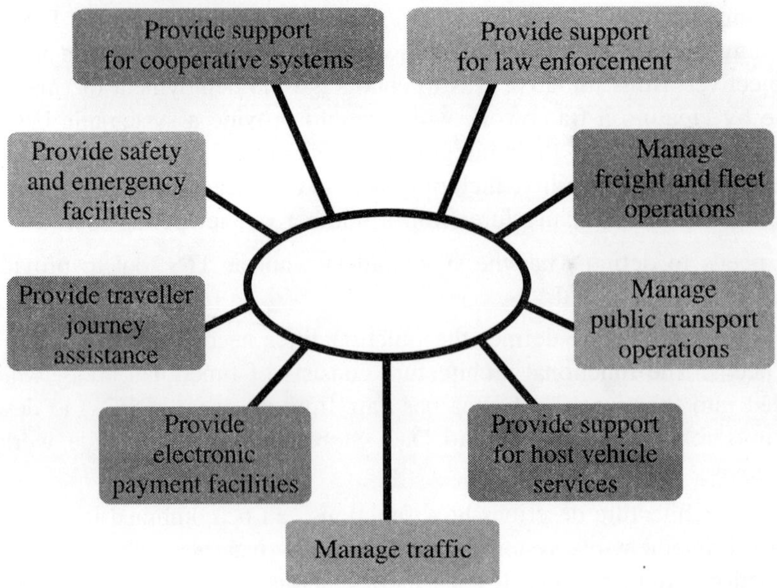

FIGURE 3.5 Requirements for FRAME Architecture of ITS in Europe (*Source*: http://frame-online.eu/ reproduced with permission).

(b) *Infrastructure planning and maintenance*

(i) Transport planning support
(ii) Infrastructure maintenance management

(c) *Law enforcement*

(i) Policing/Enforcing traffic regulations

(d) *Financial transactions*

(i) Electronic financial transactions

(e) *Emergency services*

(i) Emergency notification and personal security
(ii) Emergency vehicle management
(iii) Hazardous materials and incident notification

(f) *Travel information and guidance*

(i) Pre-trip information
(ii) On-trip driver information
(iii) Personal information services
(iv) Route guidance and navigation

(g) *Traffic, incidents and demand management*

(i) Traffic control
(ii) Incident management
(iii) Demand management
(iv) Safety enhancements for vulnerable road users
(v) Intelligent junctions and links

(h) *Intelligent vehicle systems*

- (i) Vision enhancement
- (ii) Automated vehicle operation
- (iii) Longitudinal collision avoidance
- (iv) Lateral collision avoidance
- (v) Safety readiness
- (vi) Pre-crash restraint deployment

(i) *Freight and fleet management*

- (i) Commercial vehicle pre-clearance
- (ii) Commercial vehicle administrative processes
- (iii) Automated roadside safety inspection
- (iv) Commercial vehicle on-board safety monitoring
- (v) Commercial fleet management

(j) *Public transport management*

- (i) Public transport management
- (ii) Demand responsive public transport
- (iii) Shared transport management
- (iv) On-trip public transport information
- (v) Public travel security

An ITS architecture using FRAME

The methodology for creating an ITS Architecture from the FRAME Architecture has been described in Figure 3.4. The use of particular technologies or supplier products is not prescribed in the FRAME Architecture for following reasons:

(a) The ITS Architectures created using the methodology will not become obsolete through advances in technology, or product development.

(b) It opens up the possibility for the development of new technologies to enable particular functionality to be provided.

Stakeholder aspirations: Stakeholder aspirations from ITS presented in Figure 3.6 are statements that express the expectations and desires of the various stakeholders for the services that the ITS implementation expected to provide. The stakeholders are supposed to write them, but experience has revealed that help is often needed from the architecture team. Broadly, there are four classes of stakeholder, as follows:

(a) *Want ITS:* This class comprises organisations/agencies/departments that need ITS services to enable their road networks to be used safely and efficiently. This class also includes public transport operators and freight operators where ITS can enable them to improve the efficiency with which they move people and goods.

(b) *Use ITS:* This class comprises the end users who use ITS services and/or operate the equipment. It includes travellers, commuters on a multi-modal journey as well as all classes of vehicle driver; freight shippers; public transport mangers and specialist system operators.

(c) *Rule ITS:* This class represents those who frame regulations and standards. It includes national governments, regional (state) government and the various standards making bodies.

(d) *Make ITS:* The class comprises the equipment and system manufacturers, communications providers and the system integrators.

Service providers, e.g., travel information and trip planning, may be in one or more of the Want, Use and Make ITS classes.

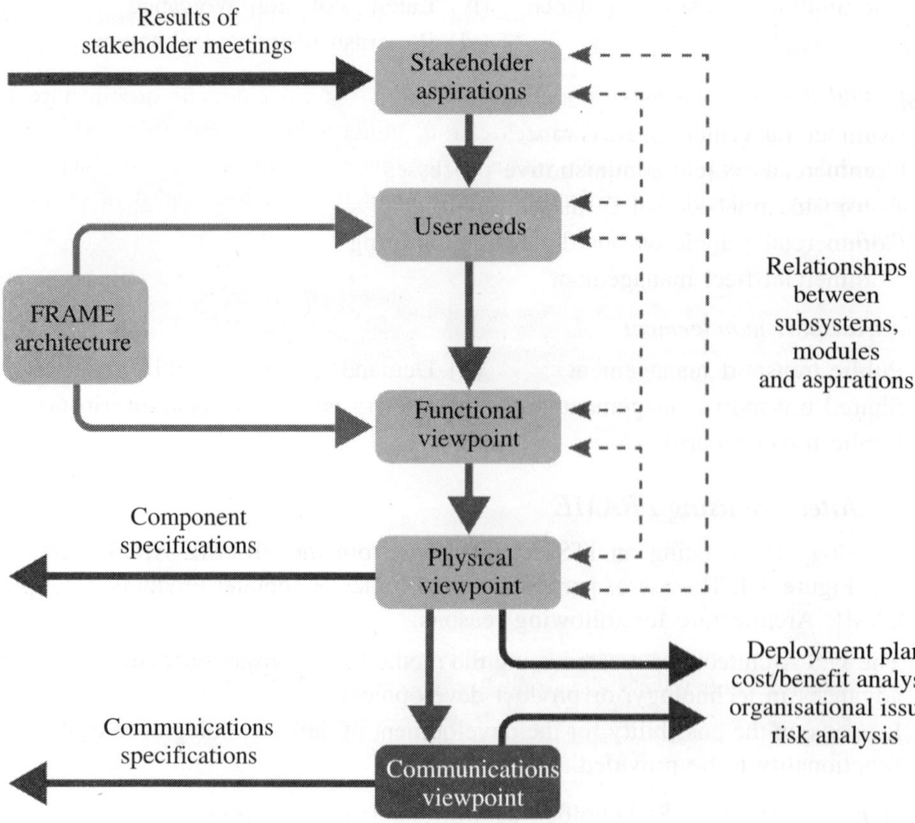

FIGURE 3.6 Stake holder's aspirations (*Source*: http://frame-online.eu/ reproduced with permission).

User needs: It is normal to find that the stakeholder aspirations will have specified formally in a variety of styles. Sometimes they can also be obscure and inconsistent with time and latest technology. Therefore, it is important to re-write them in a consistent manner which is suitable for the next stage in the process. The result is a specific and well-defined set of user needs which express the stakeholder aspirations in a consistent and judicious manner meaning is clear and whose properties are testable.

Functional viewpoint: The FRAME methodology uses the term "viewpoints" for the parts that make up the FRAME Architecture and its resultant ITS Architectures. This follows the recommendations of IEEE 1471. A functional viewpoint (also called a logical viewpoint) shows the functionality that are required to fulfil the user needs, and hence the stakeholder aspirations. While using the FRAME Architecture, the functional viewpoint is presented as data flow diagrams (DFDs) that contain functions, data stores and terminators, and the data that flows between them. Each of these is provided with its own description which, in the case of functions, includes statements explaining what they do. Since the FRAME Architecture consists of a functional viewpoint that satisfies all of its user needs, the architecture team only is required to select those parts of it that serves the user needs that have been mapped to the stakeholder aspirations. New functions are only needed for stakeholder aspirations that have required new user needs to be added.

Physical viewpoint: Once the functional viewpoint is completed, the architecture team can allocate each item of functionality to a location, either within a subsystem (Figure 3.7), or within a module that is part of a subsystem. After this, the component (subsystem or module) specifications can be developed from the definitions of the functions and data stores contained within them.

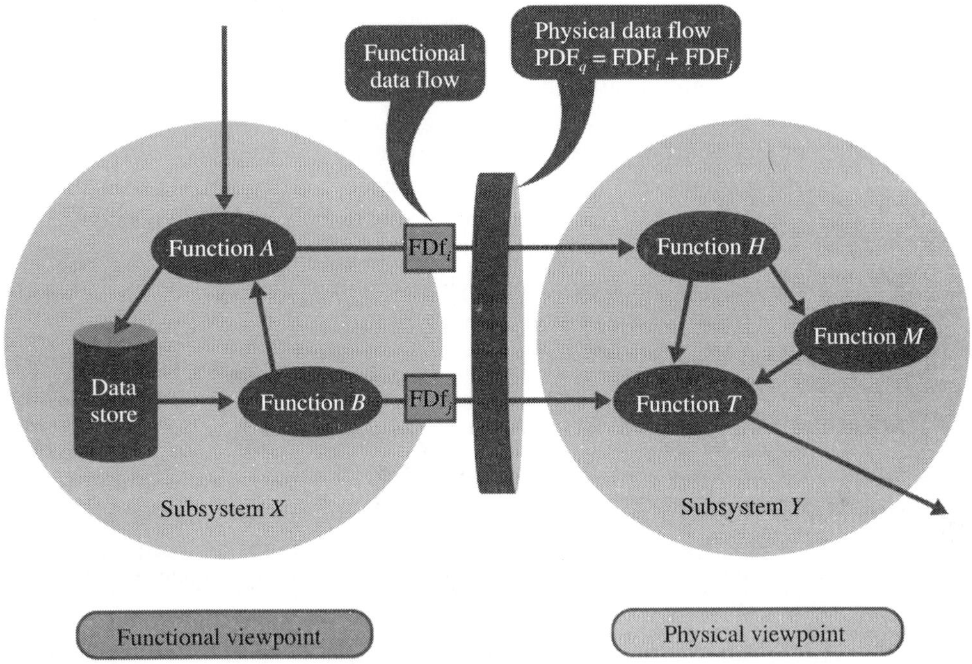

FIGURE 3.7 Physical viewpoint of ITS Architecture in Europe, (*Source*: http://frame-online.eu/ reproduced with permission).

The context diagram produced as part of the functional viewpoint also applies to the physical viewpoint. It again shows the ITS as a single item and the links needed by the functionality within it to communicate with the entities outside it.

Communications viewpoint: As evident from the diagram above, a consequence of allocating the functionality to subsystems (and modules) is that it is immediately visible which functional data flows lie within a subsystem (or module), and which pass between one subsystem and another, or between one module and another. Those which pass between subsystems or modules make up the physical data flows, and represent a communication channel between subsystems, and/or between modules.

Since subsystems by definition are located at different places (e.g., in a traffic management centre, at the roadside, in a vehicle), it is possible to develop communications specifications by analysing the contents of each physical data flow. The format of analysis in the form of desired standard helps it to work for the communications.

An analysis of the physical data flows that pass between the ITS and the terminators can also lead to development of "standard" interfaces for end users. It can play an important role to ensure that the ITS implementation can be applied in the same way, everywhere that it is deployed.

Traceability: An important aspect of the FRAME Architecture methodology is the ability to facilitate traceability all the way through the process. All of the services contained in most ITS Architectures cannot normally be deployed at the same time, due to the constraints of funds, as well for reason of dependability which means that one service needs to be established before another can be introduced. Thus those planning to implement components and communications links identified by the ITS Architecture, need consideration of any financial and dependability constraints.

A traceability matrix can be created to indicate the relationship between the stakeholder aspirations and the subsystems and modules in the physical viewpoint. This enables the ITS Architecture owners to quickly find out those components that are required to satisfy a given set of expectations, and thus meet their immediate goals. Such a matrix can also indicate whether certain aspirations can be met "for free", i.e., having identified the subsystems and/or modules needed to satisfy a given set of expectations, it may be found that it is possible to meet some other aspirations without the need for extra subsystems and/or modules.

3.9.4 Japanese ITS System Architecture

The Japanese ITS System Architecture was developed in 1999 through the joint efforts of five government ministries involved in ITS, in cooperation with VERTIS, now called ITS Japan. The objectives of the architecture are to encourage ITS applications as mentioned below:

(a) Efficient development of an integrated intelligent transportation system
(b) A maintainable and expandable intelligent transport system
(c) The development of domestic and international ITS standards

The development of the Japanese ITS architecture was guided by the following two principles:

(a) Assuring that the architecture would be able to be flexible to meet changing social needs and evolving technologies.
(b) Assuring that the architecture would lead to an ITS that was interoperable and inter connectible with other parts of Japan's advanced information and telecommunications environment.

Like any other major national system ITS architectures, the Japanese ITS architecture includes an enumeration of user services, a logical architecture, a physical architecture and areas for creating ITS standards as presented in Figure 3.8.

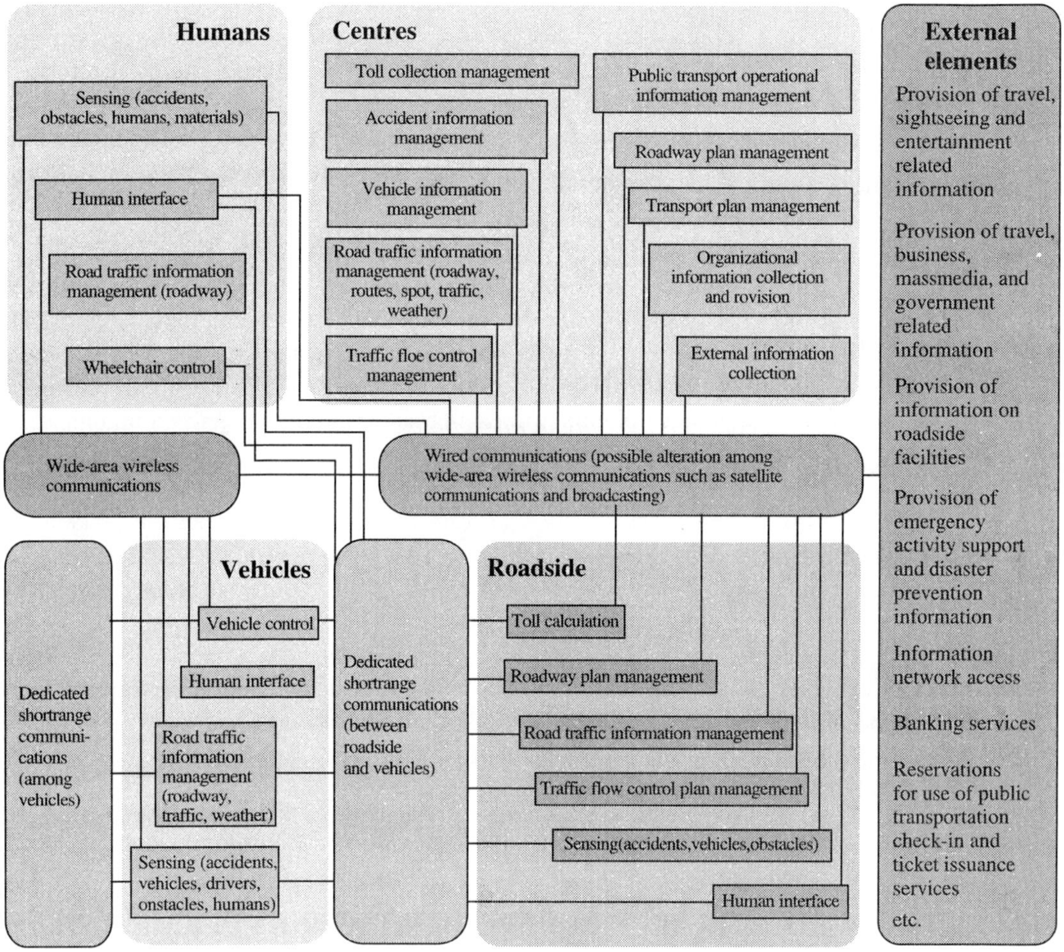

FIGURE 3.8 ITS Architecture, Japan (reproduced with permission).

Development Areas and User Services

(a) *Advances in navigation systems*

(i) Provision of destination-related information
(ii) Provision of route guidance traffic information

(b) *Electronic toll collection systems*

(i) Electronic toll collection

(c) *Assistance for safe driving*

(i) Provision of driving and road conditions information
(ii) Assistance for driving
(iii) Danger warning
(iv) Automated highway systems

(d) *Optimisation of traffic management*
   (i) Optimisation of traffic flow
   (ii) Provision of traffic restriction information in case of incident

(e) *Increasing efficiency in road management*
   (i) Improvement of maintenance operations
   (ii) Provision of roadway hazard information
   (iii) Management of specially permitted commercial vehicles

(f) *Support for public transport*
   (i) Provision for public transport information
   (ii) Assistance for public transport operations and operations management

(g) *Increasing efficiency in commercial vehicle operations*
   (i) Assistance for commercial vehicle operations management
   (ii) Automated platooning of commercial vehicles

(h) *Support for pedestrians*
   (i) Pedestrian route guidance
   (ii) Vehicle-pedestrian accident avoidance

(i) *Support for emergency vehicle operations*
   (i) Automated emergency notification
   (ii) Route guidance for emergency vehicles and support for relief activities

(j) *General*
   (i) Utilisation of information in the advanced information and telecommunications society

3.10 ADVANCED RURAL TRANSPORTATION SYSTEMS (ARTS), US

Generally, the ITS technologies are being developed for urban areas. The application of ITS to rural transportation is a recent development, rural areas encompass a significant portion of the transportation system. Rural area roads have a unique set of priorities and needs associated with the characteristics of the types of travel upon them and their maintenance and operations. The US Department of Transportation (DOT) for the first time envisaged ARTS to improve the transport system in the rural areas.

This strategic plan has been prepared by DOT for the ARTS portion of the ITS programme. The plan focuses on the Federal Government's role in developing rural ITS options and prudently managing emerging ITS technologies within rural settings from conception to viable options for implementation.

The ARTS Strategic Plan organises rural needs into seven critical programme areas (CPAs):

CPA 1: Traveller safety and security

CPA 2: Emergency services

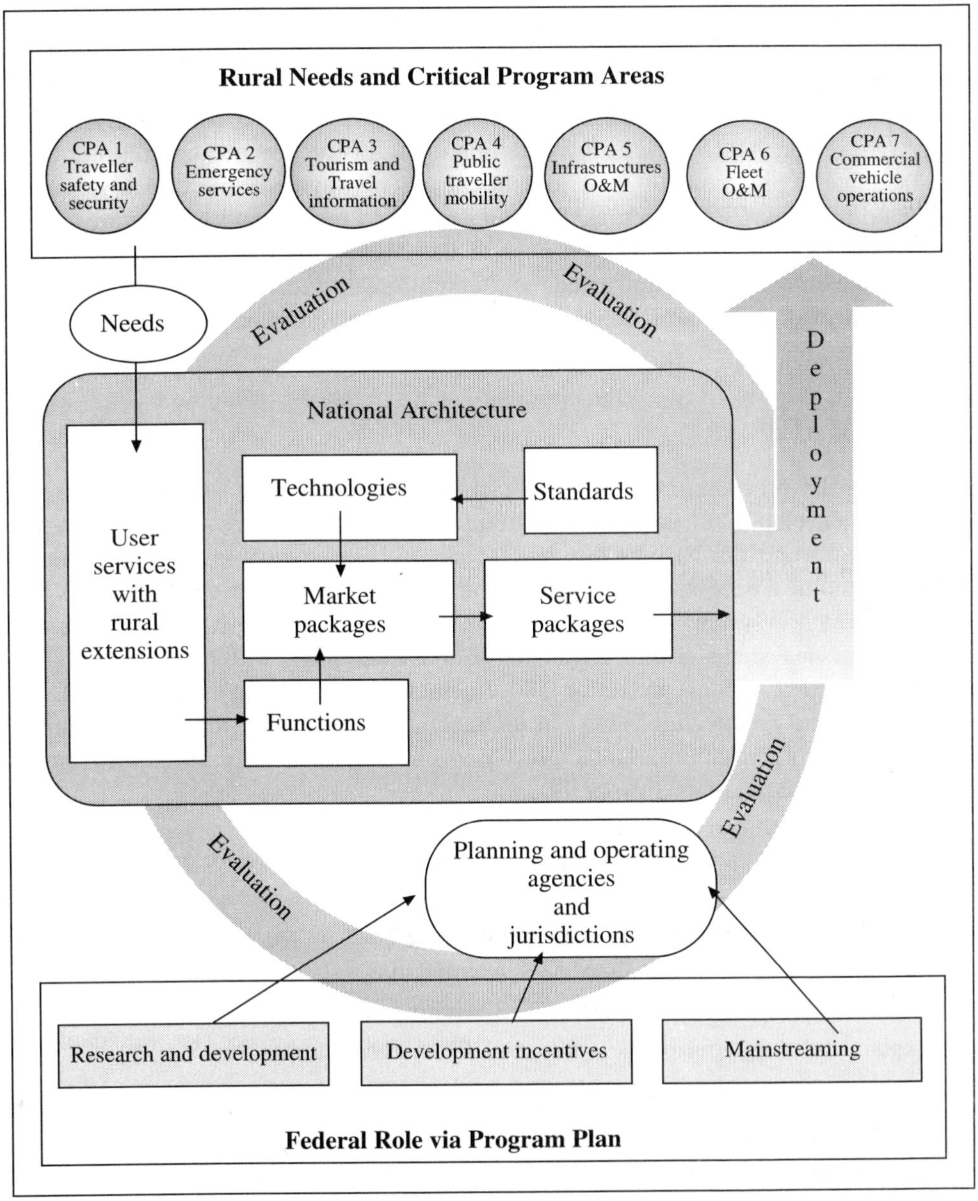

FIGURE 3.9 Advanced Rural Transportation Systems (ARTS), US.

CPA 3: Tourism and travel information services

CPA 4: Public traveller/mobility services

CPA 5: Infrastructure operations and maintenance

CPA 6: Fleet operations and maintenance

CPA 7: Commercial vehicle operations

The National ITS Architecture defines various concepts for the ARTS as a system. A layered structure consists of infrastructure that must be provided in common to support many applications that ultimately deliver services. The infrastructure will be developed largely in the public domain and will follow standards to be open to innovative and competitive applications from the private sector. ARTS activities will define the infrastructure needed to serve the CPAs, and then facilitate deployment of this infrastructure. This layered structure also emphasises issues common to many CPAs, such as communications in rugged and remote areas that must be addressed by ARTS activities. The purpose of the ARTS is not to develop a separate rural ITS, but to make sure that the rural needs and conditions are represented in what will be an interoperable, national and international system. The schematic plan of ARTS, US is shown in Figure 3.9.

3.11 SUMMARY

As discussed in the preceding section, one can appreciate the role and importance of ITS architecture. To develop a full system to regulate orderly traffic flow and safe movement of traffic on an area or corridor basis at city level, state level or country level, ITS architecture is increasingly becoming a necessity for a city committed to develop ITS programme. It is therefore important to develop a system architecture based on the linkage and stage of involvement of various functions/subsystems within a system. It is a framework within which ITS functions and traffic monitoring, incident detection and emergency support can be developed. In order to support the system architecture, physical architecture dealing with the physical subsystems and data flows and organisation architecture based on support and coordination from various organisations are to be interfaced properly.

Questions

3.1 What do you understand by ITS Architecture?

3.2 Explain primary components of ITS Architecture.

3.3 What are different components of ITS Architecture?

3.4 Explain the following in the context of ITS Architecture:

(a) Logical architecture

(b) Physical architecture

(c) Organisational architecture

3.5 What do you understand by equipment package and market package in the context of ITS architecture?

3.6 Explain the need of ITS Architecture to solve problems in urban area.

3.7 Write a short note on ITS Architecture of US.

3.8 Develop an ITS Architecture plan for your city.

Bibliography

Chen, K. and J. Miles, *ITS Handbook 2000: Recommendations from the World Road Association (PIARC)*, p. 434, Artech House Publishers, Norwood, MA, 1999.

Chowdhury, M. and A. Sadek, *Fundamentals of Intelligent Transportation Systems Planning*, p. 210, Artech House Publishers, Norwood, MA, 2003.

European Intelligent Transport Systems (ITS), Framework Architecture (http://frame-online.eu/).

Federal Highway Administration, Integrating Intelligent Transportation Systems, *Within the Transportation Planning Process: An Interim Handbook*, Report No. FHWA-SA-98-048, January 1998.

Institute of Transportation Engineers, *Transportation Planning Handbook*, 2nd ed., Washington D.C., 1999.

Intelligent Transport Systems (ITS), "Communications Architecture", ETSI EN 302 665 V1.1.1 (2009–2010), *European Standard* (Telecommunications Series), (http://www.etsi.org/deliver/etsi_en/302600_302699/302665/01.01.01_60/en_302665v010101p.pdf)

Intelligent Transportation System–I, Lecture notes in Traffic Engineering and Management, IIT Bombay, August 5, 2014.

Mathew, Tom V., "Intelligent Transportation System–I", Chapter 48, *Transportation Systems Engineering*, IIT Bombay, February 19, 2014.

Mikulski, Jerzy (Ed.), "Telematics in the Transport Environment", *12th International Conference on Transport Systems Telematics*, TST 2012, Katowice-Ustron, Poland, October 10–13, 2012.

Regional Intelligent Transportation Systems Architecture, Bloomington/Monroe County Metropolitan Planning Organisation, September 12, 2008. (https://bloomington.in.gov/media/media/application/pdf/3749.pdf)

Smadi, Mohammad, *Intelligent Transportation Systems (ITS): Architecture, Deployment, and Advances*, Advanced Traffic Analysis Centre, Upper Great Plains Transportation Institute Transportation Seminar Series, 2008. (http://www.ugpti.org/events/seminar/downloads/2008-05-07 MohammadSmadi.pdf)

The Diebold Institute for Public Policy Studies, *Transportation Infrastructures: The Development of Intelligent Transportation Systems*, p. 212, Praeger, Westport, CT, 1995.

The Frame Architecture and its Action Plan, Booklet of the E-FRAME Project, June 2011. (http://frame-online.eu/wp-content/uploads/2014/10/FRAME-ITS-Action-Plan.pdf)

Transport Association of Canada (http://tac-atc.ca/en/itsarchitecture).

US Department of Transportation, National ITS Architecture Implementation Strategy, September 1998.

Virginia Department of Transportation, Northern Virginia District Smart Travel Strategic Plan, 2000.

Williams, Bob, *Intelligent Transportation Systems Standards*, Artech House, London, 2008.

Yokota, Toshiyuki and Richard J. Weiland, "ITS System Architectures For Developing Countries", Technical Note 5, July 22, 2004. (http://siteresources.worldbank.org/EXTROADSHIGHWAYS/Resources/ITSNote5.pdf)

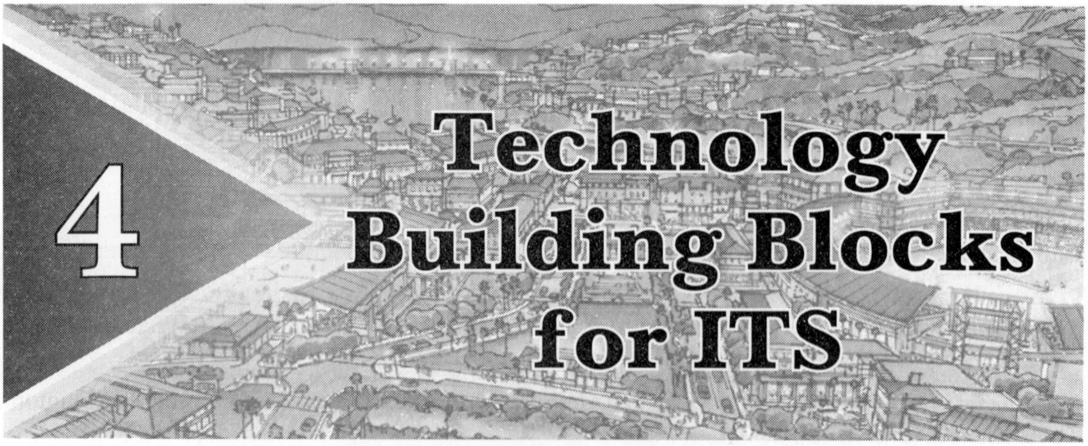

4 Technology Building Blocks for ITS

4.1 INTRODUCTION

The evolution and development of information technology (IT) form the basis of development of ITS solutions. The success of IT within a short span of time has been possible due to revolution in the data communication systems. The conversion of analog system to digital system proved to be a game changer which led very high speed, noise free communication system and development of computers. The fundamentals of information technology may be termed as building blocks of ITS. The building blocks include data, signals, waves and digital communication as shown in Figure 4.1.

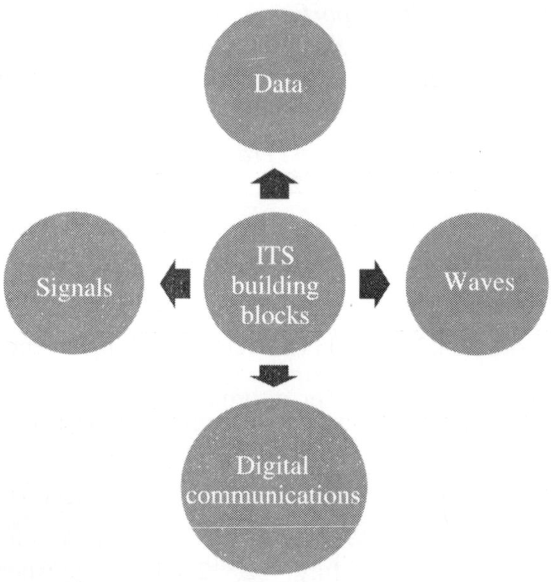

FIGURE 4.1 Fundamental building blocks of ITS.

The understanding of building blocks is essential for better appreciation of ITS technologies and their application. This chapter presents an overview of various building blocks of ITS. The building blocks of ITS are responsible from acquisition of data, analysis of data, application of various functionalities of ITS and finally communicating the information to the travellers.

4.1.1 Data Acquisition

The collection of exhaustive and accurate data also called *data acquisition* on real-time basis and rapid communication is critical for real-time monitoring and strategic planning. An effective data acquisition-management-communication system integrates tested hardware and efficient software to collect reliable data for further ITS activities. The different ITS hardwares/ equipment commonly used consist of sensors, cameras, automatic vehicle identifiers (AVI), GPS based automatic vehicle locators (AVL), and servers which can store huge amounts of data for meaningful interpretation.

4.1.2 Communication Tools

The efficiency and effectiveness of the ITS systems depend not only on the acquisition and analysis of traffic-related data, but also on fast and reliable communication, from field to traffic management centre (TMC) and information derived from data processing and models from TMC to the public. These activities involve communication between data collection centres to TMC and travel and traffic-related announcements to vehicles through on-board units and to the travellers through media like VMS, mobile applications, WhatsApp, Web pages, SMS, etc.

Dedicated short-range communication (DSRC) is mostly used to provide communications between the vehicle and the roadside at specific locations like toll plazas. DSRC is operated on radio frequencies in the industrial, scientific and medical (ISM) band. DSRC comprises road side units (RSUs) and the on-board units (OBUs) with transceivers and transponders. Wireless communications systems dedicated to ITS and road transport and traffic telematics provide network connectivity to the vehicles. Continuous air-interface long and medium (CALM) range may be used to provide continuous communications between a vehicle and the roadside units using a variety of communication media, including cellular, 5 GHz, 63 GHz and infrared links.

4.1.3 Data Analysis

The process of data analysis includes data cleaning, fusion, and analysis. The data from the collection devices like sensors, GPS, cameras, etc. that are transmitted to the TMC must be checked for consistency. Inconsistent data must be discarded and clean data has to be retained for further analysis and application. The data from different devices may be required to be combined or mixed for further analysis. The cleaned and mixed traffic data is analysed to estimate and forecast traffic states and for information to the public.

4.1.4 Traveller Information

Travel advisory system (TAS) facilities are useful for relaying transportation-related information to the vehicle users. TAS facilities include: variable message signs (VMS), highway advisory radio, Internet, short messaging service (SMS), automated cell phone messaging, mobile applications, public radio announcement, television broadcast and other modern media tools. Such systems can offer real-time information on travel times, travel speeds, delays, accidents, route closures and detours, and work zone conditions, among others. The real-time information is very useful for vehicle users for planning their journey. The concept of data flow for ITS applications from generation to the users is explained in Figure 4.2.

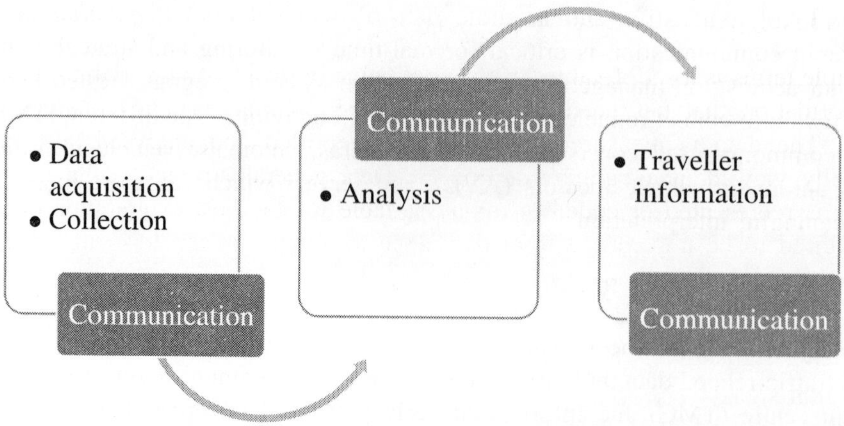

FIGURE 4.2 Concept of data flow from generation to the users.

The process of data collection to dissemination of information to the traveller deals with the four major groups of information and technologies namely, (i) Sensors and surveillance for data acquisition, (ii) Telecommunications for transfer of data, (iii) Data information processing for analysis, and (iv) Information display and delivery for travellers information. In order to perform the above tasks, there are a number of technologies associated as summarised in Table 4.1.

TABLE 4.1 Various sensor and surveillance technologies for data collection

Purpose	*Various types technologies*	*Specific technology*
Data acquisition	Sensor and surveillance	Inductive loops, passenger counters, piezo sensors, laser, radar, CCTV, laser, automatic vehicle location, smart cards, etc.
Communication	Telecommunication	Wireless application protocol, copper wire line, cellular wireless, fibre optics, DSRC, radio and TV
Data analysis	Data mining	Data warehousing, voice processing, speech recognition, Internet, customised software tools for processing transport information
Traveller information	Information dissemination and display	VMS, in vehicle information system, emergency, telephone booth, kiosks, display at bus stops, mobile

Sensor technologies are applied for use of data collection with respect to the raw data that demonstrate and characterise the current situations for the transportation network. Various equipment and products, and services are useful for providing data for multiple applications, such as traffic management, incident detection, traffic management, transit management and emergency management.

4.2 DATA

Data in simple terms is a set of values of the qualitative or quantitative variables. In computing, data is information that has been translated into a form that is more convenient to move or process. The data is measured, collected and reported and analysed, after which can be graphically viewed using graphs or images. Data generally refers to the fact that some information is represented or coded in a way suitable for the best exploitation or treatments. Data is required to be transmitted from one place to another in the form of electromagnetic signals through a transmission medium. There are following two types of electromagnetic signals:

(a) Analog signals
(b) Digital signals

An *analog signal* is a continuous wave represented by a sine wave [Figure 4.3(a)] and may vary in signal. Analog data represents information that is continuous in nature with respect to time and frequency. Analog signal has an infinite number of levels of intensity for a period of time. As the wave moves from place *A* to place *B*, it transmits a number of different magnitude of data along its trajectory. It may also said to be continuously varying that refers to a condition, physical phenomenon, or quantity such as flow, pressure, or temperature.

Examples of analog signal are: (i) analog clock (hours, minutes, seconds), (ii) human voice, etc.

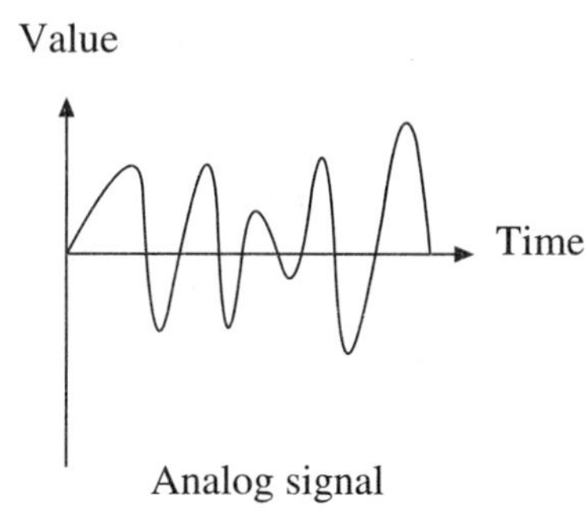

FIGURE 4.3(a) A typical analog signal.

A *digital signal* represents to an electrical signal transformed into a bit pattern. Unlike an analog signal, that carries continuous signal containing varying amounts of data over time, a digital signal provides a discrete value or data at each sampling point. The digital signal represents information that has discrete states. Digital signal has limited number of defined values. Each value can be 1 or 0.

Figure 4.3(b) demonstrates an example of a digital clock and computer memory (Data stored in binary code).

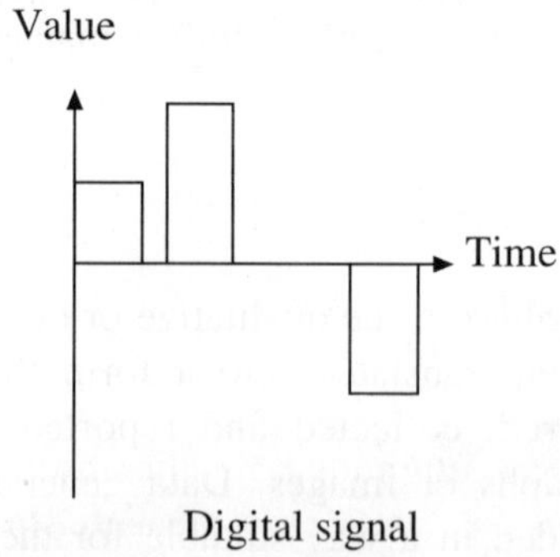

FIGURE 4.3(b) A typical digital signal.

Periodic and non-periodic signals

A periodic signal generally completes a pattern within a time frame or measurable period of time, and repeats that pattern during subsequent identical periods. It is commonly used in analog signals, because they need less bandwidth.

A non-periodic signal changes without displaying a pattern or cycle that repeats for a while. It is commonly used in digital signals, because they can represent the variation in the data.

Both analog and digital signals can be periodic or non-periodic in nature.

4.3 DIGITAL COMMUNICATIONS

Digital communication is the transfer of data (a digital bit stream or a digitised analog signal) over a point-to-point or point-to-multipoint communication channel. Examples of such channels are copper wires, optical fibres, wireless communication channels, storage media and computer buses. The data are represented as an electromagnetic signal, such as an electrical voltage, radio wave, microwave, or infrared signal.

In communications, sine wave frequency data to be transmitted is known as the *carrier wave*. The carrier wave can be in any suitable frequency to transmit data for communications systems (both wired and wireless). Therefore, a recording of the voice could be transmitted to any frequency within the electromagnetic spectrum. However, some frequencies are more efficient in performing other communications. The use of frequencies is monitored and licenced in India by Wireless Planning & Coordination (WPC) Wing, Ministry of Communications and Information Technology, Department of Telecommunications.

4.3.1 Sine Wave

A sine wave is the most fundamental form of a periodic analog signal shown in Figure 4.4. A sine wave can be represented by three parameters:

(a) Peak amplitude
(b) Frequency
(c) Phase

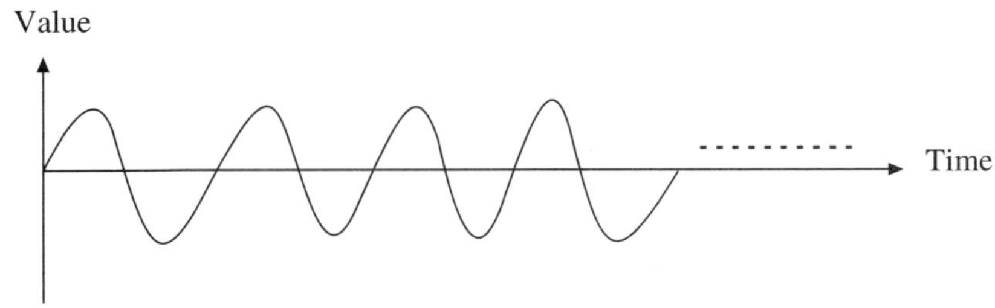

FIGURE 4.4 A typical sine wave.

Peak amplitude

The peak amplitude of a signal shown in Figure 4.5 is the absolute value of its highest intensity, proportional to the energy it carries. It is normally measured in metre or millimetre.

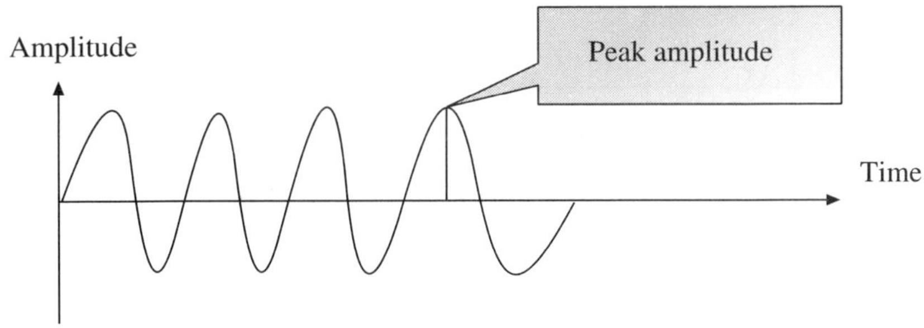

FIGURE 4.5 Peak amplitude.

Frequency

Frequency represents the number of period in 1 s as shown in Figure 4.6. It is formally expressed in hertz (Hz), which is cycle per second. Period represents the amount of time, in seconds, a signal needs to complete 1 cycle. Period (T) is the inverse of frequency (F) and vice-versa.

$$F = 1/T$$

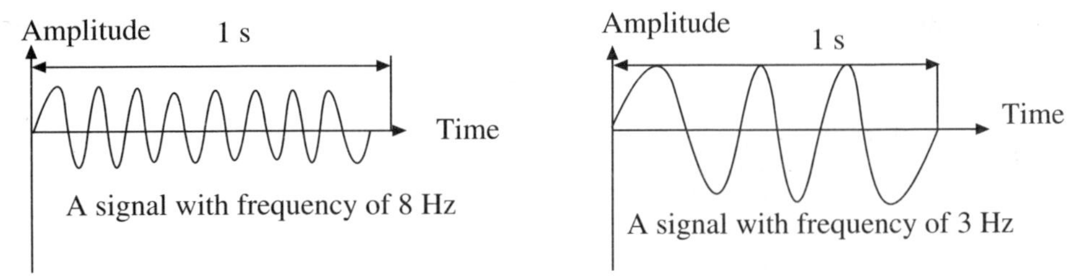

FIGURE 4.6 Frequency of sine wave.

Phase

Phase describes the position of the waveform relative to time 0 as shown in Figure 4.7. Wave that can be moved backward or forward along with time axis, phase describes the extent of that shift and is measured in degrees or radian. A phase shift of 360° corresponds to a shift of a complete period.

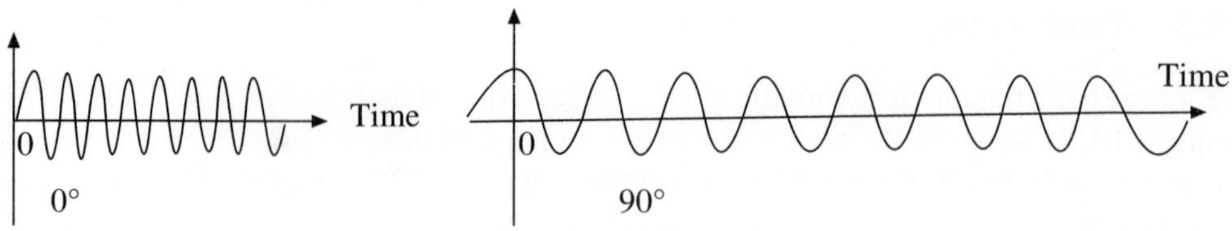

FIGURE 4.7 Phase of sine wave.

Wavelength

Wavelength as shown in Figure 4.8 takes into the period or the frequency of a simple sine wave to the propagation speed of the medium. It is the distance traversed by simple signal in one period.

Generally, *Wavelength = Propagation speed × Period*

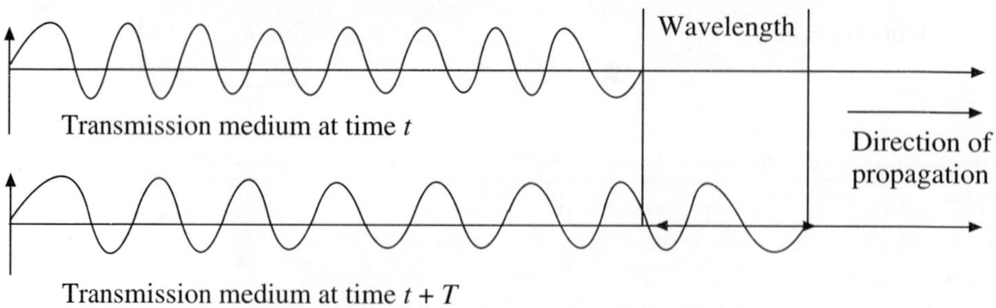

FIGURE 4.8 Wavelength of sine wave.

4.3.2 Composite Signal

Composite signal as shown in Figure 4.9 is a combination of simple sine waves with different frequencies, amplitudes and phases. It can be periodic or non-periodic. A periodic composite signal can be decompressed into a series of simple sine waves with discrete frequencies, which have integer values (1, 2, 3, and so on). A non-periodic composite signal can be decomposed into a combination of an infinite number of simple sine waves with continuous frequencies, which have real values.

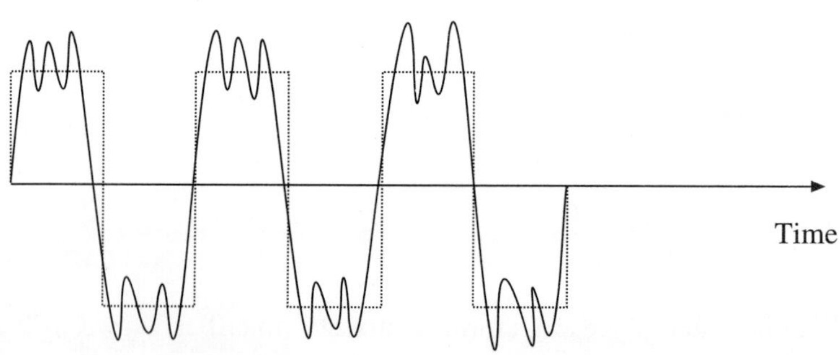

FIGURE 4.9 A composite periodic signal.

4.3.3 Bandwidth

The range of frequencies contained in a composite signal is its bandwidth as shown in Figure 4.10. It is normally a difference between two numbers. If a composite signal contains frequencies between 1000 and 5000, its bandwidth is 5000 – 1000 = 4000. The bandwidth of a composite signal is the difference between the highest and the lowest frequencies contained in that signal.

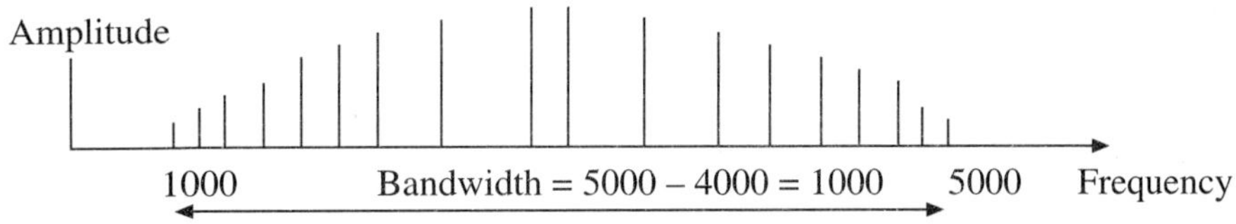

FIGURE 4.10 Typical bandwidth of a wave.

4.3.4 Electromagnetic Waves

Electromagnetic waves are form of energy wave with electric and magnetic field, which is different from mechanical waves, it can transmit energy and travel through a vacuum. Electromagnetic waves are identified according to their frequency varying from 10^4 Hz to 10^{20} Hz as shown in Figure 4.11. The different types of waves have different uses and functions in our everyday lives. This is visible light, which enables us to see. The human eye can only detect and sense only a small part of this spectrum known as *visible light*. A radio detects a different part of the spectrum, and an X-ray machine uses yet another part as shown in the figure.

An example of different types of electromagnetic wavelength is shown in Figure 4.11.

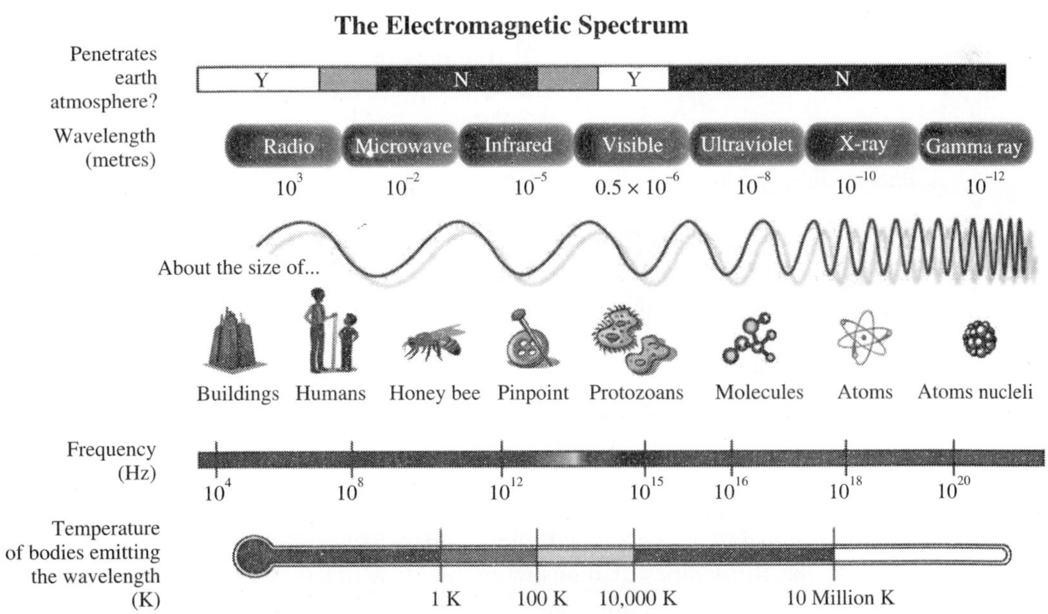

FIGURE 4.11 The electromagnetic spectrum (*Source:* https://mynasadata.larc.nasa.gov/science-practices/electromagnetic-diagram/ Free to use)

4.3.5 Bit Rate and Bit Length

The bit rate is the number of bits transmitted in 1 s, which is expressed in bit per second (bps). Bandwidth can also be referred to the number of bits per second that a channel, a link, or even a network can transmit. For example, one can say the bandwidth of a fast Ethernet network is a maximum of 100 Mbps. This means that this network can send 100 Mbps. The bit length is the distance one bit occupies on the transmission medium. It can be written as:

$$Bit\ length = Propagation\ speed \times Bit\ duration$$

Digital communications takes place through the transfer of binary data. Generally, in binary data is a representation of 1's and 0's. Its units are expressed as under:

Kilo (k) 2^{10} = 1,024 bytes
Mega (M) 2^{20} = 1,048,576 bytes
Giga (G) 2^{30} = 1,073,741,824 bytes
Tera (T) 2^{40} = 1,099,511,627,776 bytes
Kilo Hz 10^3 = 1,000 Hz
Mega Hz 10^6 = 1,000,000 Hz
Giga Hz 10^9 = 1,000,000,000 Hz

In transport sector (or any other field) one wish to transfer analog signal rather than digital. In that case these signals may be transferred as (Figure 4.12):

(a) An analog signal itself
(b) A digital representation of analog signal

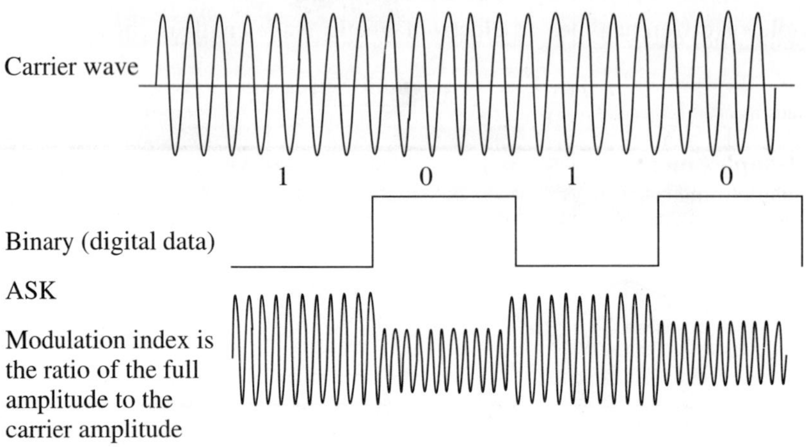

FIGURE 4.12 Conversion of analog to digital and amplitude modulation (*Source:* Technology building blocks for ITS, Phil Blythe).

Amplitude-shift keying (ASK) system can be classified into three blocks. The first block is concerned with the transmitter, the second one happens to be a linear model of the effects of the channel while the third block demonstrates the structure of the receiver as shown in Figure 4.13. The following notations are used:

(a) $h_t(f)$ is the carrier signal for the transmission

(b) $h_c(f)$ is the impulse response of the channel

(c) $n(t)$ is the noise introduced by the channel

(d) $h_r(f)$ is the filter at the receiver

(e) $s(t)$ is the signal from the transmitter

(f) $z(t)$ is the signal after filtering through $hr(f)$

(g) $z(k)$ is the signal after A/D (Analogue to digital) conversion

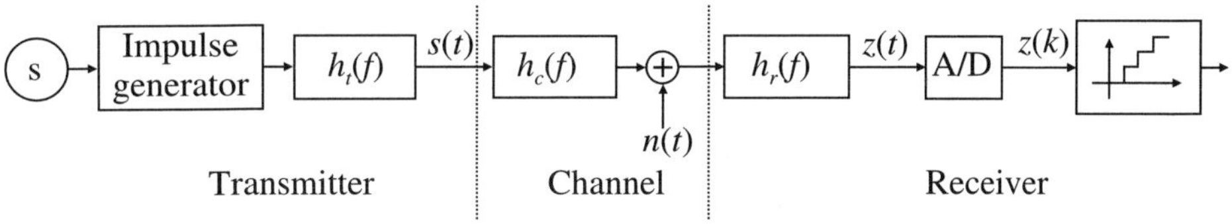

FIGURE 4.13 Amplitude-shift keying (ASK): amplitude modulation representing and transmitting digital data as variations in the amplitude of a carrier wave.

Figure 4.14 shows the transmission of signals from roadside to in-vehicle unit (IVU). Figure 4.15 shows the workability of smart card system.

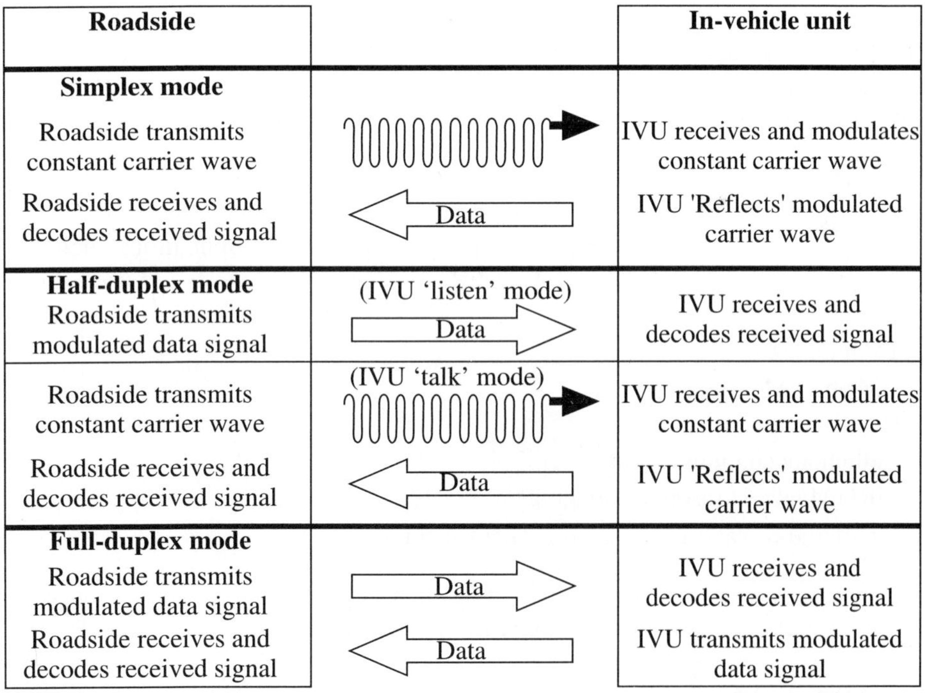

Source: Blythe, 96

FIGURE 4.14 Data transfer modes from roadside to in-vehicle unit (*Source:* Phil Blythe, reproduced with permission).

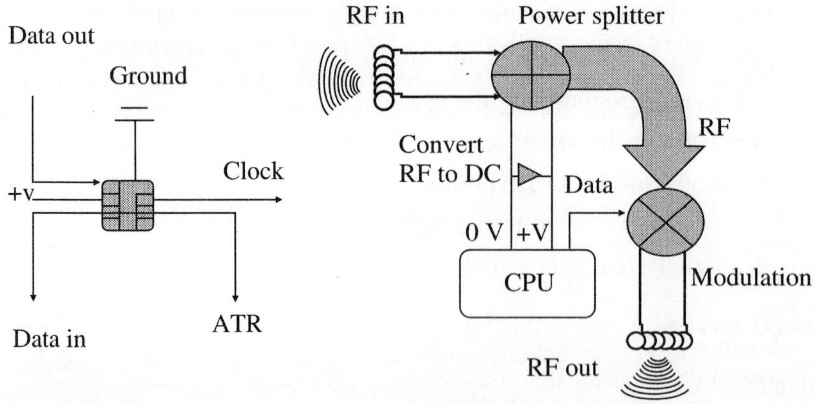

FIGURE 4.15 Flow chart of workability of a smart card (*Source:* Phil Blythe, reproduced with permission).

4.4 BUILDING BLOCKS OF COMMUNICATION TECHNOLOGIES

Communication technology offers various types of sensors ranging from point to area. The point type of sensor includes inductive loops, video image detector, piezo sensor, radar, laser and closed circuit television while area type of sensor is concerned with radio triangulation, GPS, wireless transreceiver and cellar telephone. The building blocks of communication technology addresses various types of electromagnetic waves for effective communication as a part of intelligent transport system. These are presented as follows:

4.4.1 Radio Waves

Radio waves experience the longest wavelengths of all the electromagnetic (EM) waves. They range from around one-third metre long to several kilometres long. Radio waves are often applied to transmit data and have been in use for all sorts of applications including radio, satellites, radar, and computer networks. Longest wavelength EM waves are used for following applications:

- TV broadcasting
- AM and FM broadcast radio
- Heart rate monitors
- Cell phone communication
- MRI (Magnetic resonace imaging)

An example of radio wave has been presented in Figure 4.16.

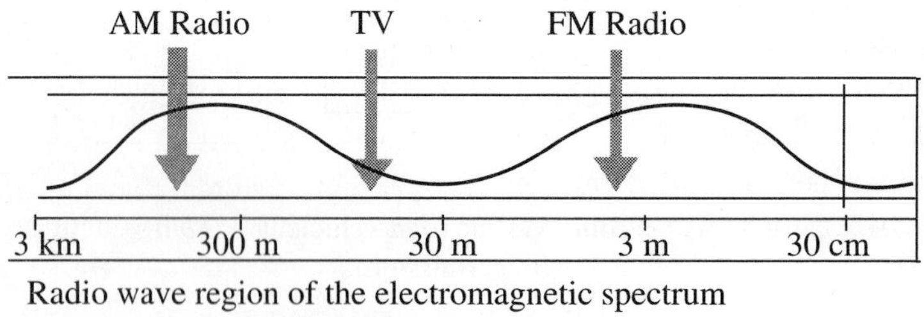

FIGURE 4.16 Radio wave.

There is an increasing use of radio waves in a number of applications in day-to-day life. One of the examples is the application of RFID where it is proposed to be used in the logistic park being developed as a part Delhi–Mumbai Freight Corridor. The logistic park will extensively use the RFID in the form of Reader, Antenna and Tag to track the movement of freight carriers. The container will tagged for transferring all necessary information with the reader while entering in to the logistic part and de-tagged after the delivering the commodity in the logistic part. The details of specifications of RFID used in Mumbal Logistics Park are available at Appendix B.

4.4.2 Microwaves

Microwaves happen to be shorter than radio waves with wavelengths measured in centimetres. The microwaves are used to cook food, transmit information, and in radar. It is useful for prediction of the weather. Microwaves are useful in communication as they can reach clouds, smoke, and light rain. The universe experiences cosmic microwave background radiation. The electromagnetic spectrum for microwave is presented in Figure 4.17. Wavelengths from 1 mm to 1 m is used for following microwave applications:

(a) Microwave ovens
(b) Bluetooth headsets
(c) Broadband wireless Internet
(d) Radar
(e) GPS

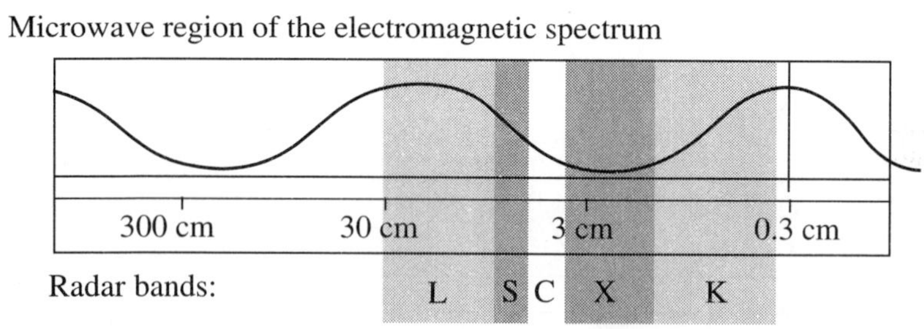

FIGURE 4.17 Electromagnetic spectrum for microwave.

4.4.3 Infrared

Infrared waves are found in between microwaves and visible light regions. Infrared waves are many a times classified as "near" infrared and "far" infrared. Near infrared waves are the waves which are near to visible light in wavelength. These are called infrared waves which are used in our TV, remote to change channels. Far infrared waves are supposed to be further away from visible light in wavelength. Far infrared waves are called thermal and radiate heat. Anything that radiate heat is called infrared waves. This includes the human body. It uses wavelengths in between microwaves and visible light. Its applications includes the following:

(a) Night vision goggles
(b) Remote controls
(c) Heat-seeking missiles

Electromagnetic spectrum for infrared wave has been presented in Figure 4.18.

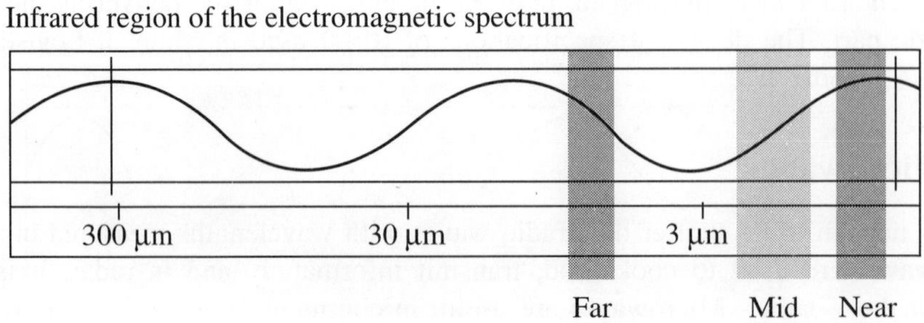

FIGURE 4.18 Infrared wave.

4.4.4 Visible Light

The visible light spectrum includes the wavelengths that can be seen by the human eye. This is the range of wavelengths from 390 to 700 nm which matches to the frequencies 430–790 THz. Only type of EM wave able to be detected by the human eye. It has Violet light which is the highest frequency light while Red light is the lowest frequency light. Figure 4.19 shows EM for visible light.

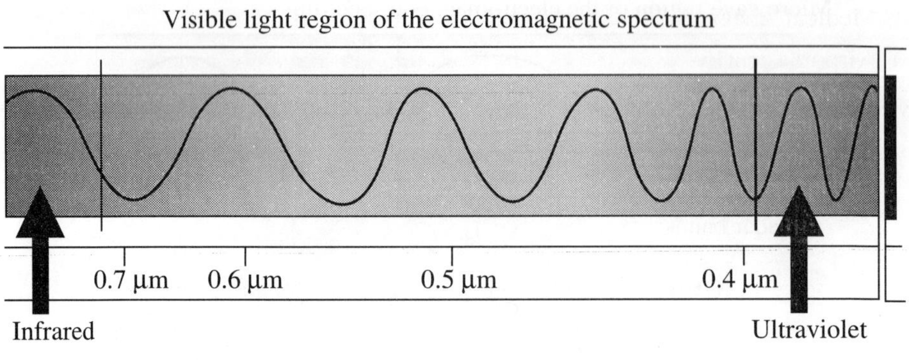

FIGURE 4.19 Visible light spectrum.

4.4.5 Ultraviolet

Ultraviolet waves experiences the next shortest wavelength after visible light. The ultraviolet rays from the Sun is responsible for sunburns. The Sun's ultraviolet rays are protected by the ozone layer. Some insects, such as bumblebees, can see ultraviolet light. Powerful telescope uses ultraviolet light like the Hubble Space Telescope to see far away stars. Its applications includes the following:

(a) Black lights
(b) Security images on money
(c) Harmful to living things
(d) Used to sterilise medical equipment
(e) Too much causes sunburn
(f) Extremely high exposure can cause skin cancer

Figure 4.20 shows electromagnetic spectrum for ultraviolet wave.

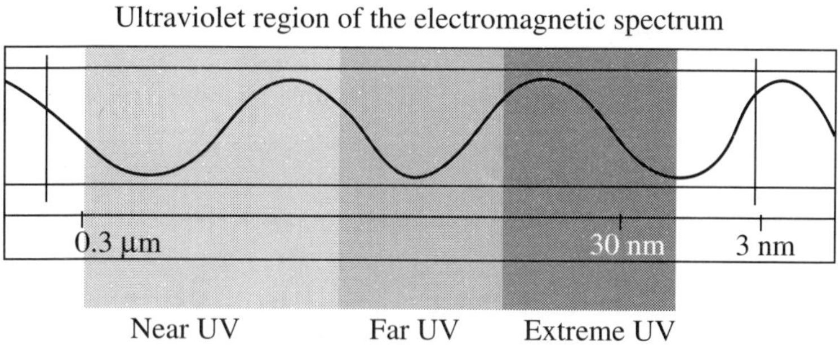

FIGURE 4.20 Electromagnetic spectrum for ultraviolet wave.

4.4.6 X-rays

X-rays has wavelengths shorter than that of ultraviolet rays. At this point in the electromagnetic spectrum, these rays appear to be more as particles than waves. They can filter on to soft tissue like skin and muscle and X-ray pictures of bones in medicine are generally taken at hospitals. It is a tiny wavelength and high energy wave. Its applications are mostly seen with the following:

(a) Medical imaging
(b) Airport security

It may damage cells with moderate dose. Figure 4.21 shows the EM spectrum for X-ray.

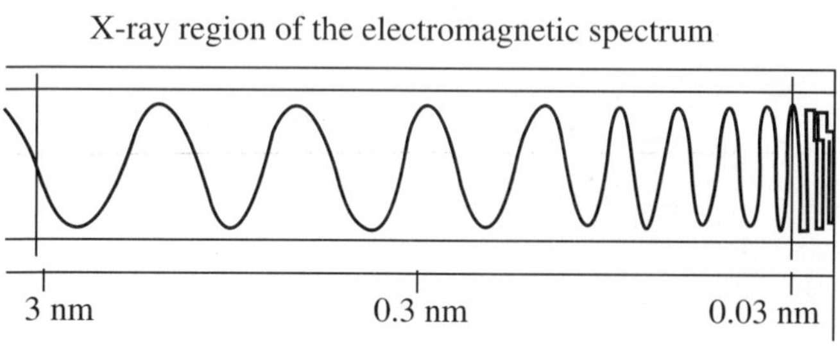

FIGURE 4.21 Electromagnetic spectrum for X-ray wave.

4.4.7 Gamma Rays

As the wavelengths of electromagnetic waves are one of the shortest, their energy gets intensified. Gamma rays are sometimes applied for treating cancer and takes detailed images for diagnostic medicine. Gamma rays are used in high energy nuclear explosions and supernovas.

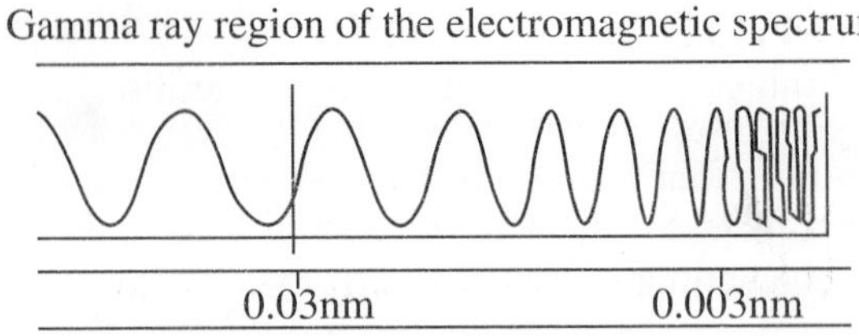

FIGURE 4.21(a) Electromagnetic spectrum for gamma ray wave.

Different popular CALM communication modes are developed as shown in Figure 4.22 and available in the market as follows:

(a) Infrared
(b) GSM (2G, 3G cellular telephone communications)
(c) DSRC 5.8-5.9 GHz
(d) Various evolutions of the IEEE 802.11 standard WiMAX, IEEE 802.16e
(e) MM-wave (63 GHz)
(f) Satellite (GNSS: GPS; GLONASS; GALILEO etc.)
(g) Bluetooth
(h) RFID

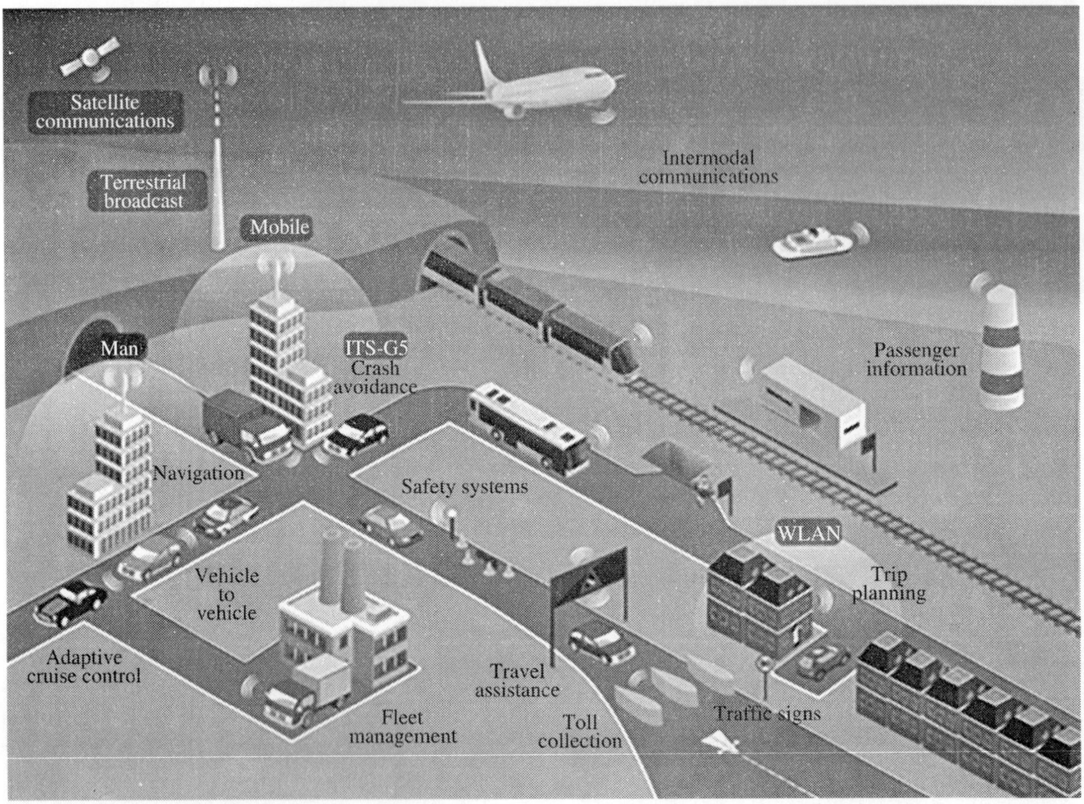

FIGURE 4.22 Application of communication technology in transport.

4.5 WIRELESS AD-HOC NETWORKS

Ad-hoc network represents to those networks which make use of random connections with other devices on cross-platforms. They are considered as new special class of communications networks. These generally help many devices to connect together to measure and sense the network and help it to fixed and mobile connections. One of the examples would be to use Bluetooth and ZigBee system as shown in Figures 4.23 and 4.24.

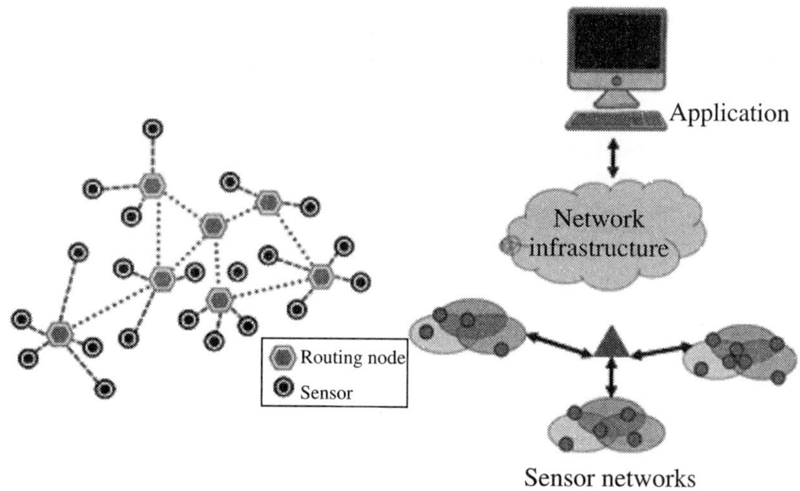

FIGURE 4.23 Wireless sensor networks (*Source*: Phil Blythe, reproduced with permission).

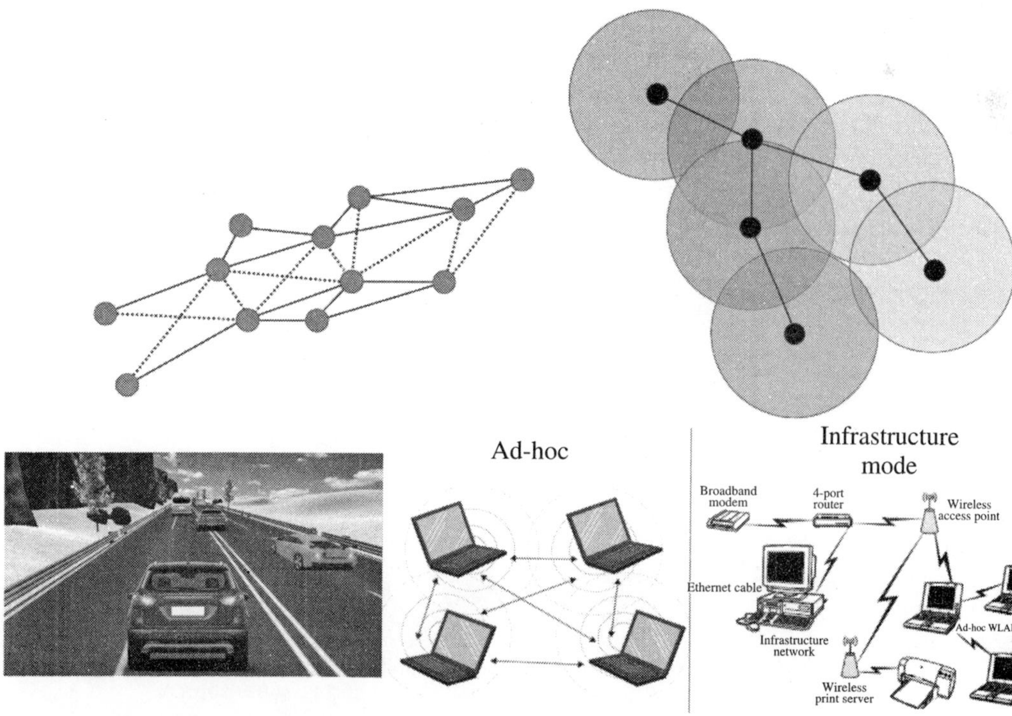

FIGURE 4.24 Wireless ad-hoc networks (*Source*: Phil Blythe, reproduced with permission).

4.5.1 Application of Ad-hoc Networks in Transportation Systems

Some aspects of mobile network technology are mentioned as follows:

(a) Potentially it helps the use of a part fixed/part mobile or a fully mobile network for mobile objects and infrastructure
(b) Potential to increase payloads to the devices
(c) A reduction in size and cost of these devices, help them to be operational pervasively in everyday objects

4.6 TELECOMMUNICATION TECHNOLOGIES

Telecommunication technologies are a mixture of media and protocols that are combined into communication techniques. The media is the physical channel used to carry the data from the origin to the destination. The protocol is the pre-defined, pre-agreed content, format, or structure of the data messages to be sent. Some of the mostly used wireless and wireline communication techniques are presented here:

4.6.1 Cellular Wireless

The mobile telephone communication is based on cellular wireless network. The technology involves wireless communications within a well-defined cell with a transceiver (mobile/cell tower) at the centre. All the communications between mobile tower and mobile phone within the cell take place using the same radio frequency. Cells are set up like a blanker across the total area to be covered by the cellular radio system as shown in Figure 4.25. A central communication and computer system links all the cell towers together and link them into the wider telephone network. This system monitors the location of each cell phone that is subscribed to the system and directs calls from the telephone network to the cell tower covering the current location of the phone. System also directs call from one cell phone to other cell phone by tracing the location of two mobile phones in the cells and transferring call from cell to cell.

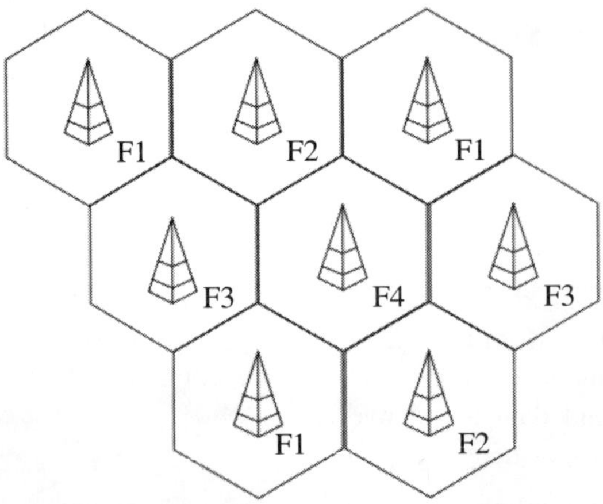

FIGURE 4.25 Cellular network for mobile telephony.

The cellular telephone technologies have been witnessing phenomenal advancement in terms of mobile technology. The first-generation analog mobile systems were introduced in early 1980 and today within a span of 35 years we have 4G advanced mobile technology on hand. A brief on evolution of mobile technology from 1G to 4G is presented in Table 4.2.

TABLE 4.2 Evolution of mobile telecommunication technologies

Type	*Period*	*Technology*	*Services*
1st Generation (1G)	Early 1980s to late 1980s	Analog signaling also called analog mobile phone systems (AMPS)	• Voice communication only • No SMS, No data
2nd Generation (2G)	Late 1980s to late 1990s	• Digital signal and the speeds up to 64 kbps • Bandwidth: 30–200 kHz	• Voice communication • SMS • Slow speed data transfer
3rd Generation (3G)	Since 2000	• Based on wide band wireless network • Peak data rates of at least 200 kbit/s • Bandwidth: 15–20 MHz	• Voice communication • SMS • High speed data transfer • Internet etc.
4th Generation (4G)	Since early 2010	• Internet protocol (IP) packet-switched networks • Peak speed of 1 GB/s • Use scaleable channel bandwidths of 700 MHz, 800 MHz, 1800 MHz, 2300 MHz and 2500 MHz	• In addition to that of the services of 3G, some additional features such as multi-media newspapers, also to watch TV pro-grammes • Very fast data transfer
5th Generation (5G)	Expected from 2019	• Based on milllimetre-wave wireless communications • Peak speed of 20 GB/s • Bandwidht: 3400 MHz to 3600 MHz	• The wide introduction of real-time services, based on sensor and video data collected from vehicle will be possible. • A host of new-age services such as machine-to-machine (M2M) comm unications, Internet of Things (IoT), connected smart cities, self-driving cars, remote control surgery to virtual reality will be possible.

GSM (Global System for Mobile Communication) and **CDMA** (Code Division Multiple Access) are two dominant technologies for mobile communication. These two technologies differ in the way calls and data travel over the mobile phone networks take place. CDMA and GSM are both *multiple access* technologies.

GSM: GSM stands for Global System for Mobile Communication. GSM networks use TDMA, which stands for Time Division Multiple Access. It's also called a "time division"

system which means that the calls take turns. Time Division Multiple Access (TDMA) allows several users to share the same frequency channel by dividing the signal into different time slots. The users transmit in rapid succession, one after the other, each using its own time slot. This allows multiple stations to share the same transmission medium (e.g. radio frequency channel) while using only a part of its channel capacity. The voice is transformed into digital data, which is given a channel and a time slot, so three calls on one channel look like this: 123123123123. On the other end, the receiver listens only to the assigned time slot and pieces the call back together. So TDMA works by assigning time slots to multiple conversation streams, alternating them in sequence and switching between each conversation in very short intervals. In order for the network to know which users are connected to the network, each phone uses a subscriber identification module card, or SIM card.

The principle of TDMA is explained in Figure 4.25(a). For example a given frequency band the digital data (voice) from a sender is assigned a time slot 1, from second sender time slot 2 and from third sender time slot 3. The data is transmitted through same frequency channel in successive time slots as 123123123…. At the receiver end, only assigned time is identified through SIM card and the receiver only listens the time slot assigned to him/her. GSM is the network provided by major companies in India like Airtel, Idea, Vodafone, Reliance, BSNL, Tata, Aircel etc.

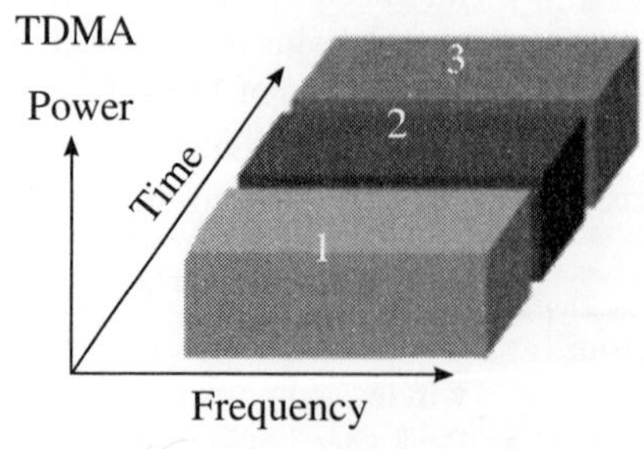

FIGURE 4.25(a) Principle of TDMA.

The GSM mobiles use **GPRS** to access internet services provided by the mobile network. General Packet Radio Service (GPRS) is a packet oriented mobile data service on the 2G and 3G cellular communication system's global system for mobile communications (GSM). GPRS enables people to enjoy advanced, feature-rich data services, such as e-mail on the move, multimedia messages, social networking and location-based services.

CDMA: CDMA stands for Code Division Multiple Access. It required a bit more processing power. It's a "code division" system. Unlike the GSM and TDMA technologies, CDMA transmits over the entire frequency range available. It does not assign a specific frequency or time slot to each user on the communications network. Because CDMA does not limit each user's frequency range, there is more bandwidth available. This allows more users to communicate on the same network at one time than if each user was allotted a specific frequency range. With CDMA technology, data and voice packets are separated using codes and then transmitted using a wide frequency range. Every call's data is encoded with a unique key, then the calls are all transmitted at once over same.

The principle of CDMA is explained in Figure 4.25(b). The digital data (voice) from the sender is assigned a unique code and the coded data is transmitted over same frequency range. The receivers each have the unique key to "divide" the combined signal into its individual calls. The receiver decodes the data with the unique key available with him/her to listen the signal (voice).

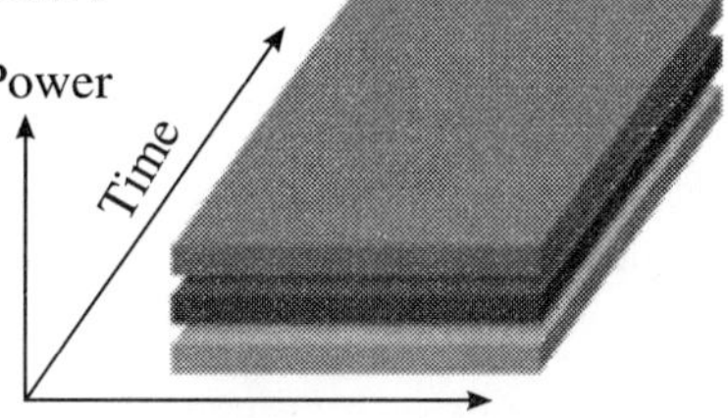

FIGURE 4.25(b) Principle of CDMA

Code division turned out to be a more powerful and flexible technology, so "3G GSM" is actually a CDMA technology, called WCDMA (wideband CDMA) or UMTS (Universal Mobile Telephone System). WCDMA requires wider channels than older CDMA systems, as the name implies, but it has more data capacity. Unlike GSM, CDMA grants users full access to the entire spectrum of bands, thus allowing more users to connect at any given time. It also encodes each user's individual conversation via a pseudo-randomized digital sequence, meaning the voice data remains protected and filtered so that only those participating in the phone call receive the data.

Phones on CDMA networks do not use SIM card. Instead, each phone is built specifically to work on that carrier's network. The mobile service providers like Reliance, Tata Indicom, MTS etc use CDMA technology.

4.6.2 Wireless Application Protocol (WAP)

To make use of cellular wireless technologies to provide mobile Internet access, the wireless application protocol (WAP) has been developed. It primarily uses the same basic principles and structure of HTML but is specially designed mobile handheld devices, such as cellular telephone, smart phones and Personal Digital Assistants (PDAs). The Web servers which store and process data for transmission across the Internet can be linked with an additional WAP server that converts the current Web pages to the WAP format. The information and service available on the Internet can be accessed on handheld mobile devices using WAP.

4.6.3 Copper Wireline Communications

Copper wireline communications also called plain old telephone system (POTS) include the ubiquitous telephone system such as Digital Subscriber Line (DSL), Integrated Services Digital Network (ISDN) and Telephone Interface (TI). In this system, a single copper wire, or a combination of copper wires is used to transfer maximum amount of data from origination to destination. For ITS, the potential applications include the transmission of sensor data from the transportation network to a central processing facility and from the facility to static information delivery devices, such as Internet terminals, kiosks, dynamic message signs, and smart bus stops.

4.6.4 Fibre-Optic Communications

The light can be reflected along a long thin fibre or strand of glass for very long distances with little loss of light or signal, through the selection and integration of suitable materials with varying refractive properties. This property of the light is used to transmit Voice video, or data at a fast speed along the optical fibres without loss of quality. Multiple telephone calls, data and video streams can be combined and transmitted along a single fibre optic cable thinner than a human hair.

4.7 DATA AND INFORMATION PROCESSING TECHNOLOGIES

The next step after collection of the raw data is to process it to produce the desired information output during information-processing and data-fusion activities for ITS application. The data is stored and retrieves utilising advanced data storage and database management techniques. The section presents a brief on techniques used for data processing.

4.7.1 Data Warehousing

The ITS applications need to collect and assemble data from multiple sources and then prepare it for extracting coherent, reliable information. Data warehousing techniques have been designed and developed specifically to meet such needs in the most cost-effective and efficient manner, enabling the end users (system operators and managers) to carry out much of the work themselves. There are a number of proprietary data warehousing techniques and standard packages available in the market which support the conversion of large quantities of data into information.

4.7.2 Online Analytical Processing (OLAP)

OLAP is a recent advance in the structured storage and analysis of data in large databases. The system makes use of relational database structures in which data is allocated to multiple tables with relationships defined between each table. This new approach adds new capability for faster response times and faster calculation of the results. Instead of the data being held in a large number of tables, it is held in a multi-dimensional array that can be accessed very quickly. Data consolidation is also carried out for commonly requested analyses, ensuring that response times to queries can be reasonably consistent.

4.7.3 Data Mining and Visualisation

This system complements OLAP and enables the user to explore and identify trends in the database through data extraction, formatting, and graphical representation. These tools support intensive analysis of the data populating the data warehouse, using data search techniques and statistical algorithms to identify patterns, correlations, and trends in the data.

4.7.4 Voice Processing

These technologies include synthesis of the human voice, known as text-to-speech technologies, and human voice-recognition technologies. These technologies can support a range of information delivery options as a text-to-speech application can be used to deliver traveller information to the driver while driving the vehicle, avoiding the distraction that could be caused by map and display alternatives. The voice recognition technologies through interactive voice response systems (IVRS) enables the user to speak navigation commands instead of pressing buttons.

4.7.5 The Internet

The Internet is a global network of interconnected computer network that provides one of the most valuable information sources and most powerful information delivery mechanisms in existence today. There is a whole range of information and communication technologies that have been developed in support of the efficient and effective operation of Internet such as:

HTTP; WWW; Browsers; TCP/IP; Internet; Extranet; HTML; Cookie.

Detailed discussion on above technologies is beyond the scope of this book.

4.7.6 Internet Protocol (IP)

An Internet Protocol address is the virtual address used for identifying a piece of network hardware. The IP address allows a device to communicate with other devices over an IP-based network like the Internet.

Most IP addresses look like this: 111.118.176.186

Like a home or business address, which identifies specific physical location, devices on a network are differentiated from one another through IP addresses.

For sending a packet to someone, the exact address of the recipient needs to be mentioned on the packet. Similarly, exact address in terms of IP address is needed to send information packet from one place to other place. Every Website has a unique IP address. Once Website address is entered into the address bar of the browser, our computer uses DNS servers to look up a hostname to find its IP address. A *DNS server* is a computer server that contains a database of public IP addresses and their associated hostnames, and in most cases it serves to resolve or translate common names to IP addresses as requested.

For example, a Website address like www.spa.ac.in is entered into the browser, the request to load that page is sent to DNS servers that look up that hostname (spa.ac.in) to find its corresponding IP address (http://111.118.176.186). Without an IP address, it will not be possible to connect to the desired Website. So IP address is used to identify the desired network, communicate with other devices, and connect to a Website in virtual cyber space.

4.7.7 Transmission Control Protocol/Internet Protocol (TCP/IP)

TCP/IP is the basic communication language or protocol for communication on Internet or private network (Intranet) of the Internet. It can also be used as a communication protocol in a private network (either an Intranet or an Extranet). When we are set up with direct access to the Internet, our computer is provided with a copy of the TCP/IP program just as every other computer that we may send messages to or get information from also has a copy of TCP/IP.

There are two layers in TCP/IP program. Transmission control protocol, the higher layer, manages the assembling of a message or file into smaller packets that are transmitted over the Internet and received by a TCP layer that reassembles the packets into the original message. Internet protocol, the lower layer, handles the address part of each packet so that it gets to the right destination. Each gateway computer on the network checks this address to see where to forward the message. Even though some packets from the same message are routed differently than others, they will be reassembled at the destination.

The client/server model of communication is used by TCP/IP. The computer user (a client) is provided a service (such as sending a Web page) by another computer (a server) in the network. TCP/IP communication is primarily point-to-point, meaning each communication is from one point (or host computer) in the network to another point or host computer. The transmission of data over Internet through TCP/IP is shown in Figure 4.26.

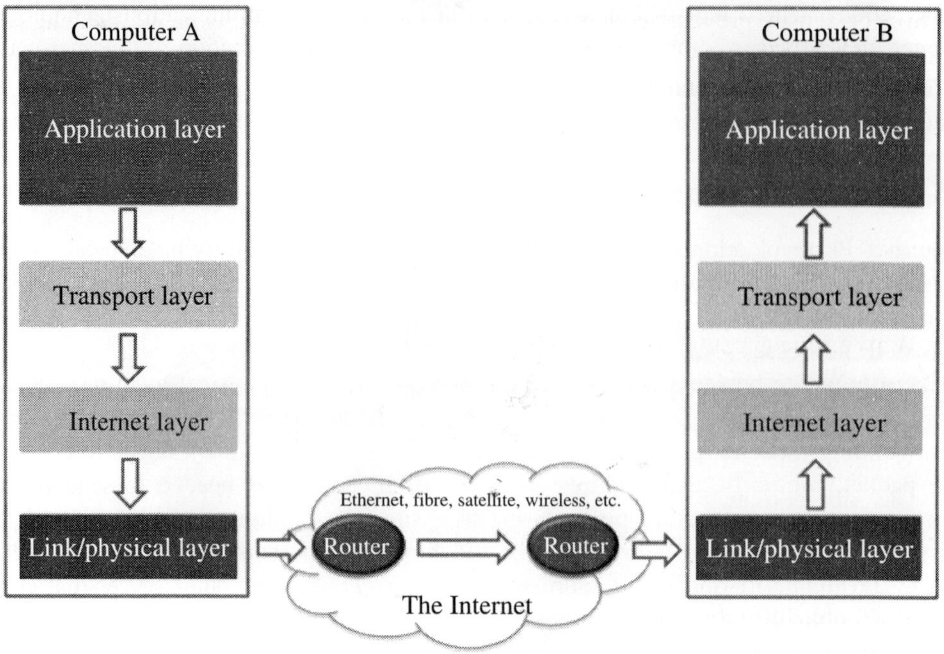

FIGURE 4.26 Data transmission over TCP/IP.

TCP/IP for ITS applications

Networked applications are critical to the success of ITS solutions. These ITS applications require the ability to handle many types of data and performance requirements. Internet is a cost-effective, high-bandwidth, intelligent technology being increasingly used in the most out of the ITS network infrastructure. Many new ITS projects are using video monitoring, which increases bandwidth demands from thousands to tens of millions of bits per second. Information is shared with many end users, including users on the Internet, generating the need for logical network partitioning and security. Transmission control protocol/Internet protocol (TCP/IP) represents an excellent end-to-end solution for these requirements.

4.8 SUMMARY

Any ITS application involves collection, communication, analysis of data to get the desired output. The building blocks of ITS include data, signals, waves and digital communication. The data is a set of information which can be measured, collected and analysed. Analog or digital signals are used to transmit data from one place to other place in the form of waves. Technological advancements in telecommunications technology along with the ultra-modern/ microchip, radio frequency identification (RFID) and detection technologies like intelligent traffic guidance have improved technical capabilities that provide intelligent transport systems safety benefits of users worldwide. Detection systems are networked systems infrastructure based on the vehicle, i.e., intelligent vehicle technologies. Infrastructure sensors are indestructible (such as reflectors on the way) devices which are installed or embedded in the road or around the road (e.g., buildings, poles and signals), as needed, that is maintained manually during road construction or sensor preventive injection equipment for rapid deployment. Vehicle systems with sensor include infrastructure deployment with respect to vehicle and vehicle for electronic beacon infrastructure for identifying communications and can also be used reading registration video technologies or detection magnetic signature of the vehicle at desired intervals to increase alertness constant vehicles operating in critical areas.

Questions

4.1 What do you understand by building blocks for ITS?

4.2 Why efficient communication tools are important for efficacy of ITS?

4.3 Differentiate between analog and digital signals.

4.4 What do you understand by digital communication?

4.5 What are the parameters which define a sine wave?

4.6 Describe composite signal.

4.7 Describe various types of electromagnetic waves used for effective communication.

4.8 Differentiate between bit rate and bit length.

4.9 What are different transmission modes of signals?

4.10 Explain different types of telecommunication technologies being used in ITS applications.

4.11 What do you understand by data and information processing in the context of ITS?

Bibliography

Automotive Council UK, "Connected and cooperative transport: for a safe, efficient and sustainable transport system", Technology Roadmaps, September 2014. (https://www.transportinnovation.se/system/files_force/dokument/roadmap_for_connected_and_cooperative_transport.docx?download=1)

Beecham Research Ltd. and GSMA, "Benefits analysis of GSMA Embedded SIM specification on the Mobile Enabled M2M Industry", September 2014. (http://www.gsma.com/newsroom/wp-content/uploads/BRL-Benefits-Analysis-GSMA-Embedded-SIM-Specification-2014.pdf)

Blythe, Phil, "Technology building blocks for ITS", Class Notes for M.Sc. in Transport Engineering, Newcastle University, 2013.

Cheung, S., S. Coleri, B. Dundar, S. Ganesh, C. Tan, and P. Varaiya, "Traffic measurement and vehicle classification with a single magnetic sensor", Paper ucb-its-pwp-2004-7, California Partners for Advanced Transit and Highways (PATH), 2004.

Forouzan, Behrouz A., *Data Communications and Networking*, 4th ed., Tata McGraw-Hill, US, 2007.

Hatem, B.A. and H. Habib, "Bus Management System Using RFID in WSN", *in the proceedings of European and Mediterranean Conference on Information Systems (EMCIS)*, Abu Dhabi, UAE, April 12–13, 2009.

Kastrinaki, V., M. Zervakis, and K. Kalaitzakis, "A survey of video processing techniques for traffic applications", *Image and Vision Computing*, 2003.

Kim No., "MSc Data Communications", in *Telecommunication Engineering*, 2015.

McKinsey, "Disruptive Technologies: Advances that will Transform life, Business, and the Global Economy", May 2013.

McQueen, Bob, Rick Schuman, and Kan Chen, *Advanced Traveller Information Systems*, Artech House, Boston MA, 2002.

Presentation on Data Communication, 2011. (https://edurev.in/studytube/DATA-COMMUNICATION/9aa34e85-72bd-4adb-b3f4-b8bfe0af1a38_p)

Sedra, Adel S. and Kenneth C. Smith, *Microelectronic Circuits*, 5th ed., Oxford, US, 2012.

Walport, M., "The Internet of Things: Making the Most of the Second Digital Revolution", A report by the UK Government Chief Scientific Adviser, 2014. (https://www.gov.uk/government/uploads/system/uploads/attachment_data/file/389315/14-1230-internet-of-things-review.pdf)

5 Various Detection, Identification and Collection Methods for ITS

5.1 INTRODUCTION

Technology is rapidly altering how transportation officials and transit service providers improve operational efficiency and safety of traffic in real-time. New and innovative technologies are increasingly being integrated into both personal vehicles and devices, improving the overall traveller experience. The success of ITS largely depends upon the technologies deployed in the application of ITS. The collection of data from field requires recording of events through detection or sensor technologies/components.

Some of the most common technologies used in various ITS applications have been discussed in this chapter.

5.2 DETECTION AND SENSING TECHNOLOGIES

Detection method or sensors are the techniques/components to detect the presence of a vehicle or pedestrians moving on the road and then to take necessary action accordingly. There are various types of detection and sensing systems for vehicles and pedestrians.

Sensing systems for ITS are vehicle and infrastructure based networked system, e.g., intelligent vehicle technologies. *Pavement loops* are used to detect the presence of vehicle at road intersections and at parking lot entrances. The presence of pedestrians waiting to cross a roadway is detected by *pressure pads*.

The presence of vehicle on road can be detected by radar and acoustic sensors. The radar transmits the radar pulses. A portion of the energy is reflected or scattered from the vehicle and roadway back toward the sensor and this energy is received and interpreted. The major benefits of radar technology are less power requirement, most accurate technology for detecting speed, traffic count accuracy and easy installation. The following section presents an overview of different sensors, detectors, radars, RFID sensors, etc. which are being used to detect/sense the presence of vehicle or pedestrians and transmit data to ITS applications.

5.2.1 Roadway Sensors

Roadway sensors are provided along with the road to detect the presence of vehicles. The road-way sensors can be divided in three categories: embedded or intrusive (i.e. embedded in the pavement), non-intrusive (i.e., installed off the pavement), and environmental.

Embedded (intrusive) detectors

There are following two types of embedded detectors:

(a) ***Inductive loop:*** An inductive loop detector is essentially a copper wire embedded in the road surface in the shape of a loop (e.g., square, rectangular, diamond, etc.). When an electrical current is passed the through the copper wire, a magnetic field is induced in the vicinity of the loop. Typically, an inductive loop consists of three to five "turns" of copper wire beneath the road surface. The loop is connected to a roadside cabinet through a lead-in cable which allows electronic equipment to monitor the magnetic field and changes in inductance as a vehicle passes over the loop as shown in Figure 5.1.

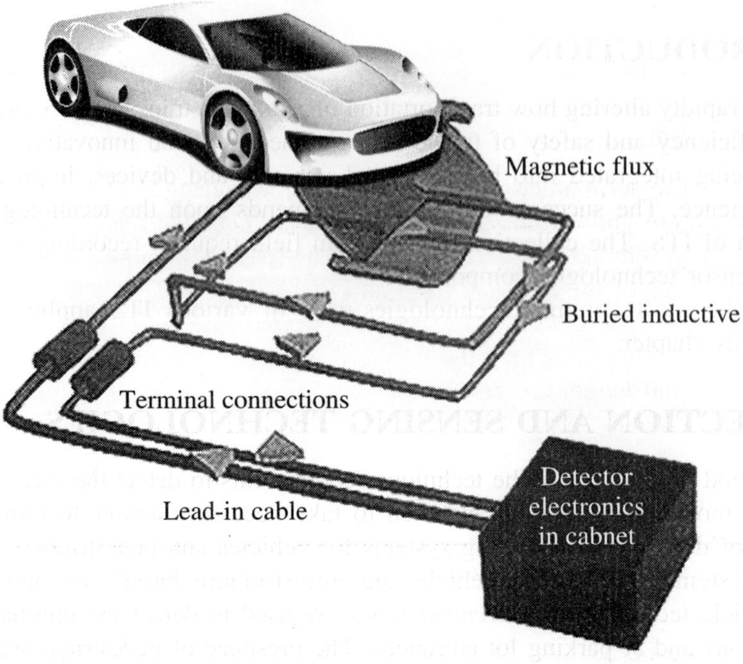

FIGURE 5.1 Inductive loop.

(b) ***Magnetometer:*** The magnetic detectors are small pencil like cylindrical probes which are placed vertically in or beneath the road surface. They measure changes in the earth's magnetic field as a vehicle passes over them to detect the presence of the vehicle.

Non-intrusive detectors

The following technologies are used for non-intrusive detectors:

(a) ***Radar:*** Radar emits radio wave signals and can detect vehicular presence and speed, depending upon the characteristics of the signal returned to them by the moving vehicles. Generally, there are two types of radar detectors: (i) *Doppler*, which measure the change of frequency between the transmitted and received signals, and (ii) *Time of Flight*, which measure the difference in time between when the signal is transmitted and received.

(b) ***Infrared (IR):*** Infrared is an electromagnetic energy at a wavelength or wavelengths which is somewhat longer than that of red light. There are two types of infrared detectors: active and passive.

The *active infrared detectors* send out invisible infrared radiation pointed at the road surface. By comparing the radiation reflected from a passing vehicle with the radiation from the road surface, the sensor can detect the presence of a vehicle.

The *passive infrared detector* measures IR light radiating from all objects with temperature above absolute zero which means that an IR device does not radiate any energy for detection purposes, instead IR light is being emitted from vehicles plying on the road and detected by infrared sensors.

IR detectors can detect presence of vehicles, speed and density of vehicles. It is easy to instal infrared sensors for detection along with good resistance even in poor weather conditions such as rain, fog, dust and snow.

(c) ***Microwave:*** The term microwave refers to the wavelength of the transmitted energy, usually between 1 cm and 30 cm. It is also called as *Doppler sensor* or radder or RF that helps detect walking, running, human targets, etc. The roadside-mounted microwave radar emits energy toward an area of the roadway from an overhead antenna. When a vehicle crosses the antenna beam, a part of the transmitted energy is reflected back towards the antenna. The return energy then enters a receiver where the detection is made and vehicle data, such as volume, speed, occupancy, and length are calculated. The microwave sensors can be effectively used for recording traffic data.

(d) ***Ultrasonic:*** Ultrasonic detectors emit continuous (Doppler) or short interval (pulsed) signals directed to the road surface where the detection of vehicle is required. The Doppler detector measures the shifts in frequency of signals reflected by passing vehicles to detect and determine their speeds. The Pulsed detector measures the time taken by reflected signals to return to the sensor.

(e) ***Acoustic:*** Acoustic detectors make use of microphones and signal processing technology to listen for sounds made by passing vehicles to determine the presence of a vehicle.

(f) ***Video image processing (VIP):*** VIP makes use of the images provided by video cameras installed near the roadway. "Machine Vision" algorithms are integrated with computerised pattern recognition software to detect passing vehicles.

(g) ***Closed Circuit Television (CCTV) Cameras:*** CCTV cameras are very useful in continuous monitoring of traffic flow conditions and verify traffic congestion and incidents. The CCTV cameras as shown in Figure 5.2 provide video images to assist the area traffic control (ATC) in determining and verifying causes of congestion, such as collisions, vehicle breakdown, load spills, construction/maintenance activities, police/fire operations, or just heavier than normal traffic.

Video vehicle detection (VVD) is one of the most widely used methods of detection used as ITS tool. Video detection is essentially an image processor which consists of a microprocessor based CPU and software that analyses video images. Using a mouse and interactive graphics, the user places virtual "detectors" on the video image displayed on a monitor. Information/ data can be progressively transmitted to a server for real-time analysis. The advanced video analytic software can be used to detect any unusual pattern, violations, incident or accident.

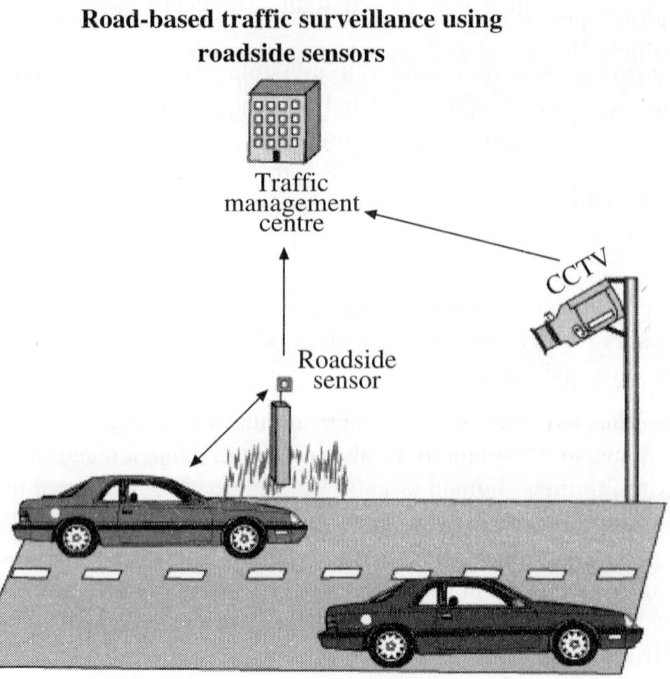

FIGURE 5.2 Video vehicle detection (*Source:* https://www.fhwa.dot.gov, reproduced with permission).

5.2.2 Environmental Sensors

(a) ***Atmospheric sensors:*** These sensors measure weather-related data, including wind speed and direction, air temperature, humidity, precipitation occurrence/accumulation and visibility. The video images can also be utilised in addition to atmospheric sensors to confirm weather conditions like rains, visibility, etc.

(b) ***Road surface sensors:*** These sensors measure the temperature as well as ascertain a variety of information with respect to the road surface such as wet or dry, presence of chemicals (e.g., anti-icing, de-icing, etc.), presence of snow, ice or frost, etc.

(c) ***Sub-surface sensors:*** These sensors measure the temperature at different depths in the substrate directly below the roadway.

5.2.3 Probe-Based Sensors

Probe-based sensors are on-board probes which are fitted on the vehicles. These on board probes monitor and measure the movement of a percentage of the vehicles in the traffic stream in order to assemble real-time traffic information. The probe-based monitoring is an effective method of data collection for wide area monitoring where large geographic areas are involved. The probe-based sensors typically measures the link speeds, link travel times, and origin and destination of vehicles. The vehicles which are equipped with on-board probes, allow the time and location of the vehicle to be tracked and monitored using automatic vehicle identification (AVI) or automatic vehicle location (AVL) devices. Examples of commonly used probe based sensors include global positioning systems (GPS) for AVL such as those used for monitoring of commercial vehicle fleets and E-Z pass transponders for AVI purposes. Cellular telephones can also be used for this purpose where the handset is located using either GPS or through information collected via the cellular transmission towers.

5.3 BLUETOOTH

Bluetooth is an accurate and inexpensive way to measure travel time and analyse origin and destination pattern of the traffic. Bluetooth with a short wavelength used wireless system to communicate between electronic devices like mobile smart phones, headsets, navigation systems, computers, etc. The bluetooth based road sensors are able to detect bluetooth MAC addresses from bluetooth devices of the passing vehicles that stretch of the road. If these sensors are interconnected, they are able to compute travel time and provide data for origin and destination matrices. Compared to other traffic measurement technologies, bluetooth measurement has some differences:

(a) It offers accurate measurement points with absolute confirmation to provide to the second travel times.

(b) It is non-intrusive, which can lead to lower-cost installations for both permanent and temporary sites.

(c) It is limited to how many bluetooth devices are broadcasting in a vehicle so counting and other applications are limited.

(d) The bluetooth based systems are generally quick to set up with little or no calibration needed.

5.4 RFID

Radio frequency identification (RFID) is a kind of wireless communication which uses radio waves to identify and track the objects. An RFID system has *readers* and *tags* that communicate with each through radio waves. The information is stored in the tags in eloctronic form. RFID tags are too small to need a battery to store information and exchange data with readers. This property of RFID tags makes it easy and cheap to apply tags to all kinds of things that people would like to identify or track.

The RFID system of detection is now widely used in industries and transportation and communication purposes. For example, RFID placed in the road infrastructure can identify and communicate with the vehicles plying on the road through the following RFID system.

The RFID system can be defined with the frequency band for operation dealing with low frequency, high frequency, and ultra-high frequency. Frequency is concerned with the size of the radio waves to communicate between RFID system components. RFID systems can work in low frequency (LF), high frequency (HF) and ultra-high frequency (UHF) bands. Radio waves differ in each of these frequencies with merits and demerits associated with using each frequency band. An RFID system operating at a lower frequency has a shorter read range and slower data read rate, but increased capabilities for reading near or on metal or liquid surfaces. An RFID system operating at higher frequency, it generally has faster data transfer rates and longer read ranges than lower frequency systems, but more sensitivity to radio wave interference caused by liquids and metals in the environment. The three types of frequency bands used for RFID system are explained in following section:

5.4.1 LF RFID

The frequency of LF band ranges from 30 kHz to 300 kHz. Typically LF RFID system works at 125 kHz. There are few frequencies that operate at 134 kHz. This frequency band covers a short read range of 10 cm only. However, it has ability to slower read speed than the higher frequencies with very low sensitive to radio wave interference. The applications of LF RFID include access control and livestock tracking. International standards for LF animal-tracking systems can be referred in ISO 14223, and ISO/IEC 18000-2. The LF spectrum is not accepted a truly universal application due to some differences in frequency and power levels.

5.4.2 HF RFID

The frequency of HF band ranges from 3 to 30 MHz. Most of the HF RFID system works at 13.56 MHz with reading distance ranges between 10 cm and 1 m. HF system offers moderate sensitivity to interference. HF RFID is commonly used for ITS applications like ticketing, payment, and data transfer.

A number of HF RFID international standards are in place, such as the ISO 15693 standard for tracking items, and can be referred from the ECMA-340 and ISO/IEC 18092 standards for near field communication (NFC)—a short range technology that is normally considered for data exchange between devices/components. Other HF international standards refer to the ISO/IEC 14443 A and ISO/IEC 14443 standards for MIFARE technology, which are related to in smart cards used in electronic money cards.

5.4.3 UHF RFID

The frequency of UHF band limits ranges from 300 MHz to 3 GHz. The systems meeting the standards of the UHF Generation 2 standard for RFID typically use frequency band of 860 to 960 MHz.

The read range of UHF RFID system is around 12 m in passive UHF systems, with faster data transfer rate than LF or HF. However, it experiences the most sensitivity to interference.

A number of UHF product manufacturers have developed designing tags, antennas, and readers to ensure desirable performance high even in difficult situations. The passive UHF tags are simpler and less costly to manufacture as compared LF and HF tags.

UHF RFID are found in the market with variety of applications, varying from retail inventory management, to pharmaceutical anti-counterfeiting, to wireless device configuration. The mass product of new RFID projects are preferring UHF over LF or HF due to higher range of UHF. The specification of UHF frequency band is guided by a single global standard called the ECP global Gen2 (ISO 18000-6C) UHF standard.

5.4.4 Passive, Active, and BAP RFID Systems

The RFID systems can further be classified as active, passive and battery assisted systems on the basis of source of power being used to transmit data. A brief of their concepts and applications are presented as follows:

Active RFID systems

The active RFID systems use battery-powered RFID tags which continuously transmit their own signal. Active tags send their own signal to transmit the information stored on their microchips. It is operated in the ultra-high frequency (UHF) band and covers a long range which may be up to 100 m. In general, active tag is fixed on large objects, such as rail coaches, wagons, containers, other objects that require to be monitored over long distances.

There are two main types of active tags: (a) transponders and (b) beacons.

Transponder: It is "activated" when they get a radio signal from a reader along with power-on and automatically respond by sending a signal back. As transponder does not radiate radio waves until reader sends signal, it conserve battery life.

Beacon: It works in most real-time locating systems (RTLS), in order to monitor, the precise location of an object continuously. Unlike transponder, beacon emits signals at pre-timed intervals. Based on desirable level of locating accuracy, the beacons can be adjusted to emit signals every few seconds, or once a day. Reader antennas receive each beacon's signal that are near the perimeter of the area being monitored. It communicates the tag's ID information and position.

Passive RFID systems

The passive RFID systems use tags with no internal power source, instead they are powered by the electromagnetic energy transmitted from an RFID reader. The reader and reader antenna transmit radio signal to the tag. The RFID tag then receives and activates the transmitted signal to power-on, and sends energy back to the reader.

The passive RFID systems uses the low frequency (LF), high frequency (HF) or ultra-high frequency (UHF) radio bands. The ranges of the passive RFID system are controlled by the power of the tags, the radio signal which is reflected from the tag back to the reader. Typically it is less than 10 m. As passive tag does not need a power source or transmitter, it needs only a tag chip and antenna. It is cost effective, smaller, and easier to manufacture than active tags.

The passive RFID can be packed in many different ways, as per the specific RFID application requirements. These tags may be sandwiched between an adhesive layer and a paper label to create smart RFID labels. The passive tags may also be placed through layers in a variety of devices or packages to protect the tag resistant from extreme temperatures or harsh chemicals.

The passive RFID can be used and applied to track packages in the supply chain, to inventory assets in the retail industry, to ensure the authenticity of the products such as pharmaceuticals, and to place RFID capability in a variety of devices. It can effectively be used in warehouses and distribution centres, though it is shorter range, by setting up readers at choke points to monitor asset movement.

Battery-assisted passive (BAP) systems

It is a type of passive tag which includes a crucial active tag feature. It uses the energy from the RFID reader's signal to power on the tag's chip and backscatter to the reader, BAP tags use an integrated power source (usually a battery) to activate the chip, therefore all of the captured energy from the reader can be used for sending the signal back. Unlike transponders, it does not offer their own transmitters.

5.5 MOBILE REPORTS

Conventionally, the emergency telephones or call boxes installed along roadside have been used to report traffic problems to a traffic operations centre. With the increasing use of mobile phones and improvement in the cellular network in recent years, the call boxes have been replaced by mobile communication. Mobile reporting from patrol vehicles and private citizen calls through toll-free hotlines has become a valuable source of information for quick detection of incidents and collection of traffic and incident information. The mobile communication system has the advantage of low start-up cost and two-way communications between the caller and the response agency. Further, the mobile applications are being developed which can be used to report any incident or cause of concern and photos of the site can also be uploaded. However, the mobile communication system can cause a high volume of calls when a major incident occurs. Moreover, sometimes it is difficult to identify the location of an incident when calls are received from private callers.

5.6 REAL-TIME TRAFFIC MONITORING USING CELLULAR NETWORK AND GPS PROBE

The real-time measurement of traffic speed and travel times are the basis of advanced travel information systems (ATIS) that provide information to roadway users to help them in making various decisions, such as route choice, departure time choice, etc. The real-time information can be made available to user through various means like mobile applications, navigation tools, Internet, roadside variable message signage, etc. The historic data of traffic flows is very useful to plan and design various components of road infrastructure.

The static means such as inductive loop detectors, CCTV cameras, RFID tags are mostly being used to collect traffic flow data. Although, these fixed tools are accurate but are expensive to deploy, operate and maintain. Therefore, these tools are mostly installed at critical locations like intersections, diversions, etc. Alternatively, floating vehicle (test vehicle) equipped with GPS can be used to measure travel time of vehicles on a particular road section, however, this method does not offer real-time information.

Mobile phone equipped with GPS system and Internet access is now being used as a sensor to collect real-time information on traffic flows. With increasing penetration of GPS enabled vehicles and smart phone users, the GPS probe data can be used for real-time travel speed estimation. The data from mobile phones of vehicle users can be used to assess the traffic flow on the road where the vehicle is moving. The mobile phone generates data packets containing its position and speed at regular intervals and send it to a designated server. On the basis of information received from several vehicles (mobile phones), the traffic estimation software can generate real-time traffic flow. The protection of privacy of data of the users has been a major challenges in using mobile phone location and speed data for traffic flow estimation. To protect the identity of the user, the information is encrypted and decrypted across multiple layers of the software so that the movement of a user cannot be tracked by the system. The example of real-time traffic monitoring system using mobile phone data as developed by University of California, Berkley in association with Nokia, NAVTEQ is presented in Box 5.1.

Box 5.1 Real-time traffic monitoring system using mobile phone data a system by University of California, Berkley

When a phone takes a measurement, it creates a data packet containing its position, speed and anything else that might be of interest. This packet is encrypted using the public key of the data processing server, but instead of going straight to that server, it goes to a proxy server that strips the packet of any identifying information. Then the packet is passed on to a virtual trip line (VTL) server that processes it and sends it to the data aggregation servers.

Reading the contents of the packet requires a decryption key. The proxy does not have the private key needed to perform the decryption, so although it knows the identity of the phone, it does not know where the data comes from. The packets that arrive at the VTL server have no identifying information. There is not a single machine that can be compromised to provide position and speed information that can be attached to a particular phone.

Preventing paths from being reconstructed was trickier and required the use of virtual trip lines (VTLs), something Nokia developed for this purpose. Instead of constantly reporting location and speed, each phone checks its current location against a downloaded database of VTL positions, and measurements are only sent when the phone crosses a VTL location. This drastically reduces the amount of data collected from any one phone, lessening the likelihood that someone could reconstruct individuals' paths from the data.

This still leaves the possibility a sequence of measurements can be processed to build up a trajectory. Nokia created an algorithm for placing the virtual trip lines in order to minimise the probability that two measurements from consecutive VTLs could be linked to the same vehicle.

Matching up measurements means taking a reading from one VTL and correctly associating it with another reading taken at the next VTL down the road. The more measurements are from the next VTL that could match the first, the harder it is to determine which belong together. The algorithm uses the number of cars on the road and their speeds to determine the best spacing to maximise the number of cars that might match going through any given VTL pair. In addition, the server that decides where to put the VTLs is separated from the one that processed the incoming data, making it less likely anyone could manipulate VTL placement to make tracking a car easier.

Finally, another layer of protection comes from randomising measurements. Instead of transmitting when crossing every VTL, the phones perform a virtual coin flip to decide whether to transmit. This makes it much harder to reconstruct individual trajectories.

The final architecture is illustrated in Figure 5.3, showing the multilayered server architecture. These precautions are not foolproof, especially in an extreme case like a single car driving down an empty road at night, but they provide a pretty stiff layer of protection.

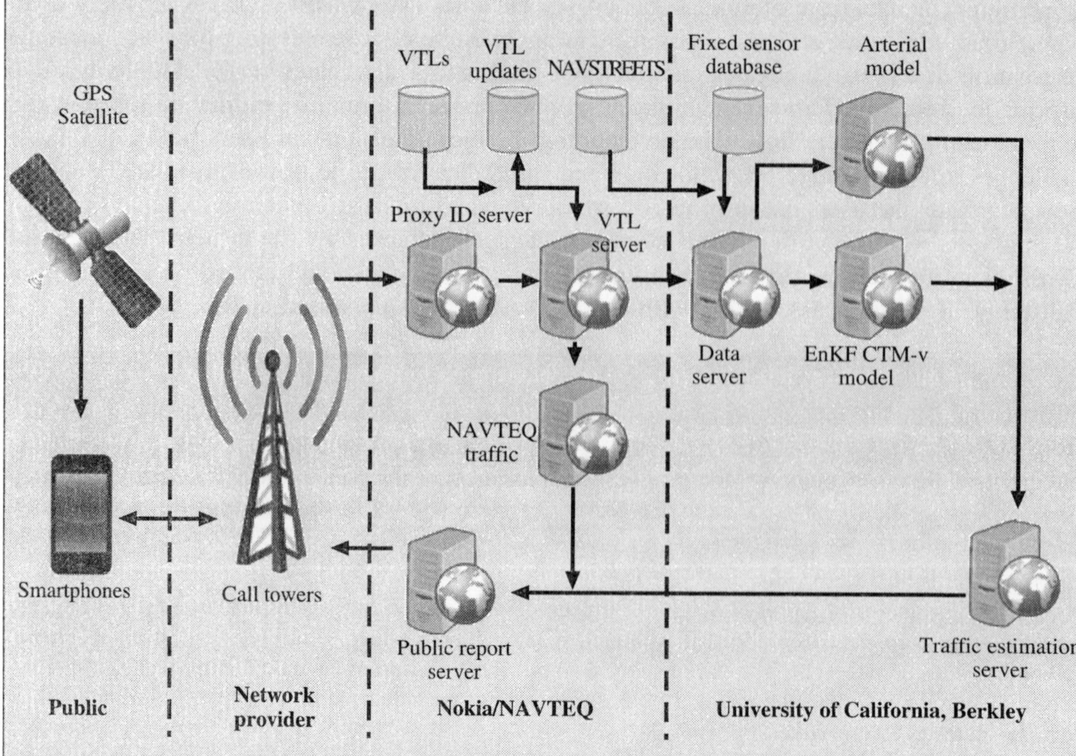

FIGURE 5.3 Architecture for cell phone network based information collection system (*Source:* https://www.wired.com/2011/03/cell-phone-networks-and-the-future-of-traffic/reproduced with permission).

The location of all the cellular users are tracked by the infrastructure may have an error of 100–500. It is feasible to this erroneous location data of vehicles collected from cellular network for edge level speed or travel time estimation, however, with some data processing to minimise the error. Chaturvedi and Srivastava (2016) has developed a model for optimal placement of ITS infrastructure by exploiting the large scale availability of the cellular network and sparse but accurate GPS probes data. The erroneous location data of vehicles collected from cellular network are processed using an innovative map matching algorithm to generate vehicle trajectories and count vehicles. The vehicle count information is used to compute vehicle flow, space occupancy and congestion estimation in real time. This enables accurate vehicle counting on all the edges and permits vehicle flow, space occupancy and congestion estimation. The edge level speed estimation is not feasible using cellular network data due to large location error. The historically collected congestion profile of edges are used to select edges for ITS infrastructure deployment. The ITS infrastructure and the historically collected GPS probes data are used in an innovative manner for edge level speed estimation in real-time.

The real-time traffic flow information can be used for travel time predictions, creation of smart roads where tolls and signals can adapt to changing conditions and provides urban planners with accurate pictures of traffic usage and its effects and improve planning.

5.6.1 Designated Spotters

These are vehicles which are designated to be utilised to act as vehicle probes and/or as spotters for traffic incidents. These vehicles can be select vehicles from a government agency fleet (e.g., road maintenance, service patrol, etc.) or vehicles of regular commuters.

5.7 SMART CARDS

Typically, the smart card comprises of microchip embedded in a plastic card. The smart card is usually of the size, 3.37 × 2.12 inches (85.60 × 53.98 mm) with approximately 0.030 inch thickness (0.76 mm). It offers properties of credit card as well as debit card. The circuitry of microchip of this card contains the program which is desgined for transferring/receiving the information for card reader. The smart cards can be used for wide range of applications ranging from payment, access control, etc. In transportation system, it may be used for electronic payment, online payment (e-challaning), driving licence, registration certificate of vehicle, permits for goods vehicles, common ticket for different modes of travel (details of smart card can be referred in the later chapter), etc. The capabilities of smart card can be categorised as memory card and microcontroller cards. Depending on the application for which the card has been developed, the cards can be of general purpose or specific purpose. The following types of smart cards are generally being used in transportation systems:

(a) Contact cards
(b) Contactless cards
(c) Hybrid cards

Contact cards: The contact type cards work as soon as it is touched or swiped or tapped at the card reader. The contact area of this card is approximately 1 cm^2 (0.16 $inch^2$) and consists of number of contact pads. As soon as the contact pad touches the card reader or inserted in card reader, the card communicates with the system. It works on same technology of passive infrared tags. The contact card does not have any inbuilt power source. It works by using the power of card reader. It is cheaper than contactless and hybrid cards. The contact cards are also not possessing inbuilt power source. It works on RF induction technology (incident radio-frequency interrogation signal) which provides relevant power to the card from the reader device.

Contactless card: In case of contactless card, the reader can read the card from a distance withourt making a physical contact. In contactless smart card, the chip communicates with the card reader through an induction technology. These cards require only close proximity to an antenna to complete a transaction. They are mostly used when transactions must be processed fast or hands-free, such as on mass transit systems, where a smart card can be used without even removing it from a wallet. The Smart Card of Delhi Metro Rail Corporation is a contacless card.

Hybrid cards: It allows the chip to work in both contact and contactless interfaces.

5.8 SUMMARY

Various detection methods explained above are increasingly becoming important to assess the magnitude and nature of traffic movement on the road. Based on the various detection methods, the violation of traffic with respect to speed, jumping the red light, type and magnitude of traffic travelling on the road, detection of presence of pedestrian at the traffic light have become extremely simpler and easier. Presently even in India, there are a number of toll booths where collection of congestion fee or toll under heavy traffic condition is not only finding this ITS application extremely beneficial but also experiencing the reduction of delay to the vehicular to a great extent.

The pace of change of the technology has been perhaps the fastest in recent days during the history of mankind. Everyday new and revolutionary inventions in information technology are created. IT has changed almost every aspect of human life beyond imagination. The adoption of IT technologies in transport sector has proved to be a game changer. This change will continue with development of advanced IT solutions almost every day.

Questions

5.1 What do you understand by automatic pedestrian detection? Explain its application.

5.2 Describe the application of red light violation detection system.

5.3 What are different sensing technologies used in ITS applications?

5.4 Differentiate between intrusive and non-intrusive sensors.

5.5 What do you understand by RFID? Describe various applications of RFID in ITS.

5.6 Write short note on following in context of ITS:

- (a) Mobile reports
- (b) Designated spotters
- (c) Bluetooth
- (d) Environmental sensors

5.7 Describe use of smart cards in ITS.

5.8 Write short note on emerging technologies for ITS.

Bibliography

Beecham Research Ltd. and GSMA, "Benefits analysis of GSMA Embedded SIM specification on the Mobile Enabled M2M Industry", September 2014. (https://www.gsma.com/iot/wp-content/uploads/2014/10/Benefits-Analysis-GSMA-Embedded-SIM-Specification.pdf).

Chaturvedi, Manish and Sanjay Srivastava, "Optimal ITS Infrastructure Deployment using Cellular Network and GPS Probes", PhD Thesis, DA-IICT, Gandhinagar, Gujarat, India, 2016.

Federal Highway Administration Research and Technology, US Department of Transportation, *Traffic Detector Handbook*, 3rd ed., Vol. I, Chapter 3, 2006. (https://www.fhwa.dot.gov/publications/research/operations/its/06108/03.cfm)

Federal Highway Administration University Course on Bicycle and Pedestrian Transportation, Lesson 10: Pedestrian Facility Signing and Pavement Markings, 2006. (https://www.fhwa.dot.gov/publications/research/safety/pedbike/05085/chapt10.cfm)

Gibson, David, Milton K. (Pete) Mills, and Doug Rekenthaler Jr., "Staying in the Loop: The Search for Improved Reliability of Traffic Sensing Systems Through Smart Test Instruments", Articles, Sept/Oct 1998. (https://www.fhwa.dot.gov/publications/publicroads/98septoct/loop.cfm)

Hatem, B.A. and H. Habib, "Bus Management System Using RFID In WSN", *in the proceedings of European and Mediterranean Conference on Information Systems (EMCIS)*, Abu Dhabi, UAE, April 12–13, 2009.

Katiyar, Vivek, Prashant Kumar, and Narottam Chand, "An Intelligent Transportation Systems Architecture using Wireless Sensor Networks", *International Journal of Computer Applications* (0975–8887), Vol. 14, No. 2, January 2011.

Mathew, Tom V., "Non-Intrusive Technologies" (Chapter 10), *Transportation Systems Engineering*, IIT Bombay, 2014.

Ontario Traffic Manual, "Advanced Traffic Management Systems", 2007. http://sudburysmap.ca/wp-content/uploads/2015/10/OTM-Book-19.pdf

Saha, Satadal, Subhadip Basu, Mita Nasipuri, and Dipak Kumar Basu, "Development of an automated Red Light Violation Detection System (RLVDS) for Indian vehicles", *IEEE National Conference on Computing and Communication Systems (COCOSYS-09)*, CS11 59. 2010.

Singh, B. and A. Gupta, "Advancements in Intelligent Transportation Systems: A Review", *Proceedings of International Conference on Advanced Trends in Engineering and Technology (ICATET)*, Jaipur, pp. 375–383, 2013.

Singh, B., A. Gupta, and S. Suman, "Framework for Development of Advanced Traveller Information System: A Case Study for Chandigarh City", *Proceedings of International Conference on Transportation Planning and Implementation Methodologies for Developing Countries 2014 (TPMDC-2014)*, Submission ID 54, 2014.

Singh, V. and P. Kumar, "Web-Based Advanced Traveller Information System for Developing Countries", *Journal of Transportation Engineering*, Vol. 136(9), pp. 836–845, 2011.

The Evolution of Automatic Pedestrian Detection Passive Activation for use with the LightGuard Systems, Inc. The "Smart Crosswalk™" Written by Michael A. Harrison – President/CEO, 2010.

Tulasi, Uppalapalli Rama and CH. Tirupathirao, "Sensor and Network Technology for Intelligent Transportation Systems", *International Journal of Professional Engineering Studies*, Vol. II, Issue 1/JAN2014, 2014.

Wang, G., D. Xiao, and J. Gu, "Design and Implementation on Software Platform of Emergency Management System for Traffic Incidents", In the proceedings of CCECE/CCGEI, Canada, 2008.

Young, Stan, "Real-Time Traffic Operations Data Using Vehicle Probe Technology", *Proceedings of the 2007 Mid-Continent Transportation Research Symposium*, Ames, Iowa, August 2007. (http://www.ctre.iastate.edu/pubs/midcon2007/youngvehicleprobe.pdf)

Zhian, L., and H. Han, "A Bus Management System Based on ZigBee and GSM/GPRS", *in the proceedings of International Conference on Computer Application and System Modelling (ICCASM 2010)*, Taiyuan, China, 2010.

6 Traffic Management System Components and Intelligent Transport System

6.1 INTRODUCTION

Traffic management is essentially adjustment of the demand and capacity of the traffic network in time and space, to better 'match' the traffic demand and supply (capacity). Traffic management measures are required to improve the safety and flow of traffic, reduce traffic emissions and utilise traffic artery capacity more efficiently. Traffic management is used to manage and control the demand for transport and decides the selection of the mode of transport, route, or the time of travel or transport. It is generally customary practice that before embarking on capital intensive transport development such as building a metro project, the city authority tries to optimise the existing transport supply with the application of traffic management techniques. Broadly, the traffic management comprises the following elements: information, control, incident management, demand management, driver support and monitoring, and fleet and transport management. It also requires reliable and real-time status information on the transport system.

Some of the examples of traffic management measures are variable settings for traffic lights, variable speed limits, variable message sign, parking guidance, dynamic lane management and dynamic route information. Many traffic management measures are increasingly using information and communication technologies to improve its effectiveness and optimise the cost. This chapter discusses the role of ITS for traffic management.

6.2 OBJECTIVES OF TRAFFIC MANAGEMENT

Traffic management is the planning, monitoring and control, influencing of traffic to travel in desired direction and advice as per formulation of management plan. It aims to:

(a) maximise the effectiveness of the use of existing infrastructure;

(b) ensure reliable and safe operation of transport;

(c) address environmental goals; and
(d) ensure fair allocation of infrastructure space (road space, rail slots, etc.) among competing users.

It is therefore an essential element in increasing the efficiency and safety of transport networks and operations.

For road transport, traffic management consists of monitoring the actual traffic situation in real-time (including volumes, speeds, incidents, etc.) and then controlling or influencing the flow using that information in order to reduce congestion, deal with incidents and improve network efficiency, safety and environmental performance, or achieve other objectives. Broadly, strategic traffic management involves managing whole networks at a macro level as well as integrating or linking different networks.

The aim of traffic management is the allocation of infrastructure (road space or network) as per strategic operational and policy goals which include efficiency, safety, environmental, economic and equity objectives. These measures include giving priority to buses, trams or other vehicles such as emergency services or high occupancy vehicles, increasing space available for pedestrians and cyclists, or providing shared road space. For rail, rules for market opening, network capacity allocation and pricing also constitute policy-level strategic management. Traffic signals and railway signalling are perhaps the earliest form of "intelligent" traffic management, aimed primarily at safety but also at managing priorities at junctions. The phasing and coordination of road traffic signals in urban areas via a control centre is still the most widespread tool for the effective management of street networks. More recent applications of traffic signalling include dynamic lane allocation (for tidal flow, or reserved lanes for buses or other authorised vehicles) and ramp metering (signalised access control to motorways).

While standard road traffic signs provide routing guidance, dynamic rerouting by road authorities on major inter-urban corridors is primarily done by variable message signs (VMS), often coordinated according to different scenarios so that a series of coherent messages on different signs can be displayed according to the location and nature of any incident. Broadcasting, Websites, text messaging services (SMS) and in-vehicle systems are also important telematic tools in this respect. The main difference is therefore that routing for rail traffic is highly prescriptive whereas for road users it is primarily informative.

6.3 TRAFFIC MANAGEMENT MEASURES

The traffic management measures used worldwide can be broadly classified under following categories:

(a) Traffic information
(b) Traffic control
(c) Incident management
(d) Demand management
(e) Driver support and monitoring
(f) Fleet and transport management

The description of each transport management measure is presented in Table 6.1.

TABLE 6.1 Transport management measures

Traffic management measure	*Description*
Traffic information	Traffic information offers real-time information for road users, both before and during travel. The information provided covers topics such as weather and road conditions, road maintenance, traffic incidents, traffic situation and alternative modes of travel.
Traffic control	Traffic is controlled per inter-section, road-section or the entire road network. Traffic control is divided into fixed and variable traffic control.
Incident management	Incident management comprises the detection, handling and elimination of traffic incidents, such as accidents, in cooperation with various authorities.
Demand management	Demand management is used to affect choices in the destination, time, mode of transport or route of travel or transport. Means of achieving this include regulating access or parking, arranging park-and-ride facilities, supporting carpooling, promoting public transport and pedestrian and bicycle traffic, providing traffic information and using peak-hour tolls and other road use charges.
Driver support and monitoring	Driver support systems comprise IT and communications technology applications, providing assistance to the driver. These include collision prevention, support for lane keeping and navigation systems. Monitoring systems include automatic speed and intersection control, monitoring transports of hazardous materials and automatic lane-use monitoring.
Fleet and transport management	Fleet management covers the planning, monitoring, control and assessment of the movement and operations of a vehicle fleet and its drivers. Transport management involves the management of transport chain operations and information flows.

Conventionally, traffic inspectors used to be deployed at major road junctions to control and manage the traffic. The manual system of traffic management has its limitations in terms of efficiency, cost and scale. With the advent of advanced traffic management systems, thanks to ITS technologies, the automatic system based traffic management systems are replacing the conventional system at a very fast pace.

6.4 ITS FOR TRAFFIC MANAGEMENT

Perhaps, traffic signals were first application of technology to replace the traffic police personnel from the road junctions. Since then, ITS has become as a key tool for traffic management and control, improving safety and user services and reducing the environmental impact of traffic, particularly at infrastructure bottlenecks. ITS applications for traffic management and control include rerouting using VMS, variable speed limits (VSL) with automated enforcement, lane control, dynamic use of expressways or access control measures such as ramp metering, as well as specific measures for freight such as information on heavy goods vehicle (HGV) parking and "stacking" of lorries in the case of disruption. Cooperative highway systems, whether vehicle-to-vehicle (V2V) or vehicle-to-infrastructure (V2I) will play an important role in traffic management and control in the future; coordination across countries and regions as well as with vehicle and equipment manufacturers, is required in order for systems to succeed in meeting traffic management objectives. Depending on the city characteristics, a subset of the following ITS based TMS tools are used all over the world for traffic management:

(a) *Traffic information*

- (i) Passenger information system—advanced traveller information system (ATIS)
- (ii) Traffic—transport information system
- (iii) Mobile Applications
- (iv) Real-time tracking systems
- (v) CCTV

(b) *Traffic control*

- (i) Intelligent signalling system
- (ii) Automatic vehicle location system
- (iii) Police patrol monitoring system
- (iv) e-challan (ticketing) system
- (v) Pedestrian/biker detection system
- (vi) Smart card based driving licence issue system
- (vii) Smart card based permit and registration certificate issue system
- (viii) Traffic control maintenance services system
- (ix) Traffic flow monitoring system
- (x) Ramp metering system
- (xi) Automatic speed detection system
- (xii) Automatic number plate recognition system
- (xiii) Automatic vehicle detection system (image processing)
- (xiv) Area traffic control system
- (xv) Red light violation detection system
- (xvi) Automatic traffic counter and classifier
- (xvii) Wrong way vehicle detection system

(c) *Incident management*

- (i) Emergency and accident response system
- (ii) Automatic incident detection system

(iii) Distress call response management system—emergency roadside telephone system, e-Calling

(iv) Smart tracking and distress alert system for automobiles

(d) *Demand management*

(i) Automatic fare collection system, electronic parking charge payment system, electronic ticketing system, and electronic road pricing system

(ii) Electronic toll collection system

(iii) Parking management and information system

(iv) Parking guidance system

(v) Intelligent parking lot management system

(e) *Driver support and monitoring*

(i) Journey time monitoring system

(ii) Car navigation system—vehicle information and communication system (VICS)

(iii) Route guidance system—dynamic route guidance

(f) *Fleet and transport management*

(i) Multi-modal transportation logistics support system

(ii) Transportation monitoring system (fleet monitoring system: public transport, taxies, jal (water) board vehicles, bank vehicles, school buses, ambulances, and government vehicles)

(iii) Goods vehicle weight monitoring system

(iv) Vehicle tracking system

(v) Towing system

These ITS tools are discussed in various parts of this book in greater details.

6.5 DEVELOPMENT OF TRAFFIC MANAGEMENT SYSTEM

It is practically feasible to implement all possible tools of ITS at once due to lack of funds, experience, and expertise. Even if funds are available, it is always advisable to progressively implement ITS tools to assess efficacy of different tools in the process. The TMS measures need to be implemented in a phased manner so that the learning from the earlier phases can be utilised in the later phases. Based on availability of funds, expertise, severity of the problem, and ground reality, the following proposed phases can be altered and adapted to suit the prevailing conditions in a city.

The following phase-wise implementation of TMS (Table 6.2) can be considered depending upon the stage of the city with respect to level of congestion, availability of fund, computerisation of vehicle registration, driving licences, etc.

TABLE 6.2 Phase-wise implementation of TMS

Phase	*Applicability*
I	City where the problem of traffic congestion is still under control, congestion is just starting to spread beyond the peak hours, and funds or trained/technical staff may also be a constraint. Smaller tier-2 cities shall start with this phase as early as possible.
II	City has developed a computerised database of vehicle owners/driving licences/ commercial vehicle permits and also has started implementing smart driving licences/ permits/registration certificates, etc.
III	Cities facing congestion and increase in road accident related fatalities. The pollution level is also increasing at a fast pace. Metropolitan cities and state capitals of India are already facing these problems.
IV	Heavy congestion is being faced. Pedestrian and bicycle traffic are important part of transportation system to access the public transport system.
V	City is ready for integrated multi-modal transport system with heavy reliance of public transport. Real-time information to passengers about public transport is key to success of this phase.

The ITS tools in different phases can be added in earlier phases or postponed for further phases as per specific need for a city. The phase-wise development plan needs to be customised to the specific needs of the city. For example, in a city which has not experienced much traffic congestion, accidents, delays, etc., public transport system may be introduced in first phase itself.

An indicative deployment of ITS tools in different phases is presented in Table 6.3 for conceptual understanding of the process. However, for every city a detailed feasibility study shall be carried out to appreciate the need and suitability of various ITS tools for the city.

The system architecture for a city depends upon the phase of deployment of ITS tools system for a city. For phase I, the ITS architecture comprises only traffic control terminals, roadside sensors, communication systems and area traffic control. The system architecture for subsequent phases will have additional ITS tools like intelligent traffic management, electronic printing, traffic rules enforcement systems, data warehouse, VMS, speed detection, red light violation, e-challan, CCTV system, VMS parking management system, passenger information systems, etc. as explained in Table 6.2. The system architecture for phase III and phase V are shown in Figures 6.1 and 6.2 for illustration. Needless to mention, the system architecture needs to be customised to the actual needs of the city after careful examination of functionalities required, availability of funds and other enabling infrastructure. The components may be added or deleted from the system architecture as shown in Figures 6.1 and 6.2 as per the ITS plan of the city.

TABLE 6.3 Phase-wise implementation of ITS tools

System	*ITS tools*	*Purpose*	*Phase*
Traffic control	• Intelligent signals/ coordinated signals • Area traffic control system (ATC)	• For smooth flow of traffic at intersection/corridor • Smooth flow of traffic in implemented network, minimum overall delay at red lights, and green signal throughout for emergency vehicles	I, II, III, IV, V
Intelligent traffic management and electronic payment	• Speed control signage, hazard warning system • Emergency roadside phone system electronic road pricing system	• To alert the vehicle drivers about speed restrictions and about any obstruction (temporarily or permanent) in traffic flow • To reduce congestion by means of introducing congestion charges	II, III, IV, V
Traffic rules enforcement system Transport authority enforcement system	• CCTV system, automatic speed detection system • Automatic number plate recognition system • Red light violation detection system • Wrong way vehicle detection system • e-challan system • Smart card based driving licences issue system • Smart card based permit and registration certificate issue system • Police patrol monitoring system • Traffic police (squad, post), traffic interceptor vans	• To catch traffic rules offender, and collect fines from them • To book the traffic rules offenders, violation of permit rules, fake driving licence holders, fake permits and RC holders, stolen vehicles in the law of court • Maintain the records of traffic rules offenders for giving severe punishment in law of court • Reduce accidents and increase safety • To collect the congestion charges	
Infrastructure maintenance and management system	• Maintain and manage roadside ITS and non-ITS based infrastructure • Maintain roads, footpath, FOBs, junctions • Maintain and manage communication system, control room • Maintain traffic signals, VMS, signage of different types	• To provide maintained road network for all road users for smooth flow of traffic and road users mobility • To provide un-interrupted platform for ITS functioning	
Data warehouse	• Automatic traffic counter and classifier	• To maintain data of traffic, vehicles for forecasting, planning, etc. • For demand management	

(Contd...)

TABLE 6.3 Phase-wise implementation of ITS tools (*Contd...*)

System	*ITS tools*	*Purpose*	*Phase*
Information dissemination system	• VMS, speed signage, hazard warning signage, SMS, Websites, social sites, e-mail, FM radio	• It helps the traveller and vehicle driver to take corrective action in their trip as well as prepare themselves as per any traffic advisory or alert is issued	
Traffic monitoring and surveillance system	• Automatic vehicle detection system (image processing) • Towing system • Journey time monitoring system	• To clear the network in time of breakdown of vehicles for un-interrupted traffic flow • To take action after detecting any incidents and reduce casualties	III, IV, V
Emergency and accident response system	• Transportation monitoring system (fleet monitoring system—public transport, ambulances) • Distress call response management system—emergency roadside telephone system	• To provide immediate relief to road users • To save life of roadside accident victims • To help road users in case of emergency	
Automatic air pollution monitoring system	• Atmospheric sensors, air quality sensors, meteorological sensors and road surface sensors	• To save the city and habitats from pollution • To book the vehicles for violation of pollution norm and collect fine and book them in law of court	
Parking management and information system	• Parking guidance system • Intelligent parking lot management system • Electronic parking charge payment system	• Guide the driver to particular parking sites among number of sites within an area • Guide the drivers to park at particular parking bay • Help the drivers to locate particular location in city through unit fitted in car • Help the drivers or road users to locate particular location in city through system available on roadside	IV, V
Pedestrian/biker detection system	• Light detection and ranging (LIDAR), RADAR, or CCTV	• To reduce and avoid pedestrian/biker causality in road accidents	
Advanced traveller information system (ATIS)	• Passenger information system	• To help the road user for information related to public transport/informal public transport (IPT) at roadside	V

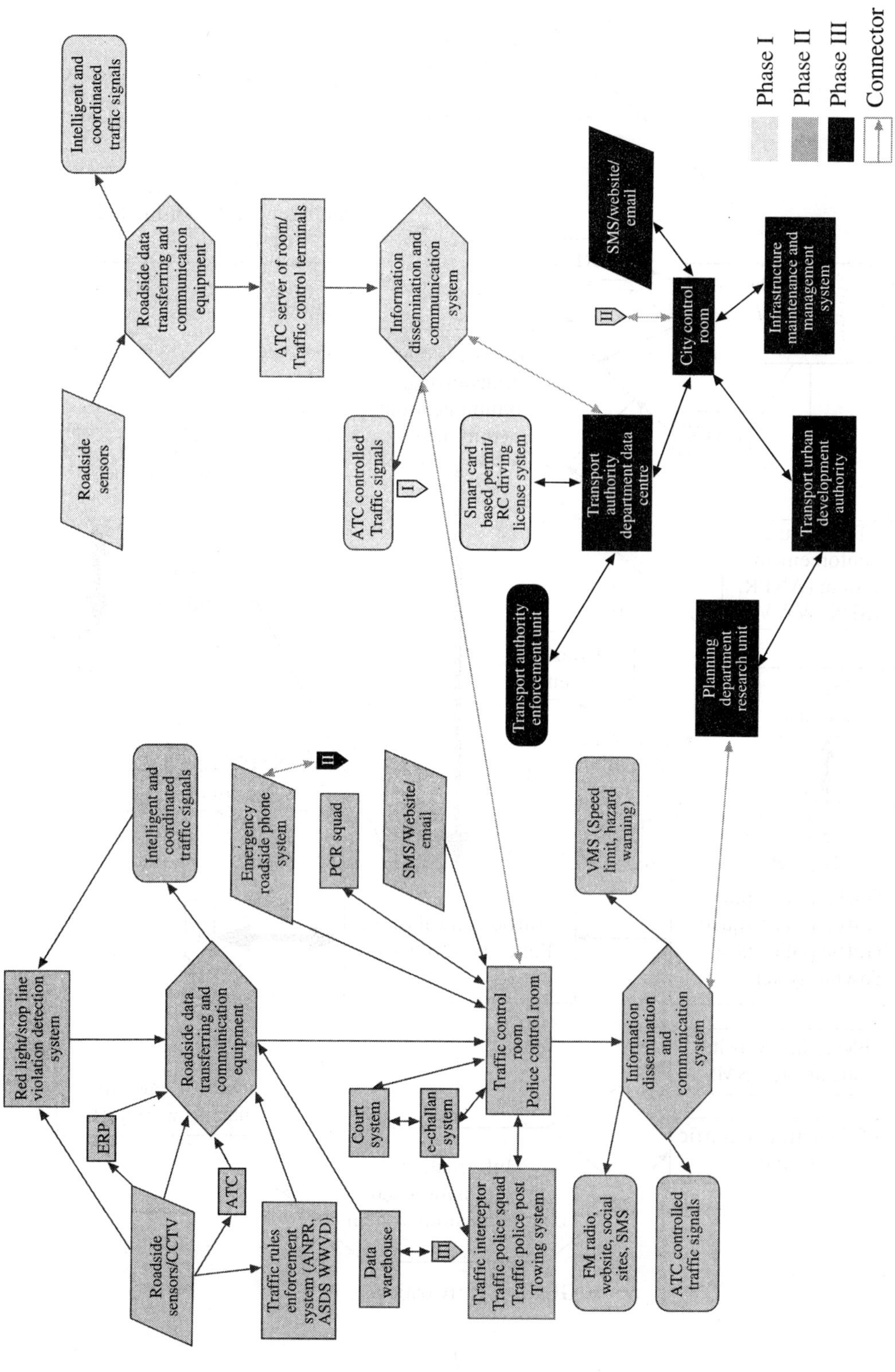

FIGURE 6.1 System architecture diagram—Phase I, II and III of ITS deployment.

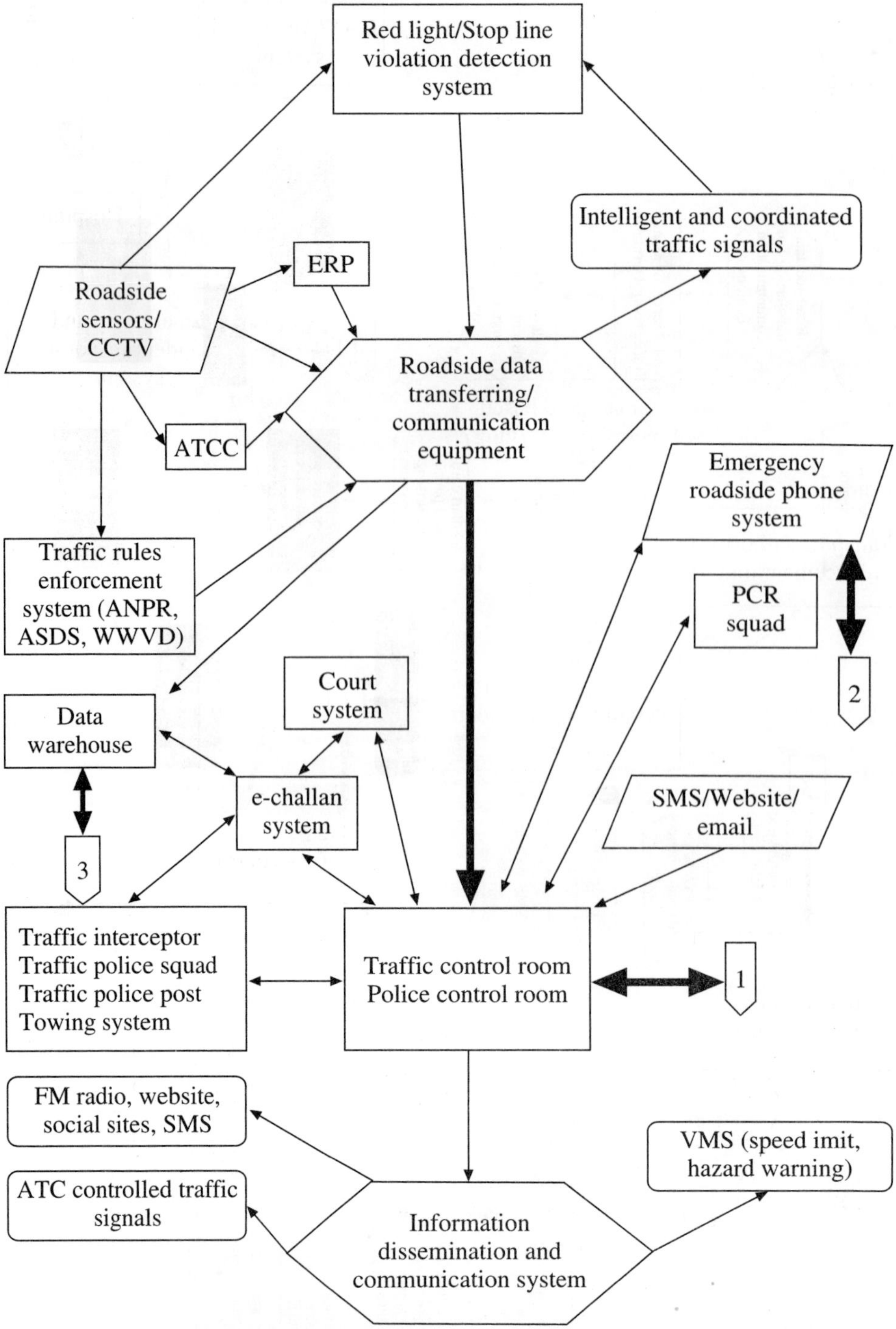

FIGURE 6.2 (Continue)

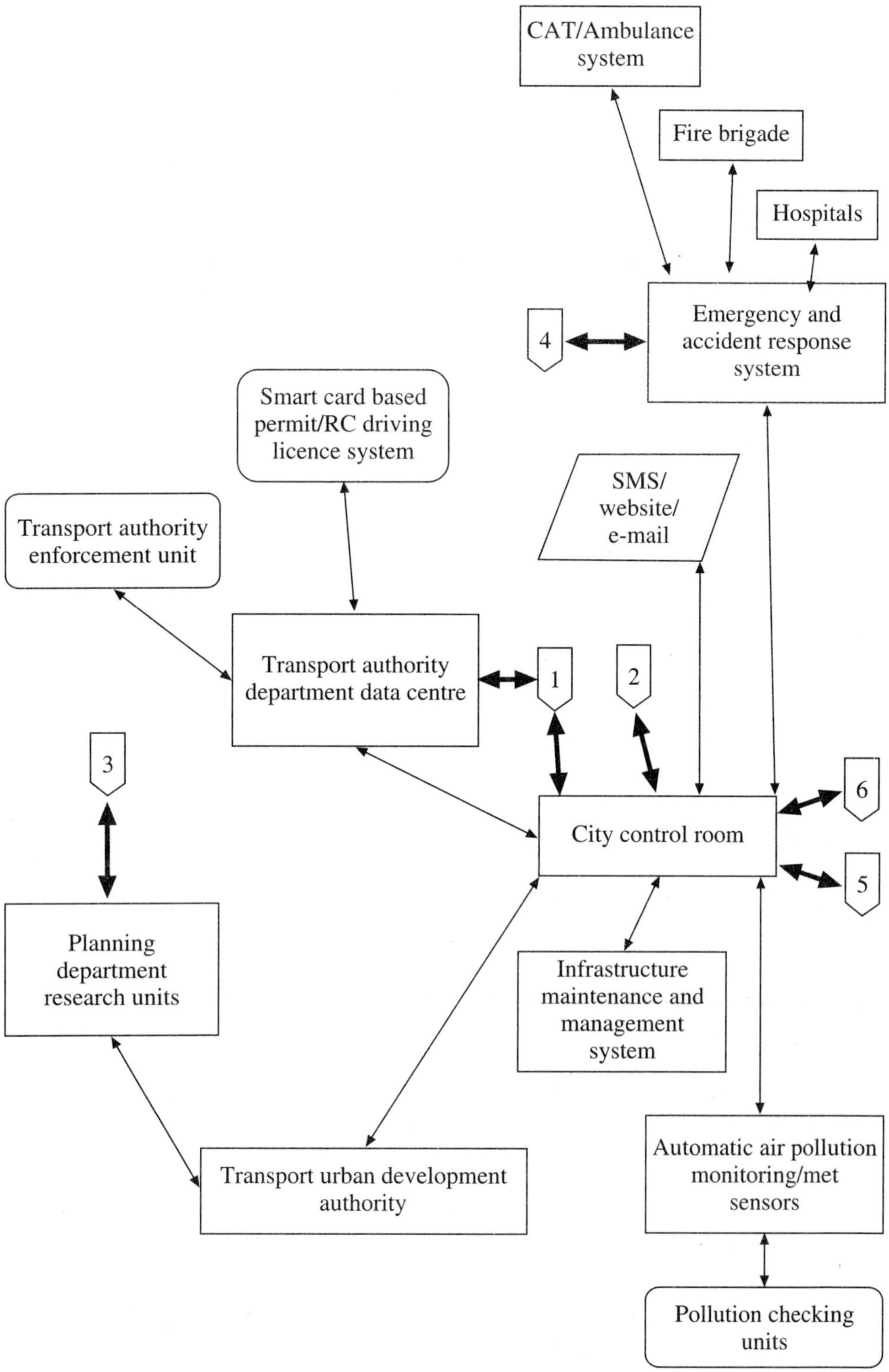

FIGURE 6.2 (Continue)

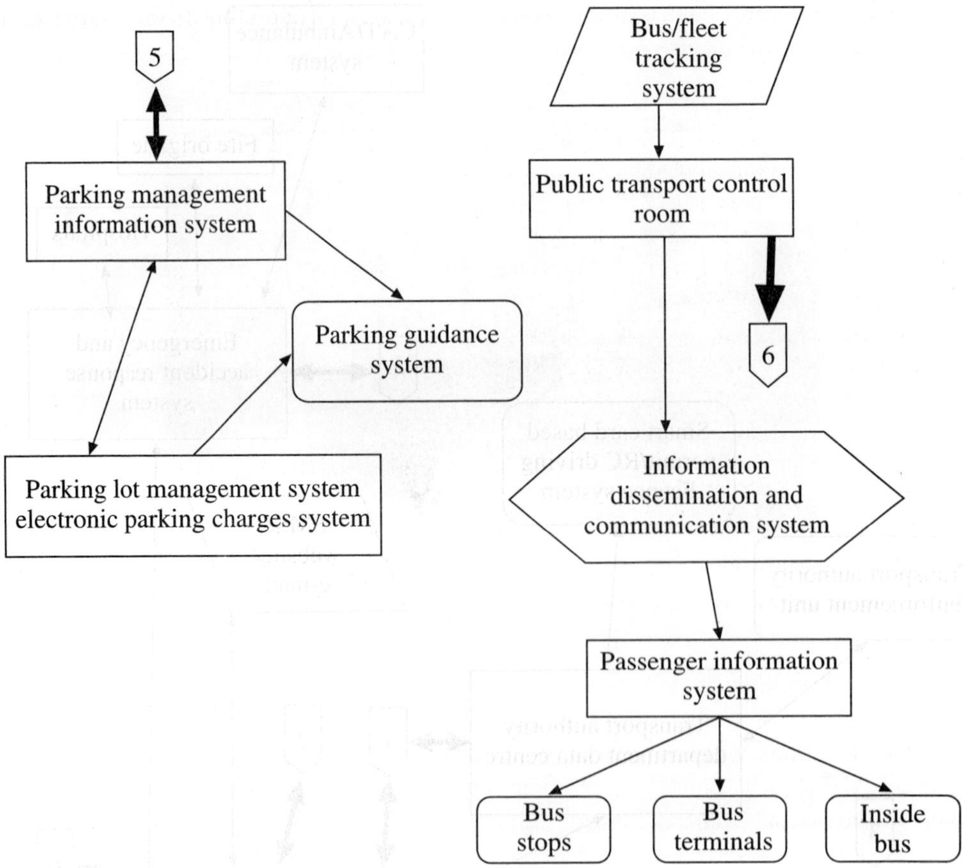

FIGURE 6.2 System architecture diagram for all phases (I, II, III, IV and V).

6.6 TRAFFIC MANAGEMENT CENTRE (TMC)

A special enabler for traffic management and ITS is the traffic management centre (also called traffic control centre). When the number of operational tasks increases, or the size and complexity of instruments and scenarios increase, a traffic management centre could become necessary. A TMC is the hub of transport management, where data is collected, and analysed and combined with other operational and control concepts to manage the complex transport network. It is the control centre for communicating transportation-related information to the media and the motoring public. TMC is a place where agencies can coordinate their responses to transportation situations and conditions. Typically, several agencies jointly manage the transport infrastructure, through a network of traffic operation centres. Often, there is a localised distribution of data and information and the centres adopt different criteria to achieve the goals of traffic management. This inter-dependent autonomy in operations and decision-making is must due to the heterogeneity of demand and performance characteristics of interacting subsystems.

For developed countries, traffic management centre plays an important role managing transport operation in urban areas. For example, city of Newcastle, UK developed its urban traffic management control centre in the Newcastle University operating from a large room as can be seen in Figure 6.3.

FIGURE 6.3 Tyne and Wear Urban Traffic Management Control (UTMC) Centre in the Newcastle University (*Source:* Ray King, Tyne and Wear UTMC specialist, Operations, Newcastle City Council, reproduced with permission).

When it is not possible to develop such a traffic management control centre due to paucity of fund or availability of space, a piecemeal traffic management measures for a single location or for a set of locations with or without any relation with between each other are commonly seen such as ANPR, red light camera, speed/detection/enforcement camera, VMS in urban areas. The following are some demonstrations as a part of ITS driven applications for traffic management. Figure 6.4 shows a concept of traffic management control centre.

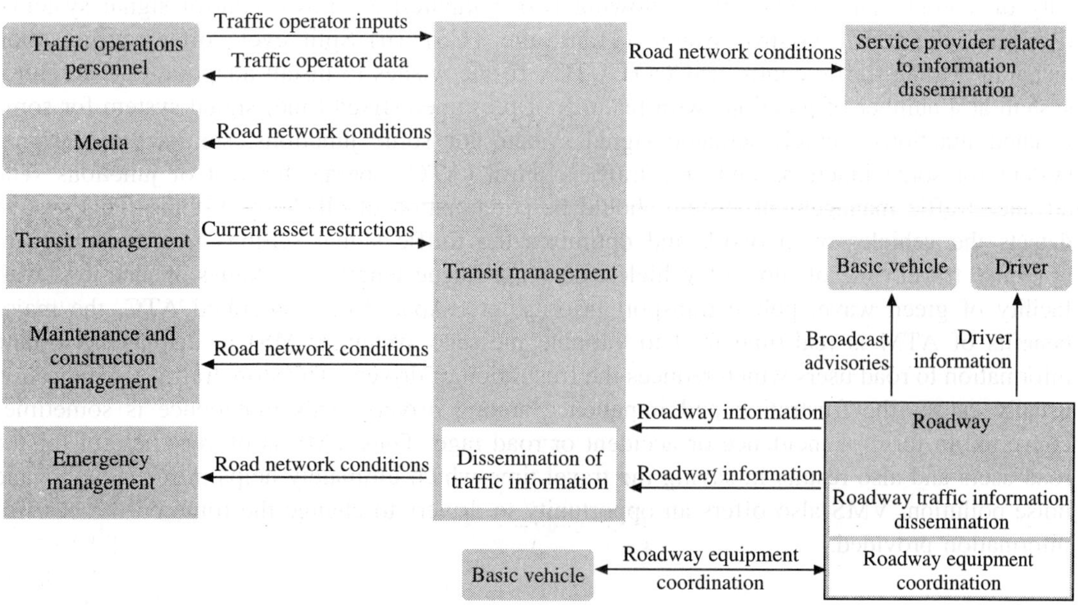

FIGURE 6.4 Concept of traffic management control centre.

6.7 ADVANCE TRAFFIC MANAGEMENT SYSTEM (ATMS)

ATMS is developed to integrate various subsystems (such as CCTV, vehicle detection, communication, variable message systems, etc.) into a coherent single interface that exhibits real-time data on traffic status as shown in Figure 6.5 and predicts traffic conditions for more efficient planning and operations.

FIGURE 6.5 Example of ATMS (https://upload.wikimedia.org/wikipedia/commons/f/f8/New_Jersey_Turnpike_Reduce_Speed_sign.jpg, Author: Ben Schumin, free to use).

One of the advance traffic management systems is a computerised traffic control signal system which is increasingly being used to analyse and manage present traffic situation, examines signal system problems and implements/modifies signal timings remotely. The City of Toronto has adopted the following two computerised traffic control signal systems: (i) TransSuite traffic control system (TransSuite TCS), (ii) Split cycle offset optimisation technique/Urban traffic control (SCOOT/UTC). If one wishes to install adaptive traffic control system at a number of junctions with features of pre-timed (fixed time) signal system for some isolated junctions, vehicle-actuated signal system for some junctions, semi-actuated signal system for some junctions and area traffic control (ATC) operate for rest of junctions. The advance traffic management system should be combination of all these systems. This system detects the vehicles on network and optimises the traffic signal settings in project area to minimise travel time of most of vehicles, decrease queue length of vehicles at junctions with facility of green wave, public transport priority, etc. Apart from integrated ATC, the major benefits of ATM is real-time feed to variable message system (VMS) to display up-to-date information to road users which reduces the frustration of drivers. The slow traffic or congestion usually causes the frustration and impatience among drivers. This impatience is sometimes converted in form of incidence or accident or road rage. Thus VMS is of very helpful for the road users and also results in saving the travel time which ultimately helps in reducing air and noise pollution. VMS also offers an opportunity to drivers to change the route on the basis of information provided.

6.8 ADVANCE TRAVELLER INFORMATION SYSTEM (ATIS)

ATIS offers to users of transportation systems, travel-related information in advance or on real-time basis to help them in decision making on route choices, estimate travel times, and avoid congestion. The following information using various technologies are very useful to meet these objectives:

(a) GPS enabled in-vehicle navigation systems (Figure 6.6)

(b) Dynamic road message signs for real-time communication of information on traffic congestions, bottlenecks, accidents and alternate route information during road closures and maintenance

(c) Website to provide a colour-coded network map showing congestion levels on highways (congestion index)

FIGURE 6.6 GPS based navigation system (*Source:* https://commons.wikimedia.org/wiki/File:Garmin_N%C3%BCvi_200.jpg, Author: Humberto Möckel, Free to use).

A typical ATIS consists of couple of information collection and processing centres and an information service provider. The information collection and processing centre linked by a combination of wireless and wireline technologies to an automatic vehicle location sensor system and passenger counters that provide an accurate location of each bus and train in the fleet and volumes of passengers boarding and alighting. This data is processed at the transit-management centre and compared to a database of scheduled times and locations. The centre also has voice and data communication links to every vehicle to support the transmission of new routes and instructions to vehicle drivers. The centre is also connected to a range of information kiosks at strategic locations, such as bus stations, railway stations, and other public places, enabling current travel information to be delivered to these locations. The centre can send information to the users directly on their smart phone, navigation tools, etc. The conceptual ATIS for a typical system is shown in Figure 6.7.

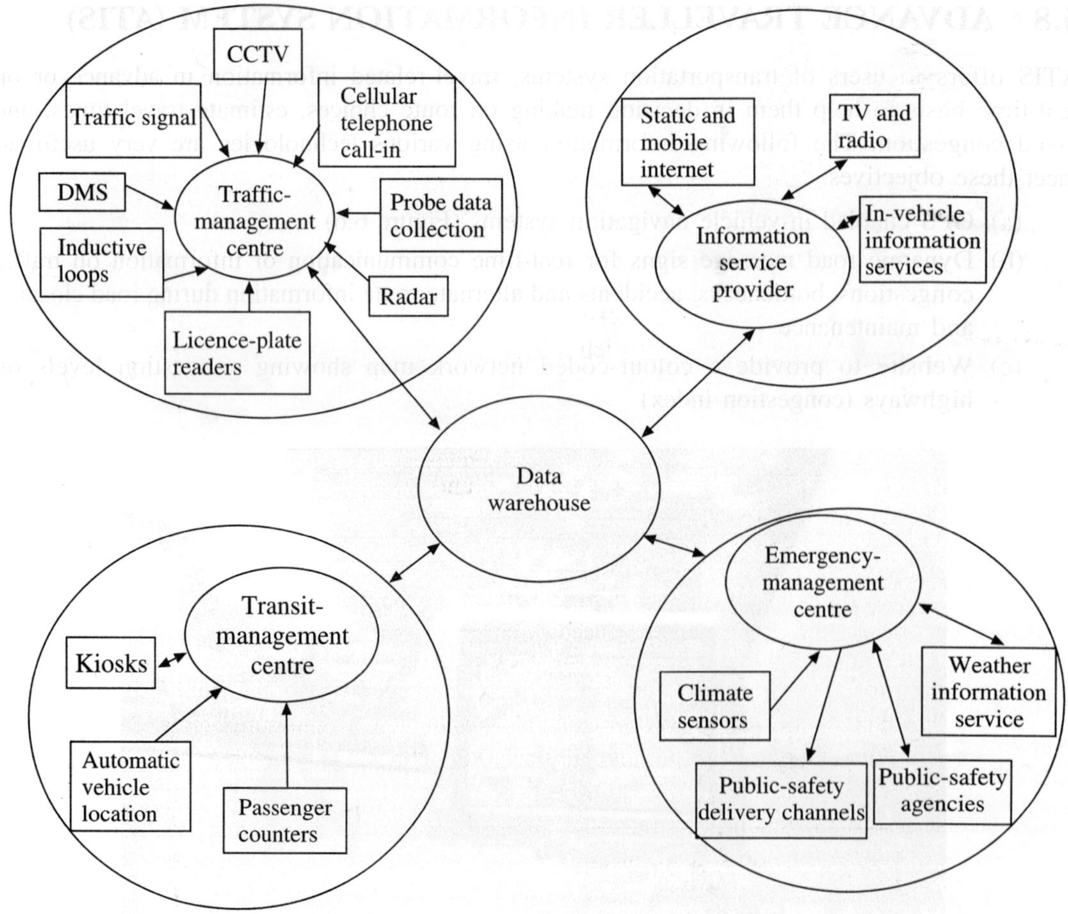

FIGURE 6.7 ATIS concept for a typical system.

6.9 ADVANCE VEHICLE CONTROL SYSTEMS (AVCS)

AVCS are tools and concepts that enhance the driver's control of the vehicle to make travel safer and more efficient. The Chapter 9 of the book is focused on Advanced Vehicle Control Systems. For example, in-vehicle collision warning systems warns the driver to a possible imminent collision (Figure 6.8). In advanced AVCS applications, the vehicle could automatically apply brakes or steer away from a collision, based on information/detection from sensors on the vehicle. Both systems are independent to the vehicle and are able to provide substantial benefits by improving safety and reducing accident induced congestion. The installation of advanced gadgets and processors in vehicles allow incorporation of software applications and artificial intelligence systems that control internal operations, ubiquitous computing, and other programs designed to be integrated into a greater transportation system.

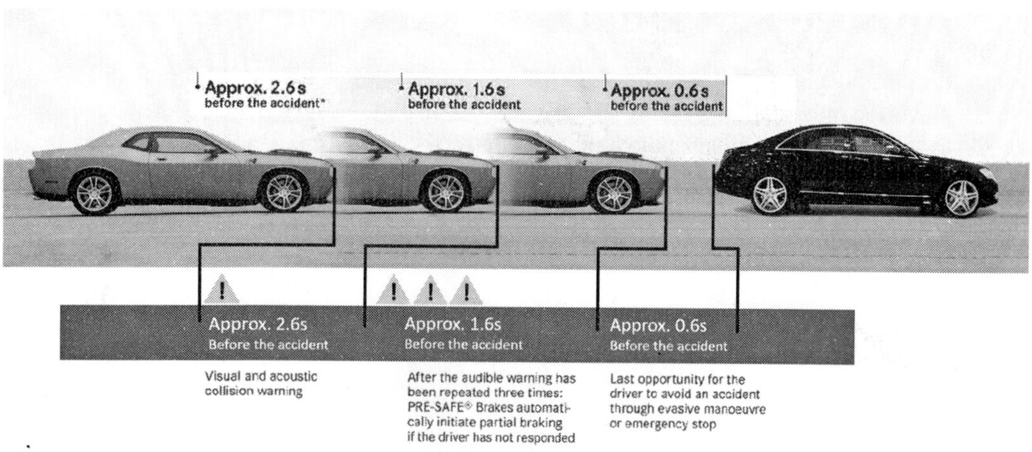

FIGURE 6.8 AVCS example.

6.10 ADVANCE PUBLIC TRANSPORT SYSTEM (APTS)

APTS applies state-of-art transportation management and information technologies to public transit systems to enhance efficiency and effectiveness of operation and improve safety of public transit. It includes real-time passenger information (RTPI) systems, automatic vehicle location (AVL) systems, bus arrival notification systems, and systems providing priority of passage to buses at signalised intersections (transit signal priority) as shown in Figure 6.9.

FIGURE 6.9 London bus countdown sign (*Source:* http://ntl.bts.gov/lib/jpodocs/repts_te/13845_files/image011.jpg, reproduced with permission).

6.11 COMMERCIAL VEHICLE OPERATION (CVO)

CVO invloves application of the technology of advanced traffic management system (ATMS), advanced traveller information system (ATIS) and advanced vehicle control and safety system (AVCSS) in commercial vehicle operation such as trucks, buses, taxis and ambulances in order to improve efficiency, safety and fleet operations. The system mainly includes automatic vehicle monitoring, fleet management, computer scheduling and electronic payment. Figure 6.10 shows the major services of commercial vehicle operations.

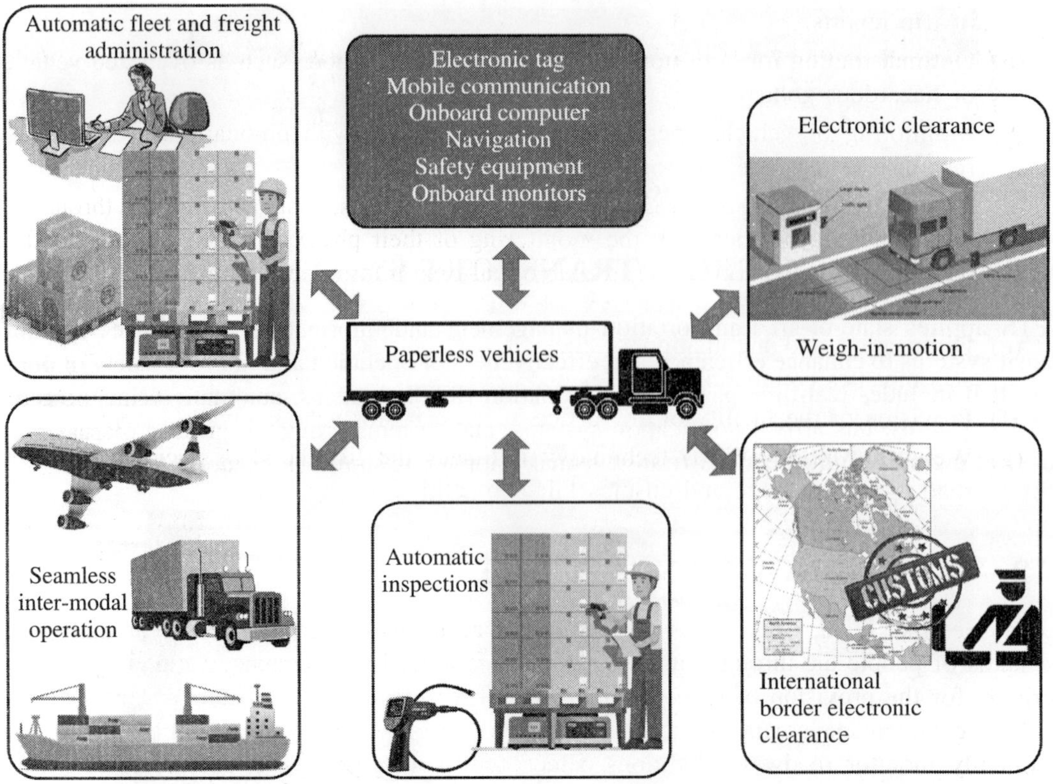

FIGURE 6.10 Major services of commercial vehicle operation (*Source:* https://www.fhwa.dot.gov/publications/publicroads/97septoct/images/p31a.gif, reproduced with permission).

The main emphasis of CVO is on the specialised operations involving long-distance heavy truck transport. Some of the main elements of CVO are weigh-in-motion, automatic vehicle identification (AVI), and automated electronic clearance.

Weigh-in-motion technology allows the trucks to be weighed while in motion without stopping, thus saving time and energy. AVI facilitates automatic identification of vehicles thus help in tracking and monitoring of the vehicle. In an automated electronic clearance system, once a truck has been cleared at the first weigh station after weighment, the data is sent to the next weight stations, where the truck is identified automatically through AVI equipment, and the truck can bypass the next weigh stations. The commercial and public fleet operators use automatic vehicle location (AVL) systems to instantly know the location of any vehicle in order to monitor their fleet more efficiently.

There are number of ITS applications available to assist drivers and fleet operators in the operations of commercial freight transport, covering both long distance freight haulage and urban goods deliveries. ITS applications can increase the efficiency and reliability of operations, and also improve the level of safety.

Some of the examples of ITS applications for CVO are as follows:

(a) Scheduling and 'matching' of vehicles, loads and drivers and the automatic generation of trip reports.

(b) Optimal routing for both normal and 'abnormal' transport, such as oversized vehicles or hazardous goods.

(c) Monitoring of vehicle operations related to safety, with on-board data storage for response to roadside interrogations.

(d) Monitoring, tracking and tracing of commercial vehicles, containers or loads throughout their journey, together with the monitoring of their physical status, e.g., for food or dangerous goods and making available real-time information to customers about their consignment.

(e) Digitisation and automation of the commercial and regulatory documentation that accompanies commercial vehicles and goods.

(f) Provision of the facility of an "office in the cab" for vehicle owner/drivers.

(g) Weigh-in-Motion (WiM) technology mitigates the likelihood of accidents, protects road against damage and ensures fair competition among carriers.

6.12 EMERGENCY MANAGEMENT

In countries like India, where there are number of accidents occurring on roads on daily basis and lakhs of people die throughout the year, there is a need of emergency support systems on corridors for the provision of timely support to the road users invloved in accidents.

Emergency management services are generally arranged by traffic control centres that continually monitor roadway conditions. Emergency management user service is essentially required to improve the management and response to emergency situations of police, fire and rescue operations. In case of an accident, the nearest emergency service vehicle is electronically located from the system and dispatched to the site. The traffic control centre then alerts other drivers in the incident area through dynamic message signs (Figure 6.11). These services help in reducing the response time, reducing the pssibility of the occurrence of secondary incidents and save lives.

ITS applications in emergency management also consist of managemet of hazardous materials movement, the deployment of emergency medical services, and large and small-scale emergency response and evacuation operations. These applications can enhance the efficiency of transportation capacity during emergencies, improve productivity for hazardous material (HAZMAT) shipping operations, and increase overall traveller safety and security.

FIGURE 6.11 Variable message sign board (*Source:* https://commons.wikimedia.org/wiki/File:MN_Changeable_Message_Sign.jpg, Free to use).

The US Department of Transport (DOT) agency have bundled different individual ITS user services into groups. The emergency management (user services bundle) user services include:

(a) Emergency notification and personal security
(b) Hazardous material planning and incident response
(c) Disaster response and management
(d) Emergency vehicle management

6.13 INCIDENT MANAGEMENT

Incident management is a corresponding user service of advanced traffic management system (ATMS). The incident management activities are closely linked to emergency management, i.e., to reduce the accidents and their severity. The following ITS services are recommended for an effective incident management system:

(a) Identification of location of accidents (video survey) and vehicle tracking
(b) Emergency telephones and call-boxes
(c) First-aid systems
(d) Real-time motorist information
(e) Variable message signs
(f) Highway advisory radio, video text, and TV systems

An incident management plan (IMP) is prepared for expressways and also for city roads. These plans also include the installation of variable message sign (VMS) boards at certain intervals to display real-time information to the road users and to avoid any mishaps to the extent possible. Implementation of various levels of traffic management plans for incident management incorporate strategic use of VMS.

The emergency communication system is essential for any kind of toll way to detect incidents as they occur. Emergency telephones are installed along the side of expressways at an average intervals of 1 km with conspicous signage regarding their availability. These emergency telephones are interfaced between drivers and personnel of traffic control centre. These phones are indispensable equipment to collect information from drivers regarding any incidents.

Mobile stations are installed in patrol cars and other vehicles operated by the expressway operator to keep a watch on the operation of expressway. These stations are connected to traffic control centre or operational offices to facilitate traffic and expressway management work. Periodical and emergency patrols are performded to grasp traffic and road conditions. When an obstruction like a fallen object is detected, the patrol personnel dispose it appropriately and promptly to avoid any mishap. The patrolling is carried out after about every two hours depending on the traffic volume. e-Calling is also being deployed as another emerging response system to inform the control centre when there are any accident happens.

6.14 URBAN ROAD SAFETY

Road safety is a multi-disciplinary multi-sector issue which is being recognised as a major socio economic concern. Urban road safety refers to measures for mitigating the risk of a person using the road network being involved in an accident causing death or injury. The road users include pedestrians, cyclists, motorists and passengers of public transport mainly buses. A large number of accidents are result of human error caused by vehicle drivers or careless on part of pedestrians. The objectives of road safety strategies are the reduction in the occurrence of an accident and the prevention of serious injury and fatalities.

The increasing number of road fatalities indicates the need of improvements required to enhance the safe design and construction of roads. Road safety is an issue of immense socio economic importance for human being as it affects people both socially and economically. An urban road safety audit is essential which focuses on reducing the number and severity of road accidents within the context of development and transport goals of local authority.

Road safety in urban areas must focus on improving urban road design so as to reduce fatalities and injuries to vulnerable road users. The safe movement of pedestrians and non-motorised transport users shall be given priority over other road users. The road safety audit shall be carried out for urban roads aimed at improving the planning, design and construction of road and operation traffic of an existing road. Furher, there is an urgent need for proper recording, collection and handling of data on accidents, which would help in creating database for policy planning.

ITS components which are being deployed for road safety are as follows:

(a) CCTV cameras
(b) Automatic number plate recognition (ANPR) cameras

(c) Speed detection cameras
(d) e-challaning
(e) Variable message sign boards
(f) e-calling
(g) Red light violation detection system
(h) Intelligent signalling system, etc.

These ITS components along with others not only help in smooth flow of traffic but also provide safety and security to the road user. The system keeps informing the road users about the real-time information on the routes, which further helps to increase the efficiency of flow of traffic.

6.15 ITS FOR INTERMODAL FREIGHT TRANSPORT

Intermodal freight transport refers to use of two or more modes to move a shipment from origin to destination. An intermodal movement consists of the physical infrastructure, goods movement and transfer, and information drivers and capabilities under a single freight bill.

There has been significant growth in intermodal freight transport especially container movement thanks to opening of markets and globalisation. This unprecedented growth in freight transport has outpaced the capacity generation and resulted into severe congestion across different modes, i.e., road, rail, sea or airways. Sharing information about congestion and operations across the intermodal freight system is critical to regain capacity and reliability. Successful intermodal freight movement requires that all of the transportation components functions as part of a large system rather than a series of interfaces. Information technology is changing the intermodal freight industry by enabling it to integrate operations across the supply chain. Advanced communication and information systems and ITS enable real-time information on intermodal freight operations and congestion on the physical transportation system.

The most preferred ITS application in the field of Intermodal freight are presented as follows:

Tracing of shipment and information systems

These systems are used to track the shipment from end to end on real-time basis to ensure visibility, reliability and security of the shipment. The integrated supply chain management system links suppliers, manufacturers, carriers, retailers and end customers.

Inventory and stowage management systems

It manages and tracks the movement of containers and trailers within port, rail and truck terminals and their placement aboard ships. These node based asset management systems are deployed to optimise the use of space in terminals; manage the stacking of containers of different sizes; make effective use of labour and equipment repair and maintenance. These systems use optimisation or expert system software and automatic equipment identification (AEI) tags. The systems are generally linked to booking and gate clearance system.

Asset location and management system (ALMS)

ALMS dispatches, locates and tracks a vehicle or container. The link-based route and fleet management systems are being used to estimate time of arrival, optimise equipment use, and improve safety and reliability. The technologies such as GPS, cellular telephone, wireless radio trans-receivers and AEI tags are used. On-board computers and sensors are also used to monitor vehicle and cargo conditions.

Traveller information system

It provides real-time information about highway congestion, road closure, incidents or accidents. The system deploys variable message signs, CCTVs, probe systems, highway advisory radio, Internet or mobile applications. These systems focus on the trip and serve the needs of drivers to decide about route and time of travel. These systems are linked to central traffic management systems.

Toll collection system

It allows electronic collection of toll for highway, bridge, tunnels, etc. The system helps to enhance the throughput, minimise queuing and delays, improve travel time, reduce fuel consumption, minimise congestion and risk of accidents at toll collection points. The system uses RFID technology. The payable toll is automatically deducted from a prepaid smart card the moment vehicles passes through the toll point.

Traffic management system (TMS)

The TMS is deployed to improve the management, operation and safety of traffic on roads and railways. Advanced traffic management systems and traffic operations centres are used to improve the flow of traffic, reduce congestion and improve the safety of road and rail operations. The systems use traffic monitoring sensors (vehicle detection loops, CCTV cameras), traffic control devices (traffic responsive and traffic adaptive systems, traffic signal systems, ramp metering) and transportation network models.

6.16 SUMMARY

With increase in vehicular trafiic in cities, the condition of traffic is worsening day-by-day. The traffic jams, severe congestion, pollution and accidents have become part of daily life of city dwellers. The application of intelligent transport system holds the key to reduce the ill effects of traffic problems. A number of ITS tools are available and are increasinly developed to address the traffic problems. The city transport planners need to devise a comprehensive traffic management plan of the city involving required ITS tools as per their city and implement the same in phased manner. The phase I may involve adoption of intelligent signals. Area traffic control may be progressively enhanced to most modern integrated advanced traffic management systems encompassing all services related to urban transport system in the city. The advanced public transport system, commercial vehicle operation for urban freight, emergency management systems and incident management systems shall be made part of the city traffic management plan. Last but not the least, the traffic management plan shall also focus on ensuring the urban road safety especially for vulnerable road users like pedestrians, cyclist, etc.

Questions

6.1 What do you understand by traffic management system? Why is it important for managing traffic in a city?

6.2 What are different ITS tools which can be used for traffic management system in a city?

6.3 What are different phases of deployment of ITS for traffic management system in a city? Why it is important to develop traffic management system in a city in phased manner?

6.4 Describe proposed systems, area to be covered and system architecture for phase I of development of traffic management system in a city.

6.5 Describe proposed systems, area to be covered and system architecture for phase II of development of traffic management system in a city.

6.6 Describe proposed systems, area to be covered and system architecture for phase III of development of traffic management system in a city.

6.7 Describe proposed systems, area to be covered and system architecture for phase IV of development of traffic management system in a city.

6.8 Describe proposed systems, area to be covered and system architecture for phase V of development of traffic management system in a city.

6.9 Explain various functionalities of traffic management control with diagram.

6.10 Write short notes on:

(a) Advanced traffic management system
(b) Advance traveller information system
(c) Advance vehicle control systems
(d) Advance public transport system
(e) Commercial vehicle operation

6.11. What do you understand by incident management and how ITS can help in effective incident management?

Bibliography

Ahn, K., H. Rakha, and J. Collura, "Evaluation of Green Extension Transit Signal Priority Strategies using Portable GPS Receivers", Presented at 85th Annual Meeting of the Transportation Research Board, Washington, D.C., 2006.

Balaji, P.G. and D. Srinivasan, "Type-2 Fuzzy Logic Based Urban Traffic Management", *Engineering Applications of Artificial Intelligence*, 2011.

Bertini, Robert L. (Department of Civil and Environmental Engineering, Portland State University) and Ahmed El-Geneidy (School of Urban Studies and Planning, Portland State University), "Advanced Traffic Management System Data", Chapter 15, 2004. (http://tram.mcgill.ca/Research/Publications/ADVANCED_TRAFFIC_MANAGEMENT_SYSTEM_Data.pdf)

Faghri, A. and K. Hamad, "Application of GPS in Traffic Management Systems", *GPS Solutions*, Vol. 5(3), pp. 52–60, 2002.

Gupta, A.K., "Transportation System Management", *National Conference on Transport System*, Chennai, at IIT, Madras, 1992.

Hall, Randolph W. and Chris Intihar, "Commercial Vehicle Operations: Government Interfaces and Intelligent Transportation Systems", California PATH Research Report, ISSN 1055-1425, 1997. (http://its.berkeley.edu/sites/default/files/publications/UCB/97/PRR/UCB-ITS-PRR-97-12.pdf)

He, Z. and Q. Zhang, "Public Transport Dispatch and Decision Support System Based on Multi-Agent", *In the proceedings of Second International Conference on Intelligent Computation Technology and Automation*, Zhangjiajie, China, 2009.

Hemily, B., "Providing Guidance for the Design and Implementation of Travel Management Coordination Center (TMCC)", Washington, DC, US Department of Transportation, ITS-Joint Programme Office, 2012.

Intelligent Transport Systems: Synthesis report and issues and challenges under Indian conditions, Centre of Excellence in Urban Transport, IIT Chennai, December 2010. (http://www.itsinternational.com/event-news/intertraffic-2012/indian-minister-explains-future-transportation-opportunities-at-intertraffic/)

Intelligent Transportation Systems for Traffic Signal Control, US Department of Transportation, (http://ntl.bts.gov/lib/jpodocs/brochure/14321_files/a1019-tsc_digital_n3.pdf)

ITS Toolkit for Traffic Management System, Institute of Urban Transport, MOUD, India.

Kadiyali, L.R., *Traffic Engineering and Transport Planning*, 3rd ed., Khanna Publishers, New Delhi, 2002.

Khisty, C.J. and B. Kent Lall, "TSM Planning: Framework", in *Traffic Engineering: An Introduction*, 3rd ed., Prentice-Hall, USA, 2002.

McQueen, Bob, Rick Schuman, Kan Chen, *Advanced Traveller Information Systems,* Artech House, Boston MA, 2002.

National Road Transport Policy, Ministry of Road Transport and Highways, Government of India.

National Urban Transport Policy, Government of India.

Qureshi, Kashif Naseer and Abdul Hanan Abdullah, "A Survey on Intelligent Transportation Systems", *Middle-East Journal of Scientific Research,* Vol. 15(5), pp. 629–642, IDOSI Publications, 2013.

Transport Research Library (TRL), UK, "Commission takes first step towards rollout of eCall system", September 9, 2011, Retrieved November 16, 2011.

US Department of Transportation, "Multi-Modal Intelligent Traffic Signal Systems (MMITSS) Impacts Assessment", Final Report, FHWA-JPO-15-238, August 19, 2015. (http://ntl.bts.gov/lib/55000/55700/55710/MMITSS_IA_REPORT_0811_v1.4.pdf)

7 Transport Demand Management

7.1 INTRODUCTION

The demand of travel for a section is measured in terms of passenger trips per hour in a direction over the section. The demand is usually high during morning and evening peak hours. Further, the demand is high in the central business district (CBD) of the city and it reduces progressively towards the outskirts of the city. The wide variation of demand across time and space causes extreme congestion in peak hours near CBD. The conventional approach to deal with congestion is to increase capacity by widening the roads, constructing flyovers, elevated roads, etc. However, it has been observed that additional capacity attracts more traffic and within short period of time, the additional capacity is offset by more vehicles on the road and consequent congestion. The enhancement in capacity to address the congestion also called supply side measure is found to be ineffective in long run. The traffic congestion in Delhi, the capital city of India, is testimony of ineffectiveness of supply side measures which remained unabated even after constructing number of flyovers, introduction of world class metro rail system. In view of limited effectiveness of supply side measures, there has been increased emphasis on controlling the travel demand itself by taking slew of measures called transport demand management tools. This chapter deals with the application of ITS tools for transport demand management.

7.2 WHAT IS TRANSPORT DEMAND MANAGEMENT (TDM)

Transport demand management is a way of influencing individual travel behaviour and providing expanded options to reduce the actual demand, or number of vehicles, placed on transportation facilities. The focus is on managing the demand side of the transportation equation rather than increasing supply by widening or building new roads. Transport demand management refers to strategies to control the traffic demand for managing the traffic within prescribed levels based on supply of available road space or other benchmarks, like environment issues, etc. TDM strategies focus on controlling traffic demand in terms of vehicles per hour per direction by encouraging a shift from single-occupant vehicle (SOV) trips to non-SOV modes, or shifting auto trips out of peak periods. It actually deals with the improvement in the modal split towards the favour of public transport uses and spreading travel demand in space and time. This demand management is mainly concerned with following focus areas:

(a) Reduce reliance or attractiveness of private transport (Push factor—Pushing people from private to public transport)
(b) Increase attractiveness of more environmentally friendly and sustainable forms of transport including public transport (Pull factor—Attract people towards public transport)
(c) Use of new strategies to reduce the need to travel into congested business areas and city centres

And these above-mentioned strategies need balanced with economic and social criteria for a viable city centre. The concept of travel demand management is explained in Figure 7.1.

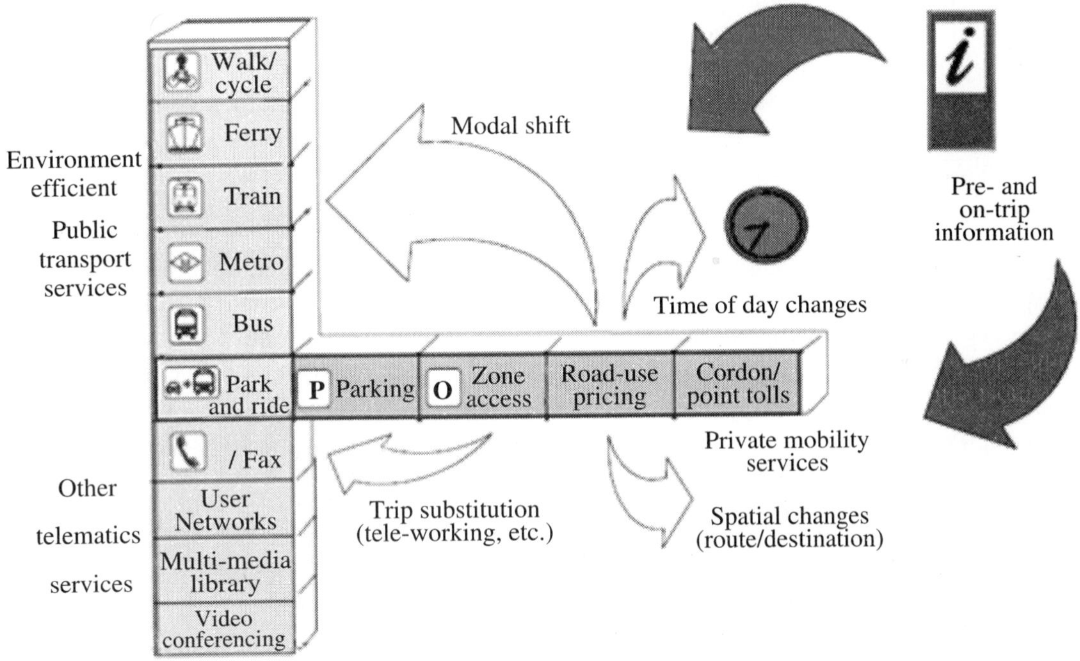

FIGURE 7.1 Demand management concept (*Source:* ITS to support the principles of demand management—Prof. Phil Blythe, reproduced with permission).

As explained in Figure 7.1, the peak transport demand can be managed through:

(a) *Modal shift:* Shift from private to public transport which reduces number of single occupancy vehicles (SOVs) on road

(b) *Spatial Changes:* It involves spatial distribution of demand by imposing restrictions on route like one way traffic during peak hours

(c) *Time of day changes:* Like staggered timings of public, private offices and institutions

(d) *Time substitution:* Like encouraging work from home, flexi timings, tele working, 5-day working instead of 6-day working per week

7.3 APPLICATION OF ITS FOR TRANSPORT DEMAND MANAGEMENT

ITS can be applied in number of ways to meet the objectives of TDM. The ITS application in TDM involves collection of reliable traffic data, analysis thereof and development of suitable TDM strategies to make them intelligently optimise the management and control of transport systems. The ITS can be practically applied in most of the TDM measures. Some of the ITS tools which are most frequently used in developed countries are as follows:

7.3.1 Promotion of Public Transport

The objective is to make public transport system attractive to the users to facilitate shifting of people from private to public transport. The people may be encouraged to use shift to public transport by improving the convenience of using public transport with the help of ITS. The ITS applications may be deployed to meet following requirements of the people in the context of public transport:

(a) To plan trips
(b) To make more informed decisions about how and when to travel
(c) To better understand the cost of a trip
(d) To influence travel behaviour
(e) To re-plan the trips 'on the move'

ITS applications such as real-time passenger information systems (RTPI), mobile application with real-time information of public transport, integrated automatic fare collection system, etc. may be quite useful for the people and may attract them towards public transport.

7.3.2 Road User Charging or Congestion Charges

The road user charges or congestion charges are collected from the users who use a particular stretch of the road (congestion zone) in the defined period (usually peak period). The objective of the congestion charges is to control the traffic demand during the peak hours by imposing a fee and thus to push people to public transport. The congestion charges is considered as one of the key instruments for reduction of traffic congestion as well as revenue generation. This is used as a part of travel demand management, as the cost which is been charged to the road user actually reflects the full external cost of travel. Figure 7.2 shows an application of electronic road pricing in Singapore.

FIGURE 7.2 Electronic road pricing (*Source:* Vinay Tiwari, Singapore, reproduced with permission).

The early age congestion pricing system was based on issue of paper ticket as permission to enter the congestion zone. The road users were supposed to display the ticket on the wind screen of the vehicle while entering the congestion zone. The violators were imposed heavy penalty. The modern ITS based congestion pricing systems has done away with the paper ticket system. The applicable congestion fee is automatically deducted the moment the vehicle enters the congestion zone.

One of the successful examples of congestion pricing is Singapore ERP (electronic road pricing) which was introduced in 1998. In Singapore a device known as an in-vehicle unit (IVU) is affixed on the lower right corner of the front windscreen within sight of the driver, in which a stored-value card, the cash card, is inserted for payment of the road user charges (Figure 7.3). It is mandatory for all Singapore-registered vehicles to be fitted with an IVU if they wish to use the priced roads. The moment vehicle passes through the ERP gantry, the applicable congestion fee is deducted from the card. The concession fee changes with the time of the day and actual congestion on the road. The applicable congestion fee is displayed at the top of the ERP gantry for information of the road users. The Singapore ERP system has been explained in greater details in Chapter 19.

FIGURE 7.3 Card to be used in-vehicle unit (*Source:* Sh. Vinay Tiwari, Singapore, reproduced with permission).

In a study conducted by Ahmedabad-based Centre for Environmental Planning and Technology (CEPT) University in February, 2016, congestion charge has been proposed to ease congestion in the Rajiv Chowk (Connaught Place) area of Delhi. The report suggests that while imposing ₹160 per trip per vehicle congestion charge could make 13% car users shift to public transport, increasing it to ₹200 would shift around 18% car users to use public transport.

7.3.3 Parking Management System

Affordable and easy availability of parking encourages use of private transport. On other hand, limited availability or costly parking encourages use of public transport and helps in reducing the traffic demand on roads. In view of significant impact of parking on travel demand, an effective parking management strategy can be one of the best transport demand management tool. Parking management refers strategies that encourage more efficient use of existing parking facilities, reduce parking demand and shift travel to non-SOV modes. Managing parking helps to reduce the undesirable impacts of parking demand on local and regional traffic levels and helps in managing transport demand in a city. At the same time, intelligent management of parking helps to ensure access to retail businesses, provides access for visitors to regional and neighbourhood attractions and supports neighbourhood vitality. The following ITS tools may be used to effectively manage parking in a city:

Demand based parking charges (DBPC)

The most effective parking strategies are pricing measures in which parking rates vary according to the demand. So as demand or occupancy of the parking spaces increase, the parking charges are also raised. This strategy controls the total parking demand, shifts travel to other modes, reduces vehicle miles travelled (VMT) and ensures a minimum number of parking spots are always available, avoiding the "circling" problem adding to congestion. A software can be deployed to estimate the parking charges as per the pricing model adopted by the authority. The parking charges can also change with time using ITS, i.e., higher in peak hours and lower in non-peak hours to spread demand over time.

Electronic parking guidance systems (EPGS)

EPGS guides the motorists from the main access roads of a defined geographic area to parking facilities which have some spare spaces. The number of spare parking spaces available in a specific car park is displayed shown on variable information signs on real-time basis. The availability of spare spaces may also be presented via phone, Internet, or in-vehicle navigation systems. This strategy controls the parking search traffic which leads to a reduction of emissions, fuel consumption and wasted time; promotes better use of existing parking facility capacity; and can direct parking traffic onto dedicated roads.

Electronic parking management system (EPMS)

The most common electronic parking management system is mainly aimed at parking access control. The main elements of a typical EPMS are:

(a) Electronic gate controlling entrances and departures to/from the parking area.
(b) Smart card based entry and exit
(c) Automated access control, based on RFID devices or ANPR (automated number plate recognition)

The EPMS system not only manages the parking lot but also gives real-time feed to variable information signs displaying the availability of space in parking. The information can also be fed to area traffic control parking, internet, mobile application, etc. for the users. The reservation of parking spaces can be managed by specific mobile apps, giving the traveller the possibility to book a parking space and therefore efficiently plan his journey. Furthermore, parking management systems could be integrated in enforcement policies, improving the security of parking area and more in general, the quality of urban life.

The ITS tools like automatic number plate recognition system, automatic collector of parking fee from smart card, integration with public transport through 'Park and Ride' may further improve the efficiency of parking system and help in transport demands management.

Conceptual system architecture for a typical parking management system

With this concept, the process of parking is tried to start from the moment the need arrives. Thus there are two types of parking users based on the moment of realisation of need of parking space. The first type of user is the one who plans the journey in advance at home only. This category of user has been provided the privilege of parking information through mobile phones, landlines, local cable television and Internet. Through the services mentioned before the parker can have the information about availability of parking space and the estimated availability of space before they even began their trip? Also the parker will have the facility of booking the parking in advance through Internet.

For the second type of users who has not planned their journey well in advance can have information by calling one centralised telephone number. The parking management system will have also VMS boards installed within 2 km of radius giving information about availability of parking space at the time of arrival at parking lot. As soon as the parker enters in this radius the parking user will be given directions through variable sign boards parking towards the parking facility. The users will also be provided information through FM radio which becomes active for the user as soon as he enters in the radius of 2 km of parking lot. By this FM service the information about the parking space as well as general directions will be conveyed time to time. By implementing these methods, the chances of loss in time as well as chances of mismanagement will be reduced. As the whole parking system will be centralised, facilitating the authorities to estimate the amount of parking space left at required time; the way people plan their journeys with concern about parking space will be completely different.

Logical architecture: The whole logic behind the parking system is about simplifying the process of parking and having a systematic and convenient parking system incorporating user's involvement at every step as explained in the Conceptual System Architecture for typical Parking Management System.

Physical architecture: From the point of entry till the point of exit, at every step central computer uses lots of physical hardware to collect, manage and to convey data. As one is approaching the parking system, VMS is easily visible as well as the FM transmission. VMS is controlled by the central control room through wireless transmission. Although power transmission cable has to be installed but by installing photovoltaic cell ensures very little use of electricity by generating its own electricity through solar power.

The second array of equipment is visible at the point of entry. As a car approaches, the automated entry stop bar opens only in case of availability. As car enters inside, a Webcam take pictures of car's driver and plate numbers, and send the information to the centralised computer system (CCS). At the same time vehicle sensor analyse the size of the vehicle determining whether it is a two-wheeler or four-wheeler. At the side of the driver there is a token button which gives the entry slip. This system here generates bar-coded slip which is easy to read and track electronically. After this a green signal is given by the LED aspects signaling the vehicle to move and automated boom barrier opens letting vehicle in the parking lot. Through all this time, the entering and the data of vehicle is being recorded by the CCS.

At the parking bays, a sensor records the presence of vehicle and sends the information to a 16-bit micro controller system which in turn activates the LED status appropriate indicators. All these information is stored in a flash memory. In case of vacancy and occupancy, the data is send to the 16-bit micro controller data concentrator which uses flash memory to store configuration. It then updates the central data server.

When the duration of parking is complete, the vehicle driver has to produce the entry slip at exit point. A barcode reader reads the token and gets information from central database and then generates receipt and prints bill. Here Webcam takes shots of the licence plate for live streaming information. At this point, there is again one signal light to indicate the movement of vehicle. Here an automated boom barrier works as access control to the parking.

The development of logical and physical architecture can be conceptualise in the form of system architecture in Indian context. As there is no concept of ITS architecture developed so far in India. Figure 7.4 shows a concept of system architecture.

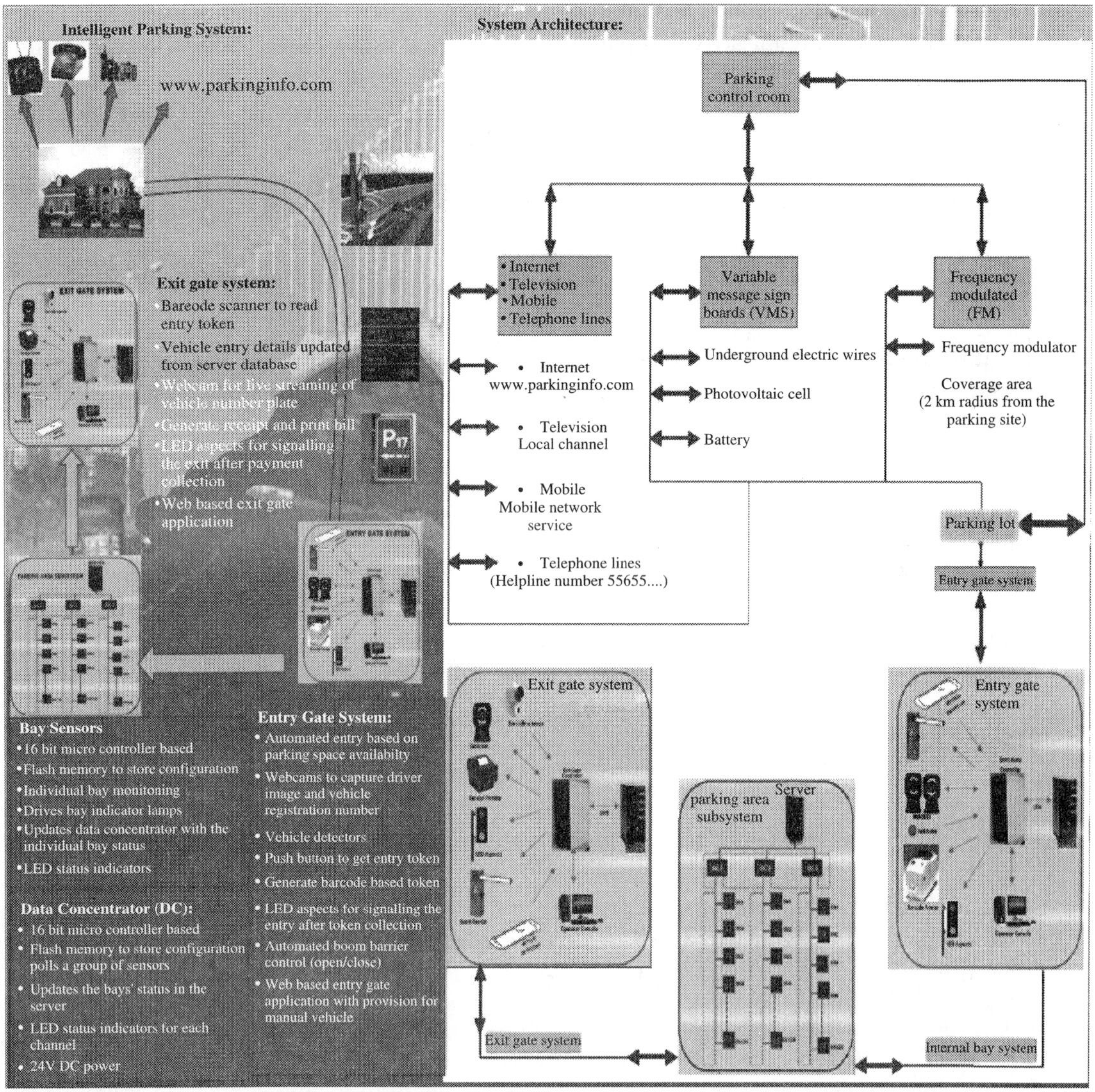

FIGURE 7.4 Concept plan and system architecture for ITS based parking management system[1] (*Source:* Unpublished Thesis work of Sh. Ankur Shivare, SPA Delhi).

Process flow chart as per user group

User group 'A' using Internet facility: This kind of user has the facility of booking parking through online portals and also can receive parking directions in advance. The user with GPS facility can find the location of parking lot on popular GPS services like Google maps, Ovi maps, etc. and get driving directions. As soon as the user following the directions with information about parking conveyed through same FM channel reaches the parking lot, he is easily able to find his way to the parking bays available using internal information system.

1 The work of Ankur Shivare, SPA Delhi is duly acknowledged.

These users can keep a tab on parking availability through the parking facility Website. These sites provide all parking data from parking availability to exit directions.

User group 'B' using mobile phone service: This kind of user receives status update through SMS and can call on one centralised number with the inbound customer care providing information about parking status. This user group has the facility of booking parking much before they reach the parking area. They can book the parking space through the centralised customer care cell. As soon as the user following the directions with information about parking conveyed through same FM channel reaches the parking lot, he is easily able to find his way to the parking bays available using internal information system.

These types of users have choice of having receiving information about range of services like monthly routine booking of service. They can also receive directions of going to parking lot as well as exiting from parking lot.

User group 'C' watching local cable channels: This kind of parking user is provided parking availability data through advertisement ribbons at the end of channel display. They also provide contact information like centralised customer care cell number and also the way of having information from VMS boards. As soon as the user following the directions with information about parking conveyed through same FM channel reaches the parking lot, he is easily able to find his way to the parking bays available using internal information system. These advertisements provide information about the Website address and other data. This kind of user has direct knowledge of availability and parking location from advertisement banners and also have a choice of booking parking space through further information is given.

User group 'D' using VMS boards: When a user in search of parking space enters in the radius of 2 km, user is able to see easy and eye catching VMS which gives the instruction to reach the parking lot from that point. Also VMS keeps displaying updated parking space status. The boards also display instructions for using the parking bays. As soon as the user following the directions with information about parking conveyed through same FM channel reaches the parking lot, he is easily able to find his way to the parking bays available using internal information system.

While exiting the parking lot, the lower portion of VMS screen area displays way out to the main roads. The parking management system (PMS) architecture presented here is typical in nature. It may change as per the user's requirements, local conditions of the place, rules and regulations, government's priority, etc. Indeed there are plenty of examples of PMS architecture worldwide and new models are being adopted with emerging innovations in ITS.

7.3.4 High Occupancy Lanes (HOL)

The high occupancy lanes are meant for vehicles having more number of passengers. Provision of dedicated lane for high occupancy vehicle is another technique of the demand management as people are encouraged to leave single occupancy vehicles (SOVs) and opt for public transport. In this system, a separate lane is provided for the vehicles with occupancy 3 or more. Based upon high occupancy the toll charges get reduced. Figure 7.5 demonstrates the application of HOL in Maryland, US (Figure 7.5).

The high occupancy lane enforcement can be done using ITS tools such as APNR, CCTV cameras, etc.

FIGURE 7.5 HOV 2+ lane on Interstate I-270 towards Frederick, Maryland, US (*Source:* https://commons.wikimedia.org/wiki/File:HOV2%2B_I270_MD_08_2009_7039.JPG, Author: Mariordo (Mario Roberto Duran Ortiz), Free to use).

7.3.5 Bicycle Rental Schemes

The bicycles can be promoted as first and last mile connectivity with public transport to encourage people to shift to public transport. The bicycle stands are spread over the city connecting public transports stations with neighbouring activity centres. A user can pick a bicycle from any stand and leave at other stand as per his convenience. The ITS tools can be used for tracking the movement of cycles, availability of the cycles at a stand, deduct cycle usage charges from the smart card of the user. The case study of bike rental scheme in France called 'Velib' has been described in Chapter 19.

7.3.6 Carpooling

The carpooling or sharing scheme encourages people to share their cars so that more than one person travels in a car. The carpooling schemes help in reducing single occupancy vehicles (SOVs) on the road and in turn manage the transport demand. For examples, carpooling by three people results in removal of two cars from the road. ITS tools like carpooling Websites, mobile applications, etc. can help in connecting people travelling on the same route. The Websites like www.carpooling.in, www.poolmycar.in, www.blablacar.in/ride-sharing/new-delhi/ etc. offered carpooling facilities in India.

7.3.7 Integrated Fare System

Integrated fare system allows use of single fare product across multiple modes of public transport. So a user is not required to purchase tickets every time he/she changes the mode of transport. The convenience of using single fare product across the modes encourages people to use public transport and helps in transport management. The ITS based automatic fare collection (AFC) systems allow a single smart card fare product to be applicable across multiple modes of transport. The More Card announced by Ministry of Urban Transport, India is an integrated fare product which can be used over all modes of public transport in India. The functionality of AFC system has been discussed in details in Chapter 16 of this book.

7.3.8 Traffic Rule Enforcement

The ITS tools can be used for effective implementation of the traffic rules to ensure transport demand management in the city. For example, the enforcement of Odd-Even scheme which was implemented in Delhi as transport demand measure can be enforced through automatic number plate recognition (ANPR) system.

Under the scheme, the vehicles with registration number ending with odd number (1, 3, 5, 7, 9) are allowed on odd days (1, 3, 5, 7, 9–29, 31) of the month and vehicles with registration number ending with even number (0, 2, 4, 6, 8) are allowed on even days (2, 4, 8, 10, 12–28, 30) of the month except on Sundays when there was no restriction.

7.3.9 Incentive Scheme to Travel in Non-peak

ITS tools can be used to launch incentive schemes to encourage people of travel in non-peak hours. In an effort to promote non-peak hour travelling in public transport, the Singapore government offers the incentives for Singapore's commuters program ("INSINC program"). Passengers automatically earn 1 point for every 1 km travelled on the train all day Monday through Friday. The INSINC points can be converted to cash rewards, which will be transferred to the commuter's card each month. Commuters are rewarded with additional points for travelling during the period preceding or following the peak hours of 7:30 am to 8:30 am.

7.3.10 Active Traffic Management

Active traffic management is one of the means of increasing peak capacity and smoothening traffic flows on busy major highways. Techniques include variable speed limits, hard-shoulder running and ramp-metering controlled by overhead variable message signs. This system has been implemented in several countries, including Germany, the United Kingdom, France, Netherlands, Belgium, Italy, Moscow in Europe, and the United States as shown in Figure 7.6. The system has proved to be highly effective in managing the peak hour transport demand.

Active traffic management (ATM) is a strategy to manage the congestion on real-time basis as per current traffic conditions. Focusing on trip reliability, the system tries to maximise the efficiency of the facility. With the help of integrated systems with new technology, it increases throughput and safety and optimises performance quickly. ATM mainly put focus on influencing travel behaviour with respect to lane and facility choices.

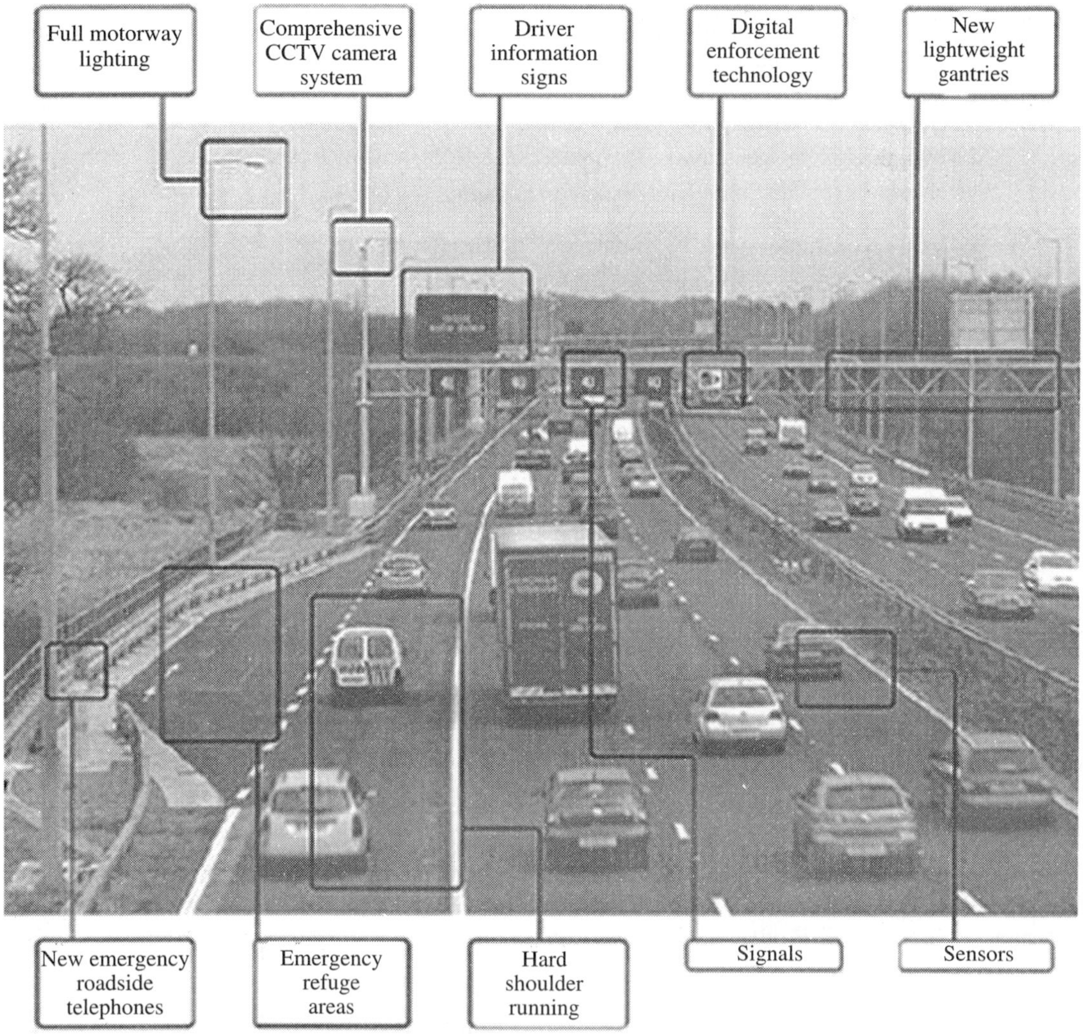

FIGURE 7.6 Active traffic management. (*Source:* Class notes—"ITS to support the principles of demand management"—Prof. Phil Blythe, Reproduced with permission.)

7.4 SUMMARY

With the increasing number of vehicles on road and limitation on increasing the capacity of the road network, travel demand management tools are increasingly being adopted to address traffic congestion in the cities. The introduction of congestion charges with the help of modern electronic payment means has emerged as the most popular TDM tools thanks to a number of success stories like Singapore, London, Seoul, etc. The results show a substantial improvement in the traffic conditions wherever congestion pricing has been efficiently implemented and enforced. The other measures like electronic parking guidance system, parking management system, cycle renting, carpooling, integrated fares, active traffic management, high occupancy vehicles, staggered office timings, etc. have also proved to be effective in controlling peak hour traffic congestion.

Questions

7.1 What do you understand by traffic demand management?

7.2 Describe different demand management strategies.

7.3 Explain application of ITS for travel demand management.

7.4 How road user charging can be used for travel demand management in a city? Give some examples of application of road user charging for travel demand management.

7.5 What do you understand by active traffic management? What are the different techniques used for active traffic management?

7.6 Describe black box PAYD insurance.

7.7 Explain use of ITS in personal travel.

Bibliography

Been, Vicki, "Building Environmentally Sustainable Communities: A Framework for Inclusivity", Furman Centre, 2010. (www.furmancenter.org/files/publications/White-Paper-Environmentally-Sustainable-Communities.pdf)

Blythe, Phil, "Transport Engineering and Operation", Class Notes, New Castle University, UK, 2013.

Menon, Gopinath, ERP in Singapore—A Perspective One Year On, *Traffic Engineering and Control*, Vol. 41, No. 2, 2000. (https://www.lta.gov.sg/ltaacademy/doc/Academic%20paper%20on%20ERP.pdf)

Myanmar business today, "ERP— A success or a failure in regulating traffic flow in Singapore", February 10, 2016.

Official Website of Land Transport Authority of Singapore, (https://www.lta.gov.sg).

The Dallas Morning News, "High Occupancy Vehicles", April 25, 2016. (http://www.dallasnews.com/)

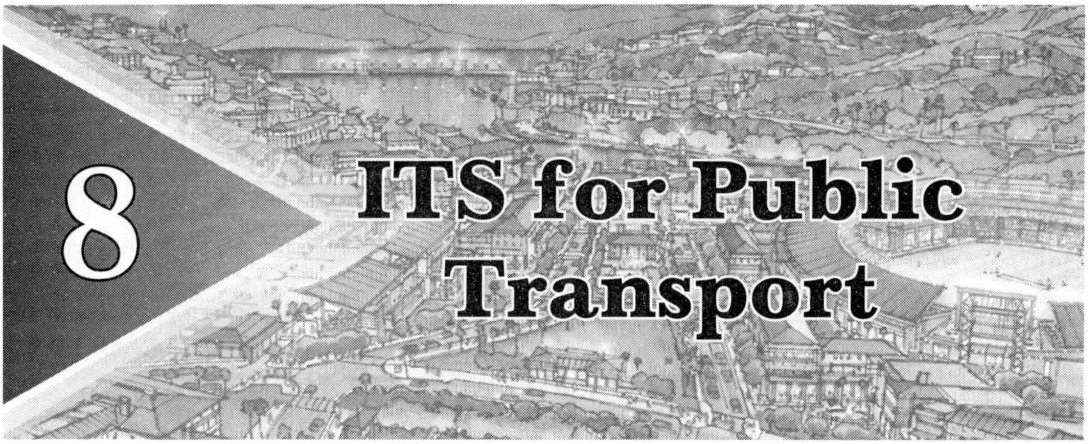

8 ITS for Public Transport

8.1 INTRODUCTION

The public transport system is a lifeline of any sustainable city. A well-organised and integrated public transport system is essential for a sustainable city. The development of a well-organise, integrated and efficient public transport system requires huge capital investment. The heavy capital cost coupled with expensive operation and maintenance of public transport systems, makes it quite challenging for public transport operator to remain financially viable without the support of public authority. A reasonable level of ridership of public transport is fundamental to financial sustainability of a public transport system for which the public transport system shall be attractive to the users and efficient in operation and maintenance. The adoption of ITS tools to manage public transport can be very useful in meeting above objectives.

Some of the general issues involved with public transport services, where ITS can help, are:

(a) Not adhering speed limits

(b) Not following bus lanes or not standing at bus stops bays

(c) Not stopping at bus stops

(d) Not completing trips

(e) Not following the route and timetable

(f) Overcrowding

(g) Coin change problem

The above issues are pretty much related at bus and at driver level. At management level, the calculation of the performance index of public transport system (like vehicle utilisation, fuel consumption per vehicle per kilometre, etc.) at aggregate level are also areas of interest. This chapter presents some of the ITS tools which can be used to make public transport attractive, and operationally efficient.

8.1.1 Use of GPS System

The GPS system can be used for tracking and monitoring of public transport vehicles. This system provides a real-time data on tracking individual vehicles. Through this technique movement of each bus, details of trip length, trip time, time spent at bus stop, change in route, complete movement of the bus on a route for every minute, crisis circumstances, log of total trips, etc. are recorded. This strategy additionally helps in creating passenger information system at every bus stop. Through control room, one can watch, coordinate and speak with every bus of this system. This is valuable in order to handle emergencies also.

GPS system can also help in controlling the eve teasing or pickpocketing as when any incident happens then driver can essentially press one button and control room may make prompt move by advising the police control room or PCR vans about the incident along definite area of bus. This again will help in controlling law and order problem.

In this system, every bus is equipped with GPS and all these buses are connected with a server in the control room. The principle server records the movement of all buses. According to the assessment from continued monitoring of buses, the control room can take fundamental activities in case of any deviation or infringement. Further, the control room can transmit the message to checking staff in the field to make fundamental action at the spot against those buses which violate the traffic rule (speed limit), etc. In the event of this, passenger information system is provided at bus stops, the expected arrival of bus at that specific bus stop will likewise be communicated. This certainly helps in improving the reliability of public transport system and trust of people in the system.

This technique in India must be effective if condition of art control room is developed and this control room is controlled by top of the line experts. Through this system one can have speed details, halted details, route diversion or deficient route can be assembled. It is helpful to check the speed and to see that bus stops at bus stops or not.

Hence, this technique can be utilised for checking the speed, issues concern to not stopping or waiting at bus stops, not completing the trips, not following the route and not adhering to timetable. The system also provides the traveller information at every stop which is very common in developed countries. From management point of view, the system gives complete details for vehicle utilisation, fuel consumption kilometre wise, passenger wise, depot level, route level and the details of every bus.

8.1.2 Automatic Passenger Counter (APC)

The automatic passenger counter is the technology which is used to record the number of travellers using the public transport with time and the location of every stop as passengers get on and off transport. The technology can collect the information records physically or with a hand-held device, with a reduction in time, cost, and effort. The infrared bars at the doors or weight delicate mats on the steps can also be used as APC which correctly records the time, region, and the number of travellers as they enter or leave the vehicle. With the information provided by the APCs, the public transport operator can improve the plans that better serve the transportation needs of the commuters.

8.1.3 Automatic Vehicle Location (AVL) and Automatic Vehicle Identification (AVI) Systems

Automatic vehicle location (AVL) technology is used to track the location of transit vehicles in real-time through the use of GPS devices or other location-monitoring methods.

Automatic vehicle identification (AVI) is basically an identification technique used for moving objects such as vehicles. This is usually accomplished with the help of radio frequency identification (RFID) as the object passes a RFID reader.

Information about the vehicle location is communicated to control centre in either raw data format or as processed data.

There are several benefits associated with this technology, which can be used in BRT vehicles:

(a) This system improves the control to ensure efficient operations.
(b) This system can be calibrated to distribute service times and covers the signal priority and control centre.
(c) This technology improves the bus safety in an emergency condition by showing the vehicle location.
(d) This also improves quality of service by notifying the location of next bus and expected arrival time.
(e) This improves the system integration by providing the better bus scheduling and knowing the location of each vehicles.
(f) It also reduces need of voice communications.

8.1.4 Traffic Signal Priority (TSP)

Traffic signal priority is a technique that accords priority to public transport at signaled intersections by keeping a traffic light green or turning it green earlier to facilitate the movement of vehicle without stopping at the intersection. Conventionally signal priority was only implemented for emergency vehicles, but now is being used for public transport as a measure to encourage people to use public transport.

The TSP system is combined with AVL technology, which allows the system to provide priority only when needed (i.e., when a public transport vehicle is running behind schedule). The signal priority system keeps transit vehicles on schedule and improves on-time performance. As a result, the reliability of the public transport improves and fewer vehicles are necessary for a route which results into a large savings for a transit agency.

8.1.5 Geographic Information System (GIS)

A geographic information system (GIS) is a system designed to capture, store, manipulate, analyse, manage, and present spatial or geographical data. It is widely used in transportation planning agencies, especially among metropolitan transportation organisations. A geographic information system (GIS) technology facilitates the public transport operator to assemble, store, examine and show information by geographic location. With GIS programming, a spatial analysis of data is possible which can be helpful in identifying the areas of economic activities where demand of transportation is high. Travel coordinators further can use the information given by GIS to alter, enhance, or include new administrations.

8.1.6 Real-Time Passenger Information (RTPI)

RTPI is a public information system which provides real-time passenger information regarding public transport system. RTPI may include both estimated arrival and departure times, as well as information about the nature and causes of disruptions. This technology provides information related to the status of transit service to the passengers including information needed before making a trip, during the trip and at the termination of a trip on real-time basis. The information may be communicated to the users physically within a transportation hub and remotely using a Web browser or mobile device. The AVL or GPS technologies may be used to generate necessary data to provide real-time information. RTPI improves the overall experience of the public transport and promotes the use of public transport.

8.1.7 Fare Collection

Fare collection is the way in which travellers pay their fares, either on the bus or at the station or ahead of time. The use of magnetic strip or RFID based electronic fare collection has changed the way fares are collected in a public transport system. The RFID based smart cards are increasingly in use as integrated fare media applicable across all modes of public transport systems in a city. More details on Intelligent Ticketing System are available in Chapter 16.

8.1.8 Lane Control Technologies

This technology is used for dedicating a lane for public transport (bus usually) for a short of time. Intermittent bus lane can also be called a moving bus lane. This concept consists of using a general-purpose lane that can be changed to a bus-only lane just for the duration of time needed for the bus to pass. Afterward, the lane reverts back to a general-purpose lane until another approaching bus needs the lane for its movement. By using this technology speeds and reliability are improved and bus can move independently from the general traffic. Moving bus lane does not require expensive capital cost because it uses the existing roadway infrastructure and takes time to move separately from general traffic. This allows the bus lane to be used for general traffic majority of time. It has low implementation costs and uses well-known, widely used and proven traffic signal management technologies and products.

8.1.9 Passenger Wi-Fi

Wi-Fi technology facilitates customers to connect the Internet using mobile devices, such as laptops, cell phones, smart watches, etc. The public access to Wi-Fi is provided through hotspots either free of charge or for a fee (pay-per-use or subscriptions models are possible). These wireless hotspots can be set up both on-board transit vehicles and at stations. At stations, wireless service can be provided with off-the shelf routers and modems (cable or DSL). For on-board Wi-Fi, two options exist: installing routers on each vehicle connected to satellite networks, or creating a continuous wireless mesh along the corridor through a number of overlapping fixed access points. Wireless Internet provision is not an essential transit component; however, it offers a high-quality customer amenity. One of the many potential benefits to customers is an increased utility of transit as it can allow people to make use of their time in transit for entertainment or work, which could have the effect of decreasing their perceived transit travel times. This is especially important for customers with long journeys. Moreover, the availability of Wi-Fi can improves customer perception about the public transport services and generates a feeling of "premium" levels of service. This may increase the attractiveness of transit to non-traditional market segments.

8.1.10 Surveillance/CCTV/Security Systems

Security systems in public transport include both surveillance CCTV and alarms, both of which can be located on-board vehicles as well as off-board at stations or along guide ways.

Surveillance devices primarily consist of closed-circuit television (CCTV) cameras, generally equipped with microphones. The CCTV network facilitates the control centre to remotely monitor vehicles, stations and guide ways. These system gives the advantage of improvement in emergency reaction to both the operator and public. Cameras and microphones can allow central control to continuously monitor and assess the installed and off-board circumstance to decide suitable help by transit security, law authorisation or different types of emergency response (i.e., medical aid, Hazmat, terror threat).

8.1.11 Request-Centric Wireless Bus Information Management System

A distributed bus information management system is used to manage the bus operations on real-time basis as per the information provided by the on-board passengers, passengers waiting at the bus stops, and occupancy of the bus. The system integrates passenger requests and manages information exchange between a bus and a bus-stop. On the basis of bus management signals, the bus-stop management subsystem obtains the information about the bus route identification, the number of accommodation available (based on occupancy and carrying capacity), need to stop or not, etc.

The following steps are involved in the process:

(a) The passengers waiting at the bus-stop enter the desired route/destination information to the bus-stop via wireless option such as GPRS, WiFi, etc.

(b) The bus-stop transmits the requested information to the bus.

(c) The on-board passengers in the bus enters stop requests or transfer information to the bus.

(d) The decision of stopping at the bus-stop is made by the bus-stop management system on the basis of the stop request and availability of the accommodation in the bus. If there is no request to stop the bus or bus is full (as per carrying capacity including seating and standing), the bus management system shall decide not to stop the bus at that bus-stop and transmits Non-stop passing signal to the bus-stop. However, if there is any request to stop the bus at the bus-stop, the stopping signal is transmitted to the bus-stop.

(e) The updated transfer information is transmitted to the passengers waiting at the bus-stop by the bus-stop management system.
(f) The bus-stop transmits the updated transfer information to the bus.
(g) The bus transmits the updated transfer information to the on-board passengers in the bus.

There is bi-direction communication between bus and the bus-stop. Through bi-direction communication, the bus management subsystem can determine whether the bus-stop management signals should be processed. On the basis of the signal system, control system and communication system, a bus-stop and a bus work as a group to dynamically perform request analysis and service control. Therefore, passengers on the bus or waiting at the bus-stop can obtain sufficient information to flexibly schedule their travel plan.

8.2 ITS OPERATIONS FOR PUBLIC TRANSPORT

The three major implications of ITS integration with public transport are:

(a) Fleet operation and management
(b) Passengers information
(c) Electronic fare collection

8.2.1 Fleet Management and Operations

The fleet management provides the fleet manager a complete overview of all public transport vehicles. A fleet management system records performance parameters defined by the fleet manager with the help of ITS technology. Fleet management and operation includes five different technologies, as follows: automatic passenger counters (APC), automatic vehicle location (AVL), geographic information systems (GIS), scheduling and dispatching (S&D), transit signal priority (TSP) and lane control signals (LCS). These separate technologies are often integrated in various software packages, which allow for the integration of many different transit functions. The computer applications improve the resource utilisation to meet service demands, which help in making public transportation more attractive to customers.

Use of APC for fleet management

The APC automatically keeps a log of the number of passengers, time and location of each stop as passengers get on and off the bus. This technology also records on-board transit vehicles, boardings and alighting at each stop and keeping a running total of passengers on-board the vehicle. The APC units include sensors (typically infrared) at door of the vehicle to monitor traveller movements on and off a vehicle. An APC system prepares a digital record at every bus stop, typically including stop location, stop date and time, time of entryway opening and closing, and number of travellers boarding and alighting. APC information can be downloaded manually through a portable PC or wireless data via a local area network, on real-time basis. The key benefit of an APC system is its ability to provide ridership data along the transit network with time of the day. This data can be constructively used for effective planning and scheduling of public transport as per the demand. The manual ridership surveys historically has been labour extensive and has provided only a snapshot of ridership conditions to transit agencies, given that only a sampling of bus routes generally can be surveyed. The APC has drastically improved the reliability of the data collected which is being used to effectively manage the precious fleet of public transport (Figure 8.1).

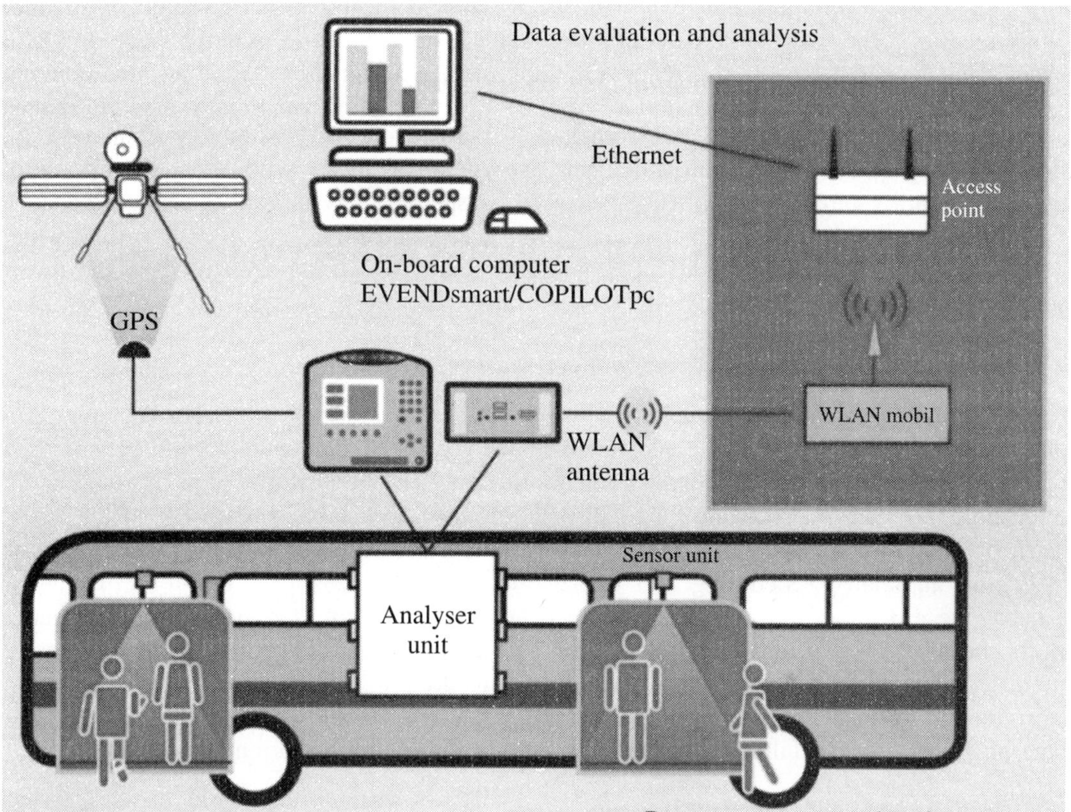

FIGURE 8.1 Automatic passenger counter for fleet management (*Source:* http://www.init-ka.de/gfx_content/products/APC_55_large.jpg, reproduced with permission).

AVL For fleet management

Automatic vehicle location system is useful for monitoring and tracking of vehicles on real-time basis which can be used to enhance operational efficiency and utilisation of the fleet. Satellite based geo-positioning technology is one of the most common AVL systems. With AVL, the controllers or despatchers in the control centre can find out at any given time, the location of buses in their fleet that are equipped with the technology.

Access to this type of information, can be especially important in an emergency situation, when either passenger or driver safety may be threatened. In addition, bus location information can be shared with the customers to assist them in planning an itinerary based on real-time information. Moreover, this information can be used to assess the running of the buses whether they are following the schedule and if necessary, buses can be diverted to avoid the problem areas.

Many transit agencies have found that AVL has been improving their punctuality performance while at the same time it has been reducing the fleet size. For example, through the use of AVL in Kansas City, the transit system was able to reduce the number of buses needed for its routes by 9%.

Geographic information systems for fleet management

A geographic information system (GIS) enables a transit agency to collect, store, analyse and display data by location. GIS can provide transportation agencies with specific information regarding mobility problems. With GIS software, a spatial analysis can determine the areas of high demand and help operations planning manager to create a fleet operation plan which is in line with the travel demand. The technology helps in optimal utilisation of the public transport fleet.

Scheduling and dispatch software for fleet management

Scheduling and dispatch software is to generate timetable schedule for the public transport. The software also be used to route, schedule, and dispatch vehicles in demand response operations. It is generally integrated with GIS and AVL to coordinate different transit functions. A combination of technologies such as, software based dispatching and AVL can increase the efficiency of transit operations, enhance safety, improve service, and optimise costs. For example, systems integrating automated scheduling and dispatching and AVL enable a controller/dispatcher to know the exact location and status of each bus under control. This real-time information allows the dispatcher to tackle any issue with service or to respond to any incident. Further, automated dispatching software and AVL allow the integration of services among many separate transportation agencies to bridge transportation gaps. The software can also help in integrated route planning and scheduling of multiple modes of public transportation in a city.

Transit signal priority for fleet management

Transit signal priority (TSP) is the process of altering traffic signal timing at intersections to give a priority to public transport (Figure 8.2). TSP can be activated by public transport vehicles working in their own particular right-of-path or in blended movement along a road (known as *mainline* priority), or operating in an auxiliary lane at an intersection (known as a *queue jump*).

With mainline TSP, the typical treatment is to increase the green aspect of the signal as or curtail the red signal to permit need for BRT vehicles, thus reducing the delay of public transport at intersection. With a queue jumping type TSP, the public transport vehicle gets a separate green phase to go through the intersection before the adjacent through traffic. In either case, the signal aspect timing are altered in accordance to protect the sign cycle length and thus keep the signal system in coordination. TSP is different from signal pre-emption, which interferes with typical sign operation and changes the sign cycle length to accommodate special events, such as a train approaching a railroad grade crossing adjacent to a signal or an emergency vehicle responding to an emergency call.

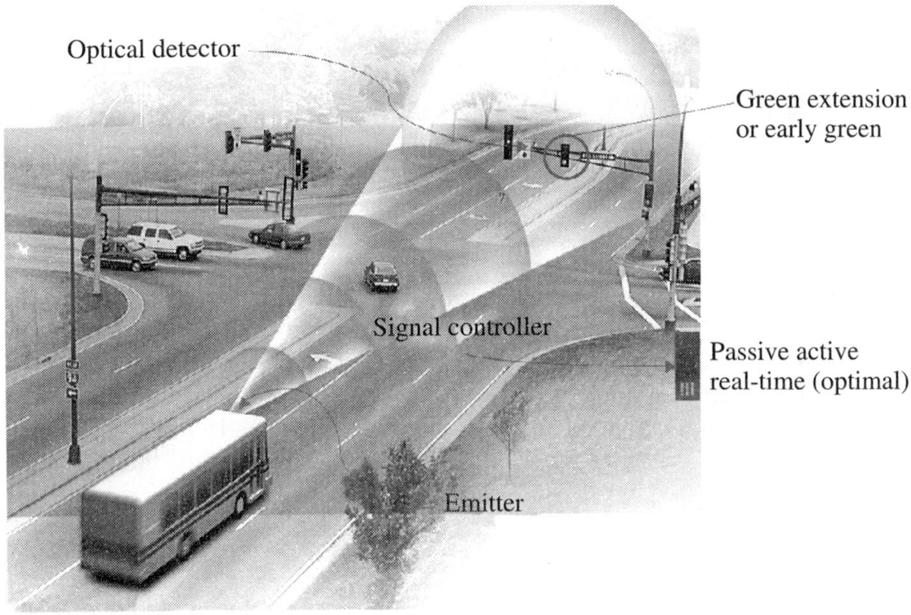

FIGURE 8.2 Transit signal priority for fleet management.

Lane control signals

Lane control signals (LCS) have been or are currently being installed on highways (freeways) in several major metropolitan areas in the United States. The LCS consists of overhead lane specific signals denoting a red "X" if travel in the lane is prohibited, and a green arrow if travel is permitted. The LCS symbolically displays the current status of each lane to inform the vehicle drivers about the condition of the lanes ahead so that they may take appropriate action. The functionality of LCS is shown in Figure 8.3.

Some of the benefits associated with LCS implementation include:

(a) Safety—reduction in the number and severity of primary and secondary accidents
(b) Traffic management—improved traffic flow stability (less variation in vehicle speeds) headway optimisation, and increased vehicle throughput
(c) Reduced travel time, energy consumption, and pollutant emission
(d) Reduced queue lengths
(e) Improved safety and efficiency for roadway maintenance operations
(f) Improved emergency vehicle movement

8.2.2 Traveller Information Systems

Traveller information system provides passengers information for planning their trips and to keep them updated during their trips. There are broadly two types of transit information—static, such as route maps, dynamic such as route delays and other real-time information. The information can be accessed by the passengers from differing locations such as home, work,

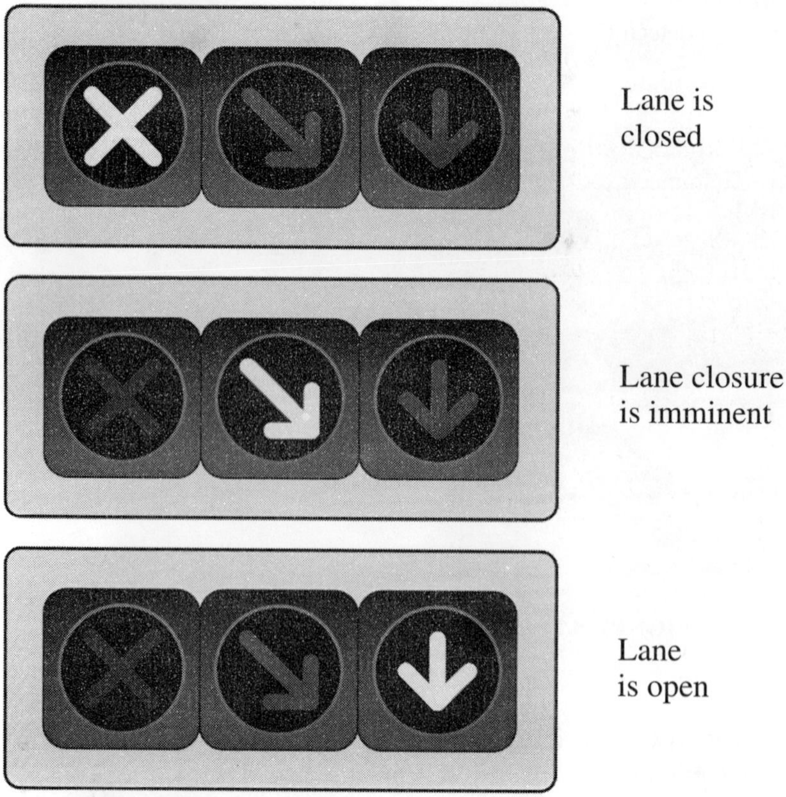

FIGURE 8.3 Lane control signal (LCS) (*Source:* https://commons.wikimedia.org/wiki/File:Lane_signals_en.svg, Author: De Nelson, Free to use).

stops, wayside areas, and on-board vehicles. Information can be accessed through number of means, such as automated trip itineraries, in-vehicle display/announcers, variable message signs, monitors, and interactive kiosks. These technologies help the users to better plan their trips and reduce the anxiety in users by providing needed information in case of delays.

Automated trip itineraries

Automated trip itineraries facilitate the users with the information necessary to plan a trip from origin to destination. Automated trip itineraries consist of a broad range of data, such as modes of travel, travel time, transfers, schedules, fares, tourist information. When automated trip itineraries are integrated with AVL technology, real-time information concerning traffic congestion and possible delays can be reported. Since automated trip itineraries are electronic, they can be accessed by various means, such as touch-tone telephones, personal computers, pagers, hand-held devices, kiosks, and Internet, fax machines, cable and interactive television. The automated trip itineraries provide accurate and timely information for customers through a variety of means which makes it very convenient for the customers to plan their journey.

In-vehicle information system

In-vehicle information system usually provides desired information through audio and visual means. It is used en-route, to provide passengers with next bus stop information (Figure 8.4). A display system (usually LCD screen) is placed in the front, and sometimes in the middle of each vehicle which displays next stop information along with a pre-recorded message simultaneously announces the same information. In-vehicle announcers help passengers with either visual or hearing impairment to know about their stop and get ready to get down at the stop. The in-vehicle information system also helps new customers like tourists, foreigners, etc. unfamiliar with a route to find their stop.

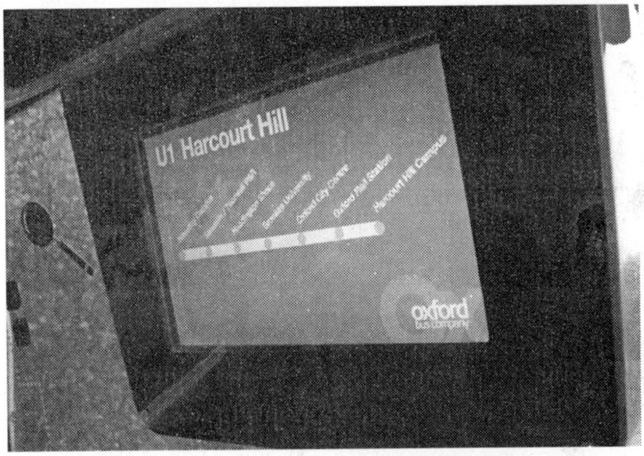

FIGURE 8.4 In-vehicle information system (*Source:* https://commons.wikimedia.org/wiki File:Passenger_Information_Display_inside_Oxford_Bus_Company_612_(15567002205).jpg, Author: Au Morandarte from London, Middlesex, England, Free to use).

Variable message sign (VMS)

VMS is an electronic message sign with systematically arranged cluster of light emitting diodes (LEDs) which are wired together and electronically driven and controlled. VMSs are used on roadways to provide travellers information about special events such as traffic congestion, accidents, incidents, roadwork zones, or speed limits on a specific highway segment. Timely information helps the travellers in planning their journey efficiently. A typical VMS proposed by DIMTS in Delhi is shown in Figure 8.5.

FIGURE 8.5 Variable message sign (*Source:* http://www.dimts.in/Services_Transportation_Intelligent_Transport_System.aspx, reproduced with permission).

Interactive kiosks

Interactive kiosks can be generally touch screen computer-like terminals or larger machines that provide travel information at the click of screen by the user. Found in malls, hotels, airports, businesses, and transit centres, interactive kiosks are accessed easily by a combination of computer-mouse, touch screen, or keyboard. These kiosks can provide a wide range of travel information, such as schedules, routes of public transport, fares schemes, traffic congestion, and weather condition in the city. When integrated with automatic vehicle location information, kiosks can facilitate customers with real-time information, such as the on time status of their transit vehicle. Interactive kiosks can promote transit services by providing easy access to information about a wide range of services for both new and existing customers. These kiosks contribute to greater customer convenience and satisfaction, and for the transit agency, the potential for increased revenues from both enhanced ridership as well as advertising on these kiosks. An interactive kiosk available at one of the metro station of New York is shown in Figure 8.6.

FIGURE 8.6 Interactive kiosk (*Source:* https://commons.wikimedia.org/wiki/File:NYCT_6401_(6167580014).jpg Author: Metropolitan Transportation Authority of the State of New York, Free to use).

8.2.3 Electronic Fare Collection

An electronic fare collection (Figure 8.7) is a system in which electronic media (cards or tokens) are used in place of paper tickets, to pay for transit fares. The objective of an electronic fare collection system is to offer convenience to the customers as they are not required to purchase the ticket for every journey. Electronic fare collection system also reduces the expense of handling and protecting transit revenues. In an automated fare system, different types of media can be used in a variety of ways including shopping at retail outlets. The types of electronic media include magnetic stripe cards, credit or debit cards, or smart cards which can be contact or contactless. For example, some systems may use these media for transit, retail purchases, and banking, while other systems may use them for transit only. The benefit of an electronic fare system is a reduction in the cost of handling, fare processing and arresting revenue leakage. The details of electronic fare collection system have been discussed in Chapter 16.

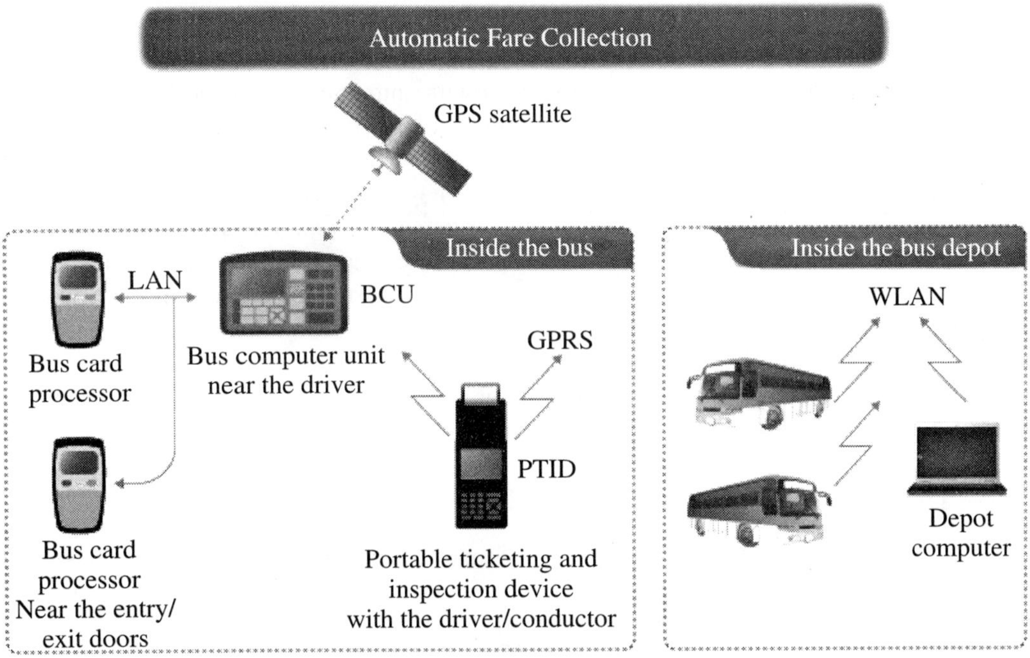

FIGURE 8.7 Automatic fare collection system (*Source:* http://www.prodigyelectronics.com/wp-content/uploads/2014/07/afc01.gif, reproduced with permission).

[Prodigy Electronics, is an Original Design Manufacturer (ODM) of custom high technology embedded electronic components and products. It is based in Hong Kong.]

8.3 TRANSPORT INTEGRATED MANAGEMENT SOLUTIONS

There are software tools available in the market for efficient operations of the public transport fleets. These software tools are increasingly being adopted by public transport operators to cut costs, boost efficiency and improve service excellence. A typical transport integrated management software tool has following four modules:

(a) ***Roster:*** It allows to create, maintain, and modify templates for common rosters. It also offers a detailed view of drivers and vehicles available for all shift dispatch points and allows us to impose scheduling conditions.

(b) ***Dispatch:*** It collects data from the roster and performance modules to a full view of the operational day, all on a single screen. It allows to find and assign vehicles, manage drivers, cover shifts, and allocate to multiple drivers.

(c) ***Timesheet:*** It allows to generate timesheets. It links with the dispatch tool to update automatically if any changes are made to shifts.

(d) ***Performance monitor:*** It is an all-in-one tool that measures staff and company performance, as well as customer satisfaction. It integrates with the dispatch tool and allows to track events in real-time and correct them accordingly.

The actual functionalities of the tool may vary as per the requirements envisaged by the public transport operator. The software tool can also be integrated with the ITS system of the city to exchange real-time data with operation control centre for traffic management in the city.

8.4 BEST PRACTICES IN PUBLIC TRANSPORT

A number of ITS applications have been developed by various organisations across the world and customised to offer transportation solution as per their own specific needs. In developed countries, ITS is increasingly used not only for mitigation of congestion and demand management, but also for road safety and improved infrastructure.

Some noteworthy implementations of ITS for public transport systems around the world are described in the following sections.

8.4.1 United Kingdom—Smart Card

In Greater London, UK, the oyster card acts as a smart mobility card for electronic ticketing on public transport (Figure 8.6). The oyster card can be used on all public transport modes across London which includes London Underground, London Buses, the Docklands Light Railway (DLR), London Overground, trams, some river boat services, and most National Rail services within the London fare zones. This system was started by Transport for London (TfL) which is responsible for operation and management of all modes of public transport in London.

FIGURE 8.8 Oyster card, London.

A standard oyster card can hold single tickets, period tickets and travel permits that should be added to the card before travel. Passengers swipe the card on an electronic reader while entering and exiting the transport system so as to validate it or deduct fare for travel. Cards may be topped up (recharged) by recurring payment authority, by online purchase, at credit card terminals or by cash, at stations or ticket offices. The card has been developed to reduce the number of cash transactions at ticket offices and the number of paper tickets. Usage is encouraged by offering substantially cheaper fares than with cash though the acceptance of cash is being phased out. On London buses, cash is no longer accepted.

It was first issued to the public in the year 2003 with a very limited features and further functions have been added progressively in phased manner. By July 2015, over 56 million oyster cards had been issued and more than 80% of all journeys on public transport in London were made using the card.

Technology and architecture

The card operates as a RFID system which are activated only when they are in an electromagnetic field. The readers read information from the cards, calculate whether to permit travel, assess any fare is due and write back information on the card.

FIGURE 8.9 Oyster card reader, London (*Source:* https://commons.wikimedia.org/wiki/File:Oyster-Reader.jpg, Author: Tom Page from London, Free to use).

Travellers swipe the card on a distinctive yellow circular reader (a cubic tri-reader) on the automated barriers at London Underground stations to 'touch in' and 'touch out' at the start and end of a journey.

In buses, the users must touch the oyster card only once at the point of boarding: as London buses have a single flat fare of £1.50, there is no need to calculate an end point of the journey (Figure 8.10).

FIGURE 8.10 Use of oyster card inside buses, London (*Source:* https://oyster.tfl.gov.uk/oyster/entry.do, Reproduced with permission).

8.4.2 Electronic Toll Collection (ETC) System—Florida

ETC system offers a facility of cashless tolling which aims to eliminate the delay on toll roads by collecting tolls electronically. ETC determines whether the vehicles passing through the toll plaza are enrolled in the program, and electronically debits the accounts of registered car owners without requiring them to stop. In case the vehicle is not registered under cashless tolling schemes, it alerts enforcers for penalising the violators (Figure 8.11). The ETC system allows easy and quick collection of toll and saves a lot of time of the vehicle users who had to wait earlier in case of manual collection of toll.

Electronic toll collection systems rely on four major components: automated vehicle identification, automated vehicle classification, transaction processing, and violation enforcement.

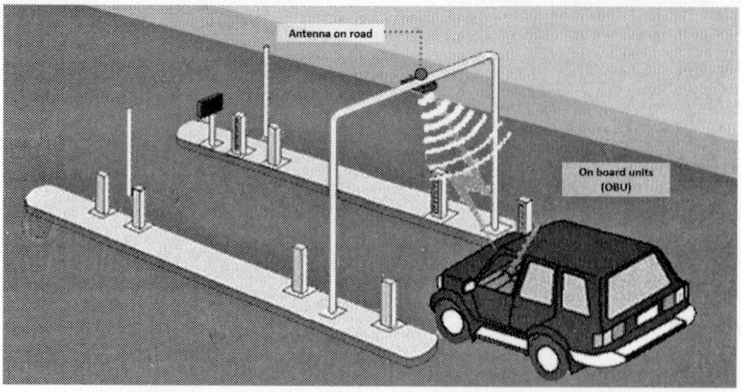

FIGURE 8.11 Electronic toll collection system—Florida.

Automated vehicle identification is the process of identifying the vehicle subject to tolls. The majority of toll facilities detect the passage of vehicles through a limited number of toll gates. At such toll gates, the task is then to identify the vehicle in the gate area.

Automated vehicle classification is closely related to automated vehicle identification which determines the type of the vehicle. Most toll facilities charge different rates from different types of vehicles which makes it necessary to distinguish the vehicles passing through the toll facility.

Transaction processing functions deals with maintaining customer accounts, posting toll transactions and customer payments to the accounts, and handling customer inquiries. The transaction processing component of some systems is referred to as a "customer service centre". In many respects, the transaction processing function resembles like banking services, and several toll agencies have contracted out transaction processing to a bank.

A violation enforcement system (VES) is useful in controlling unpaid tolls and thus arrest the revenue leakage. A number of technologies can be used to discourage toll violators which include identification of number plate of the vehicle and sending challan to the person under whose name the vehicle has been registered. Police patrols at toll gates are found to be highly effective. However, the heavy cost of police patrols makes their use on a continuous basis financially unviable.

Indian Government has also launched an ambitious plan to adopt ETC at all toll plazas available on National Highways. The project is spearheaded by National Highway Authority of India (NHAI). The details of ETC project of India are presented in Appendix F.

8.4.3 Seoul Public Transportation

Seoul has established itself model city for public transportation in terms of convenience, safety, reliability, punctuality, and economic efficiency. Public transportation has improved considerably since 2004 when the city integrated its entire public transportation system and carried out public transportation reforms.

The population of Seoul has increased sharply between 1960s and 1980s and the number of vehicles increased considerably in 1980s thanks to enhanced purchasing power of citizens. As the supply of public transit reached its limit, Seoul bowed its responsiveness to manage the demand for public transportation. Such change in the policy direction led to public transportation reforms in Seoul from 2004.

In 2004, the Seoul Metropolitan Government (SMG) carried out wide spread reforms in its public transportation system, through which Seoul has emerged as a model city for public transportation. As a result, modal share of public transport reached around 65% as of 2014 (with bus and subway accounting for 27% and 38%, respectively).

Seoul public transport policies

In 2004, SMG made following unprecedented reforms in its transport policies as listed below:

(a) ***Rationalisation of the bus route system:*** In 2004, SMG rationalised its bus route system by replacing the some of the arbitrary bus routes with a Hub-and-Spoke based dual system of trunk and feeder lines additional to some problematic bus routes having too many curves or redundant long-range services. So the rationalisation of the bus services was made to maximise operation efficiency and enhance connectivity. The earlier numbering system were meaningless to the citizens; however, the new bus numbers clearly indicate where buses originate from and end their trips. Earlier, the bus routes were divided into city, express, and circular. In rationalised system, the buses are classified into trunk, feeder, inter-regional, and circular lines and colour-coded so that citizens can easily identify the desired route. After rationalisation of the bus route system, bus operations in Seoul have improved considerably in both mobility and accessibility, marking turnaround from the downhill spiral in ridership.

(b) ***Median bus lane system:*** SMG introduced the dedicated median bus lane system along with rationalisation in bus routes in 2004. The aim being the optimisation of bus operations and hence focused on reliability, punctuality, faster service, and maximum convenience for the citizens. These dedicated median bus lane system also meant that buses were given priority over private modes on the roads.

(c) ***Quasi-public bus operation system:*** The private bus operators provided services on selective profitable routes which was a concern as it was not in the public interest and waned the quality of the bus service. To address this concern, a quasi-public bus operation system was introduced in which SMG manages the bus routes and revenues whereas the private companies operate the buses.

(d) ***Integrated transit-fare card (dubbed T-Money) system:*** Through the reforms of 2004, public transport fare collection systems was integrated into one for entire Seoul metropolitan area. Moreover, in the new system the fares are charged on the basis of actual distance travelled rather than number of trips which was the case earlier. This integrated fare product reduced citizens' burden of transport costs substantially. Further, SMG expanded its integrated fare system to all operating transport modes in the Seoul metropolitan area making it very convenient for the users.

(e) ***Improvements in bus vehicles:*** To control the pollution in Seoul from fossil fuels, SMG replaced its diesel buses with CNG buses. Further, SMG is working with local bus manufacturers to develop electric buses which are free of exhaust or noise and also introduce low-floor buses to help those with disabilities as well as children and seniors.

These reforms resulted in very positive outcome which included rapid growth in bus ridership, improvements in the average speeds of buses due to dedicated median bus lanes, lower cost burden for various transportation services, improved air quality, reduced fare burdens, grander aesthetic value of Seoul. Due to these continuous improvements, Seoul now brags a well-established integrated public transportation system which is environmentally friendly and well-suited to serve the needs of the citizens.

TOPIS, the centre of the world's advanced transportation

It is Seoul Metropolitan Government's integrated transportation management centre which collects information from bus operations, road traffic, transport fare media and provides information to city's road traffic management system, bus operation management system, unmanned enforcement systems, traffic broadcasting system and Seoul Metropolitan Police Agency and exerts comprehensive control and management of traffic situations in Seoul (Figure 8.12).

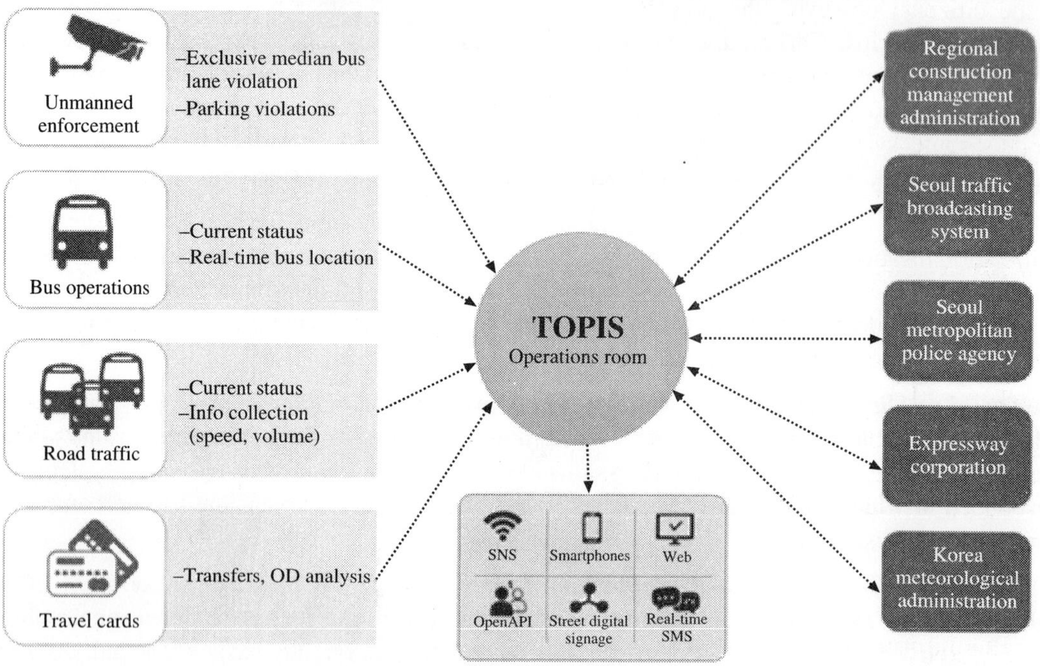

FIGURE 8.12 TOPIS, Seoul.

Functions of TOPIS

(a) Manage real-time traffic flow
(b) Supply information on traffic congestion
(c) Real-time management of bus operations
(d) Monitor traffic situations and supply congestion information promptly
(e) Supply information on bus operations
(f) Offer bus detour and assignment orders
(g) Support scientific transport administration
(h) Bus operation support and bus operation planning
(i) Improve traffic flow and surface transport planning
(j) Operate vehicle enforcement systems—Crackdown on violations of exclusive median bus lanes and illegal parking

Major systems: bus management system (BMS) and bus information system (BIS)

BMS is designed to identify optimum bus operation headways and improve the punctuality of bus operations based on GPS and wireless communications. The major function of BMS is to control the bus operations, monitor for any emergency and support bus dispatch, i.e., it helps policy analysis, origin/destination (O/D) analysis per route, analysis of boarding and alighting passengers per bus stop, and demand analysis per route.

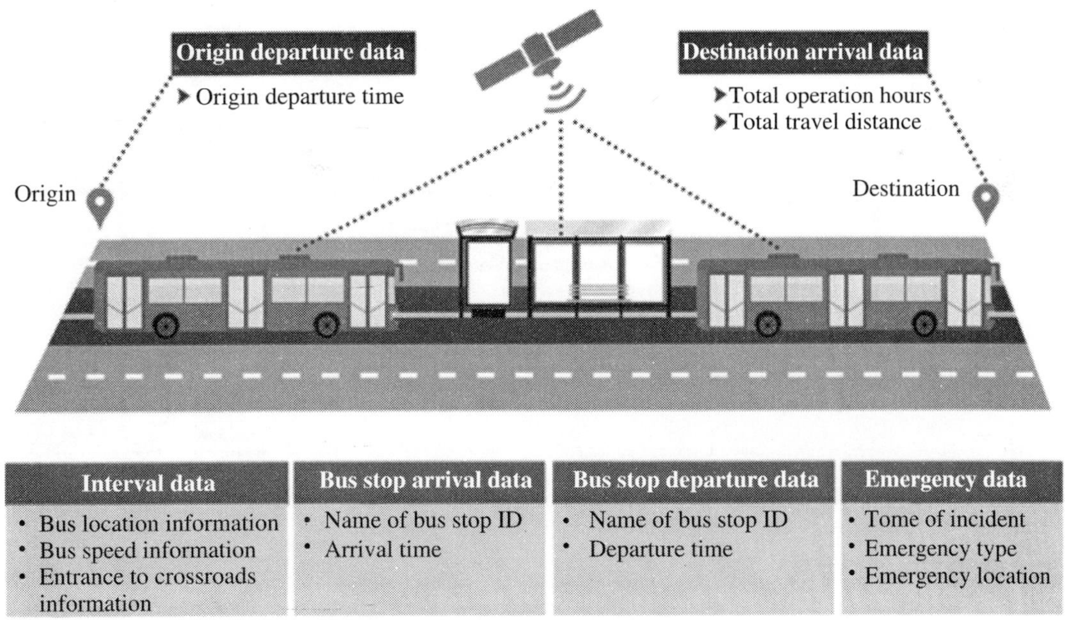

FIGURE 8.13 BMS and BIS.

Bus information system (BIS) provides public transport information collected from BMS to the users (passengers) on real-time basis. People have access to real-time information on buses, metro trains leading to a more convenient use of public transport and in turn promotion of public transport. All types of information about public transport including current location, estimated arrival time, first and last operation time of the day, operation status are made available to the users via smartphone application, TOPIS Website, and others (Figure 8.13).

Vehicle enforcement systems (VES)

SMG is using a wide network of CCTV system to enforce parking regulations and the median bus lane regulation. The VES has been of great help to improve efficiency of bus operations in Seoul city. Further, SMG has also been running on-board CCTVs in its fleet of vehicles for on-board surveillance successfully.

Smart city management

SMG has integrated its TOPIS functions with emergency response functions executed by the Seoul safety control centre smart urban management (SUM) to offer following functionalities (Figure 8.14):

(a) Integrated round-the-clock monitoring of transportation, disaster and emergency drills
(b) Supply of real-time transportation and emergency information digital signage boards
(c) First transport simulation program that anticipates emergencies based on speed of vehicles

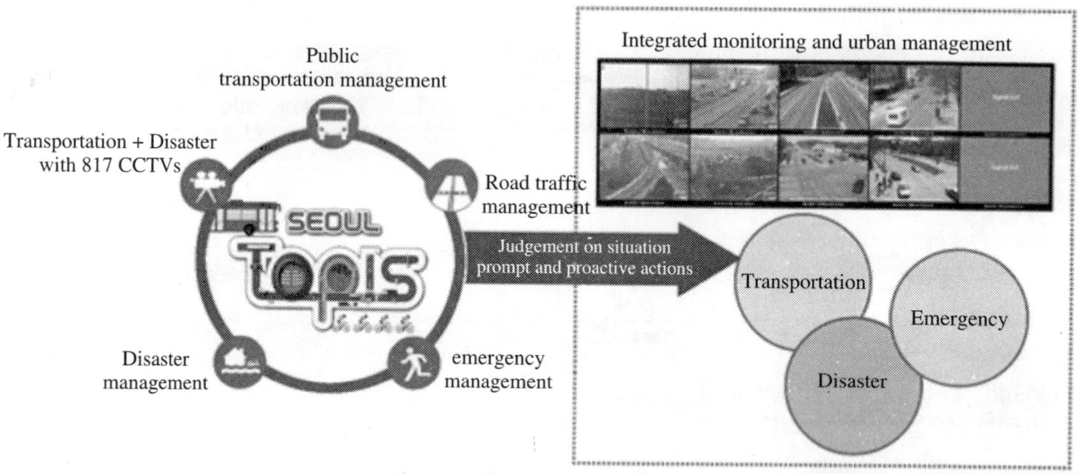

FIGURE 8.14 Smart city management using TOPIS, Seoul.

8.5 SUMMARY

The promotion of public transport to achieve a modal share in favour of public transport has been a top priority for city management. The two pronged approach of 'push and pull' has been recommended by the experts as strategy to promote public transport in a city. The push approach forces people to leave private transport and opt for public transport by making private transport prohibitively costly or inaccessible. The congestion charges in CBD of the city, restriction on car ownership are some of the push measures. Making public transport very efficient, reliable, punctual and customer friendly attracts (pulls) the people to use public transport. ITS plays an important role in both push and pull measures. Electronic road pricing (ERP) to efficiently collect congestion charges has proved to be a game changer in push measures. Similarly, the ITS technologies such as real-time passenger information systems, smart card based integrated fares, automatic vehicle location systems, Wi-Fi on-board and at stops, etc. have largely contributed to make public transport customer friendly and attractive to use. ITS can be effectively used to create a public transport oriented cities.

Questions

8.1 Describe role of ITS in public transport planning.

8.2 How GPS can be used for public transport management?

8.3 What is the use of automatic passenger counter for public transport management?

8.4 What are the benefits of automatic vehicle location systems?

8.5 Describe application of traffic signal priority.

8.6 Write short notes on the following:
- (a) Geographic information system
- (b) Real-time passenger information
- (c) Lane control technologies
- (d) Passenger Wi-Fi
- (e) Surveillance/CCTV/security systems

8.7 How fare collection system can be used to promote public transport in a city?

8.8 What are major applications of ITS in integration with public transport?

8.9 Describe some of the best practices of ITS application for public transport.

Bibliography

Auckland Transport, Auckland Regional Public Transport Plan, 2013. (https://at.govt.nz/media/308538/RPTP-2013-updateNov29-13.pdf)

Booz, Allen and Hamilton, "Intelligent Transportation Systems Field Operational Test Cross-Cutting Study", Advanced Traveller Information Systems, Prepared for U.S. Department of Transportation, Federal Highway Administration Highway and Vehicle Technology Group Washington, D.C. McLean, Virginia, September 1998. (https://www.fhwa.dot.gov/publications/research/operations/its/jpo99038/fotatis.pdf)

Chen, Ying-Chih et al., "Request-Centric Wireless Bus Information Management System" published in *Inventions,* 2016.

Faghri, A. and K. Hamad, "Application of GPS in Traffic Management Systems", *GPS Solutions*, Vol. 5(3), pp. 52–60, 2002.

Guidelines on Advance Passenger Information (API), WCO/IATA/ICAO, 2014. (http://www.wcoomd.org/en/topics/facilitation/resources/~/media/95C80C194B634618A9FB56423BEB95AE.ashx)

Kanarachos, Stratis, "A new method for computing optimal obstacle avoidance steering manoeuvres of vehicles", *Journal: Int. J. of Vehicle Autonomous Systems*, Vol.7, No.1/2, pp.73–95, 2009.

Kimpel, Thomas J. and James G. Strathman, "Automatic Passenger Counter Evaluation: Implications for National Transit Database Reporting", Centre for Urban Studies College of Urban and Public Affairs, Portland State University Portland, OR 97207 (503) 725–4020, 2002.

Lin, W.H. and J. Zeng, "Experimental study of real-time bus arrival time prediction with GPS data", Transportation Research Record, (1666):101–109, 1999.

Oyster Card, Transport for London, Retrieved December 22, 2013. (https://oyster.tfl.gov.uk/oyster/entry.do)

Passenger Counting Systems, TCRP Synthesis 77, Transit Cooperative Research Program, 2008. (http://onlinepubs.trb.org/onlinepubs/tcrp/tcrp_syn_77.pdf)

Portillo, Daniel , "Automated Vehicle Location using Global Positioning Systems for First Responders", Institute for Information Technology Applications, Technical Report Series, 2008. (http://www.usafa.edu/df/iita/Technical%20Reports/Automated%20Vehicle%20Locator%20Using%20Geoposition%20Positioning%20System%20for%20First%20Responders.pdf)

Seoul, ready to share with the world! Seoul Public Transportation, a report by Seoul Metropolitan Government, 2004. (http://citynet-ap.org/wp-content/uploads/2014/06/Seoul-Public-Transportation-English.pdf)

Shi, Qi, "Urban Expressway Safety and Efficiency Evaluation and Improvement using Big Data", University of Central Florida, 2014.

Takahashi, Masami, Satoshi Katagata, Jun Ando, Shin-ya Shiroto and Masahiro Saito, "Automatic Fare Collection System with Contactless IC (Smart) Card", East Japan Railway Company 2-2-2 Yoyogi Shibuya-ku, Tokyo Japan, 2001. (www.uic.org/cdrom/2001/wcrr2001/pdf/sessions/4_5/238.pdf).

Wang, Yinhai, "Comprehensive Evaluation of Transit Signal Priority System Impacts Using Field Observed Traffic Data", Final Technical Report Agreement T2695, Task 60 SS-RTSP Evaluation, 2008.

9 Advanced Vehicle Control and Safety Service and Emergency Management Systems

9.1 INTRODUCTION

Most of the road accidents are attributed to human error. If the human factor can be eliminated or reduced by use of technology, a number of fatal accidents could be avoided. The recent developments in sensor technology in the vehicle industry and vehicle control system have led to a significant progress in technology based vehicle safety systems. Over the time, as the microprocessors and sensors are continuously shrinking in size and getting cheaper in cost, the use of vehicle control systems has now become technically and economically feasible and common. The Vehicle automation programs have proved remarkable potential, such as autonomous cars (driverless) driving on their own based on inputs such as video images from on-board cameras fitted in the vehicle and satellite-based positioning data of the vehicle. The future of the personal transportation lies in autonomous (driverless) vehicles. Despite of all these advancement and technology, however, the complete shift from the conventional to fully-automated highways is still at the nascent stage with very few noteworthy examples. While a wide acceptance of advanced vehicle control and safety (AVCS) towards driverless cars may still take some time considering the solution of all constraints and issues of automated highway system, the existing AVCS technology is being leveraged for comfortable applications and safety needs that demonstrates short-term benefits and widely encourage acceptance of vehicle automation in the world.

Emergency management systems consist of automatic alert or alarm system in case of any accident and monitoring threats, and automated security system support in secure areas. The system can also activate wide area alerts to notify public for emergency situation like road accidents, severe weather conditions, disasters, security threats, etc.

9.2 ADVANCED VEHICLE CONTROL AND SAFETY (AVCS) SYSTEMS

The aim of the advanced vehicle control and safety systems is to improve the transportation system safety by supplementing drivers' abilities for maintaining vigilance and control of the vehicle by enhancing the accident avoidance capabilities of vehicles. Following AVCS services are included in this group:

(a) ***Longitudinal collision avoidance:*** By the use of this, the user service can provide assistance to the driver in avoiding collisions to the front and/or rear of the vehicle longitudinally. This is achieved by implementing rear-end collision control and warning, adaptive cruise control (ACC), head-on collision control and warning, and backing collision warning to the drivers. Autonomous cruise control (ACC) is an optional cruise control system for road vehicles that automatically controls speed of the vehicle to maintain a safe distance from vehicles ahead based on sensor information from on-board sensors.

Box 9.1: Collision Avoidance System (CAS)—USA

A collision avoidance system is a safety system fitted in automobile which is designed to reduce the severity of an accident. CAS uses radar and sometimes laser and camera to detect an imminent crash. Once detected, these systems either warn to the driver when there is an imminent collision or apply automatic brakes to stop the vehicle (Figure 9.1). Cars with collision avoidance may also be provided with adaptive cruise control, and use the same forward-looking sensors.

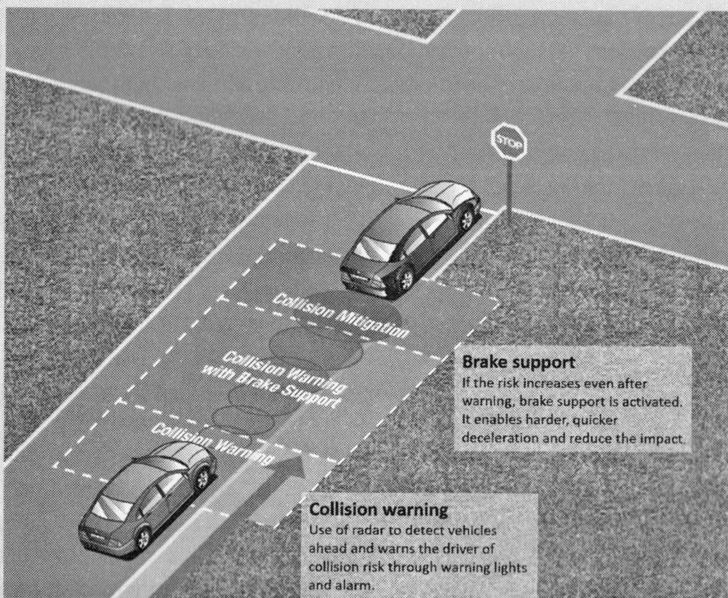

FIGURE 9.1 Collision Avoidance System—USA.

Collision avoidances system generally starts with a system called *adaptive cruise control* which is like a regular cruise control, but with the addition of radar sensors that can "observe" the traffic ahead and slow down the car to maintain a safe distance from the vehicle ahead.

Collision avoidance systems can also generate audio alarms or flash warnings on the windshield, the system senses that the vehicle is getting too close to another vehicle. The system will then apply brake pressure perhaps if driver does not react in time. If the computers and sensors determine that a crash is not avoidable, they tighten up seat-belts, adjust headrests, or close the power windows and sunroof to make the car safer in the collision. Collision avoidance may also include other related technology, like lane departure warning and blind spot monitoring systems. This technology works by using a camera to "observe" the road and alert the driver in case the car begins to move out of a lane, or if another car approaches from the side. Lane departure warning systems may generate sound warning tones or vibrate the steering wheel, and some may even help steer the car to keep it headed straight. Blind spot monitoring systems generally display warning lights in the driver's peripheral vision and may also generate alarms. Some systems use radar instead of cameras with same end results. The generic design of a Distributed Collision Warning System is shown in Figure 9.2.

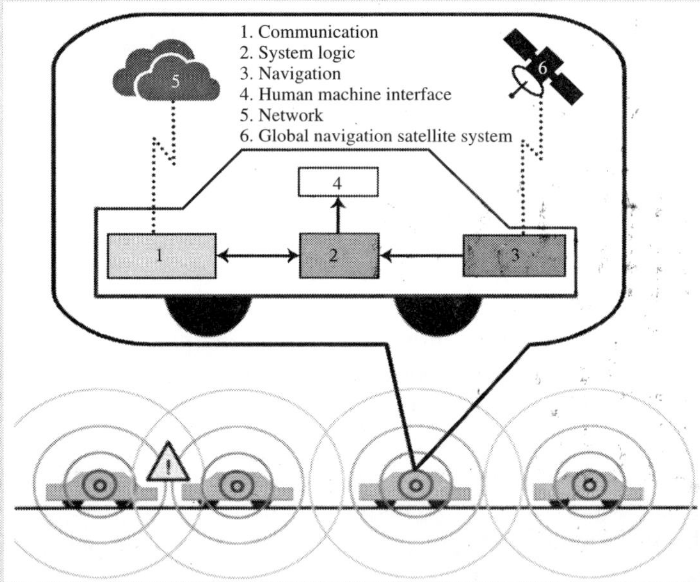

FIGURE 9.2 Generic design of distributed collision warning system.

(b) ***Lateral collision avoidance:*** This functionality assists drivers in avoiding side collision type accidents that result when a vehicle goes out of lane of travel, by warning drivers and by assuming some control of the vehicle. This service assists to the drivers the blind spot situation display/lane change, and lane departure warning and control and collision warning control.

(c) ***Intersection collision avoidance:*** The system provides assistance to the driver in avoiding collisions at intersections. With this system, the position of vehicles within the intersection area can be tracked through the use of vehicle-to-vehicle communications or vehicle to infrastructure communications.

(d) ***Vision enhancement for crash avoidance:*** This system assists the driver during periods of bad visibility by in-vehicle sensors capable of capturing an image of driving environment and providing a graphical display of the image to the drivers. The system thus helps in reducing the number of vehicle accidents that occur during the bad weather.

(e) ***Safety readiness:*** With the help of this system, warnings/alerts regarding driver's driving performance, the condition of the vehicle, and the condition of the roadway as sensed from the vehicle can be provided to drivers.

(f) ***Pre-crash restraint deployment:*** This system helps in reducing the number and severity of injuries caused by vehicle collisions by anticipating an imminent collision and by activating passenger safety systems prior to the actual impact.

(g) ***Automated vehicle operations (AVO):*** This service provides a fully automated vehicle-highway system in which instrumented vehicles operate on instrumented roadways without operator intervention. This system facilitates driverless operation (autonomous) of the vehicles.

9.3 EMERGENCY MANAGEMENT

Emergency management consists of the following two user services:

(a) Emergency notification and personal security
(b) Emergency vehicle management

First one is related to the driver and personal security and the later one is to minimise the severity of the accident. This service enables the travellers to inform appropriate emergency response handling personnel regarding the need for assistance due to emergency or non-emergency situations. The traveller can provide information manually, or it could be activated from vehicle automatically on the occurrence of any incident such as collision. The service also provides functionality for monitoring threat alerts and automated security system support in secure areas. The secure areas consist of physical areas connected to travel which include remote areas and related transportation infrastructure. Further, the service can generate the wide area alert to notify the travelling public about emergency conditions such as bandhs, severe weather conditions and warnings, terrorist attack, natural and man-made disasters, military activities, and emergencies. The second one is related with efficient management and assignment of emergency vehicles like ambulances, police vans, etc. Each user service is discussed in detail in the following sections.

9.3.1 Emergency Notification and Personal Security

This user service consists of two subparts, i.e., (a) driver and personal security (driver-initiated distress signals), and (b) automated collision notification (automated information to emergency personnel of the location, nature, and severity of the serious crashes).

Driver and personnel security: In this subservice, GPS transmits the location of the vehicle for help and assistance to the emergency personnel, repair service or the vehicle towing centre. In these days, the vehicles which are equipped with the GPS trackers and sensors and connected to cloud data can easily adopt this system.

Automated collision notification: In this subservice, the information is automatically conveyed to the emergency helpline by sensors installed in the vehicle about the location of the vehicle where it met with accident. The additional information about the condition and severity of the accident can also be transmitted based on the technology/application used in the system. This emergency notification requires crash-sensing device and GPS technology.

The performance requirements for emergency notification and personal security include:

(a) Continuous real-time transmission of status from appropriately equipped vehicles to at least one response unit from any location nationwide

(b) Minimum time taken for transmission of information

(c) GPS tracker should be very precise in providing information about the location

This system should be capable of operating in both manual and automatic modes depending upon the severity of the crash or the requirement of the user. The system should check and monitor conditions of the components of the vehicle from inside the vehicle every few seconds and for critical safety or emergency condition cases, it sends the appropriate distress signal automatically.

The emergency notification and personal security system requires in-vehicle GPS device for vehicle position/location, a two-way communications link between the vehicle and the emergency control centre using commercially available smart phone service, and control centre equipment for processing vehicle location, type of assistance required, and routing of the request to the appropriate response agency. One such system called 'Safe Drive' has recently been introduced in Kerala state of India. The details of 'Safe Drive' are presented in Box 9.2 below:

Box 9.2 Safe Drive

"Safe Drive" is an innovative smart car gadget that turns vehicles into "safe, smart and connected". The gadget has been designed, developed and manufactured by Elsys Intelligent Devices, a Kerala based start up. The gadgret is easy-to-install and compatible with most cars in India. The gadget is equipped with crash sensors, GPS module, microphone, speaker and cellular connectivity. It offers an "In-Vehicle Emergency Call" feature that automatically notifies the emergency centre for help in the event of a severe collision—even when the victims are unable to respond due to injury or shock.

Upon detecting a serious impact, Safe Drive initiates a voice call and sends out a "Minimum Set of Data" that includes the exact location of the incident and speed of the vehicle. This helps the emergency centre assess the situation, speak with the victims, notify their family/friends and coordinate emergency assistance.

It also has other features such as a one-touch assistance button, hands-free two-way calling, live vehicle tracking, trip history, smart alerts, roadside assistance, driving analysis and score.

Source: http://tech.economictimes.indiatimes.com/news/technology/innovative-smart-car-gadget-safedrive-launched/60109634

9.3.2 Emergency Vehicle Management (EVM)

The EVM service is required for more efficient management and assignment of emergency response vehicles such as ambulances, police vans, fire tenders, towing vehicles, cranes, etc. The objective of EVM is to minimise the amount of time between a public safety operator's notification of an incident and the arrival of emergency personnel at the accident site. This service is provided by public safety agencies having fleet management capabilities, route guidance, and signal priority and pre-arrangement for emergency vehicles.

This service consists of three subdivisions:

(a) Emergency vehicle fleet management

(b) Route guidance

(c) Signal priority

The primary users of all EVM services are police, fire, and hospitals. This system requires real-time reliable transportation and traffic information in order to provide travel alternatives for the arrival of the emergency vehicle at the accident site. The information about exact location of the incident and of emergency vehicles in the fleet is essential at the dispatch centre. Emergency vehicle management identifies the closest and most appropriate response vehicle, and transfer complete, accurate information on the nature and location of the incident to it so that it can reach the site in minimum possible time with all the necessary equipment. The system should also provide real-time traffic signal prioritisation by maintaining current information on signal timing, vehicle's location which is in emergency, and its routing.

9.4 INFORMATION MANAGEMENT SYSTEM

ITS information management system is used to collect and analyse data generated by other ITS applications and enables other ITS applications that make use of the collected information. ITS information management system enables ITS applications such as decision support system (DSS), predictive information, performance monitoring, and it can also help in transportation planning, research and safety management activities. Timely collection and analysis of data are very important for efficient management of any incident.

9.4.1 How Information Can Be Used?

(a) Arterial traffic conditions are made available to transit management system. These conditions help in adjusting transit vehicle route and their schedule. Transit management system can use the data collected from information management system to plan management strategies.

(b) The operating conditions of the expressway are provided to the expressway management system. These conditions are communicated to transit management system. The transit management system can use this information to plan and adjust transit routes and schedules.

(c) Transit management systems are also made available the information about incident location, severity, and duration which assist transit management system in adjusting the transit routes and schedules.

(d) Data collected by electronic payment systems can be handy in assessing ridership characteristics and this data can be also used by transit management system to plan transit routes and schedules as per the travel demand.

(e) Information about weather condition is recorded by road weather management system, which is communicated to transit management system. The transit management system can use the information provided by road management system to plan and adjust transit routes and schedules in response to severe weather condition.

9.4.2 Uses

(a) Incident locations are regularly reviewed by incident management system to plan a patrol routes, and frequencies for service patrol vehicles.

(b) The data of past traffic can be used to predict travel time in different traffic conditions.

(c) Effective schedules and routes can be developed from data collected through automatic vehicle location, computer aided dispatch system and automatic passenger counting systems. This data may be reviewed by transit agencies to manage schedule operations more effectively and efficiently as per the demand.

(d) As information management and data collection systems evolve over the time, they collect information from one source to many more complex sources. To provide support for regional operations across jurisdictional and agency boundaries, it is required to merge the data from multiple sources, integrate real-time information and collected information.

9.4.3 Benefits

Data archiving has been helping in improving ITS integration and provides coordinated regional decision making. The major sources of the archived data available to urban planners and researchers are traffic surveillance system, commercial vehicle operations, electronic payment system, transit systems, and road weather information system. The collected data provides information which was previously not available and enables analysis of problems and their solutions which were possible with conventional data. As more advanced data analysis techniques are being developed and efficiency of data reporting systems is improving, more examples of the effectiveness of information system are becoming available. The methods to assess the benefits of information management that must be developed to appreciate the efficacy of the system. Some of the successful examples of information management are presented in Table 9.1.

TABLE 9.1 ITS goals and findings

ITS goal	*Selected findings*
Customer satisfaction	In Virginia, a Web-based data management system (DMS) is being used to provide decision makers and transportation professional data about traffic, incidents and weather, needed for planning and traffic analysis. A study of Website activity reveals that 80% of the Website usage was in downloading data files that was required to create maps and graphics. Overall, the users are satisfied by the system as it provided variety of data but users wanted more information like traffic counts, turning movements, work zones and as well as on broader coverage.
Efficiency	In Portland, the Tri-met Transit Agency is using archived AVL data to construct running time distributions and provides more information to operators and dispatchers. The evaluation of the data helped in reducing variation in run times, enhanced schedule efficiency and maximised the use of resources.
Productivity	The Iowa Department of Transportation (DOT) has initiated a project to make data reporting and analysis tools available to local law enforcement organisations which has resulted in an expansion in generated crash reports received electronically. It has allowed the agency to provide state wide crash data on a quarterly basis.

9.5 CASE STUDY—REGIONAL INTEGRATED TRANSPORTATION INFORMATION SYSTEM

A major traffic accident in the Washington, D.C. capital Beltway can cause traffic backups and delays, as well as secondary incidents for hours. There are four major transportation agencies and countless emergency management groups that can respond to traffic incidents, but in early 2003 there was only limited automated data sharing. Maryland, Virginia and D.C. transportation agencies and the metro transit system approached the University of Maryland's Centre for Advanced Transportation Technology Laboratory (CATT Lab) for help coordinating traffic around the Beltway. Accordingly RITIS was developed by CATT Lab.

The regional integrated transportation information system (RITIS) is an automated data sharing, dissemination, and archiving system that includes many performance measure, dashboard, and visual analytics tools that help agencies to gain situational awareness, measure performance, and communicate information between agencies and to the public. RITIS automatically fuses, translates, and standardises data obtained from multiple agencies in order to provide an enhanced overall view of the transportation network.

Participating agencies are able to view transportation and related emergency management information through innovative visualisations and use it to improve their operations and emergency preparedness. RITIS also uses regional standardised data to provide information to third parties, the media, and other traveller information resources including Websites and paging systems. There are three main RITIS components as shown in Figure 9.3 including, real-time data feeds, real-time situational awareness tools, and archived data analysis tools.

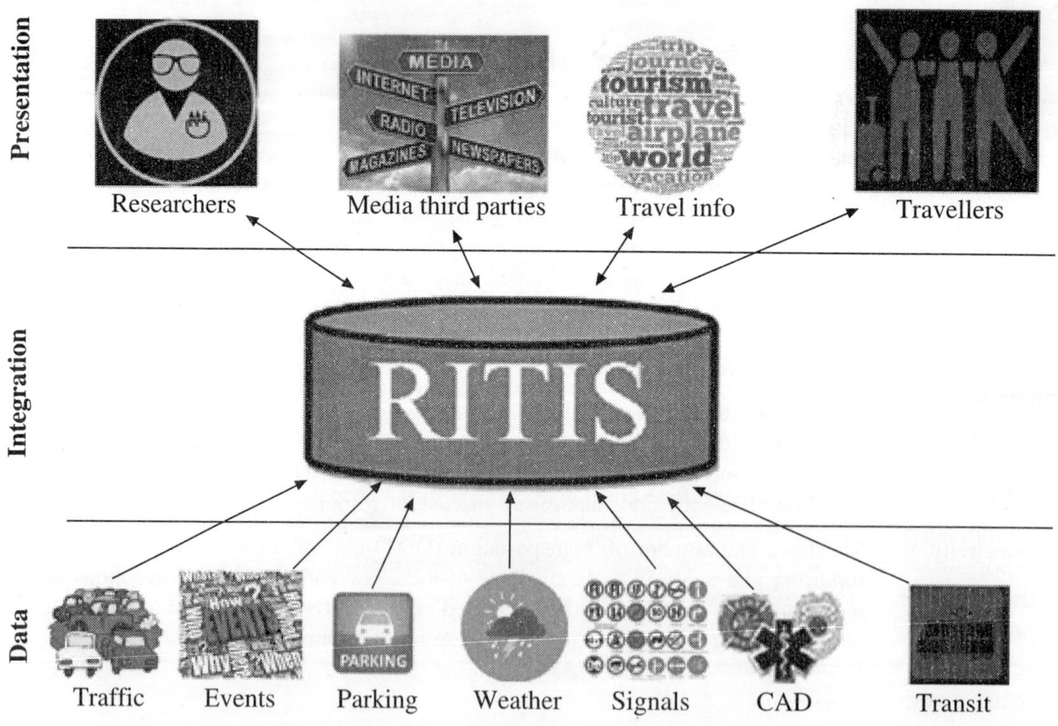

FIGURE 9.3 RITIS system implementation.

The system offers the following:

(a) Give everyone a real reason to collect data and support your programs.

(b) Provide data fusion of information from different sources and systems so users have a more complete picture of the transportation system than they would have just using their own resources or archives of individual systems.

(c) Provide easy, free access to all the data (or as much as you legally can) to everyone.

(d) Develop interesting, fun, useful applications for the data that make people aware of what you are doing.

9.6 COMMUNICATIONS AND INFORMATION MANAGEMENT

Effective communication, information management and information and intelligence sharing are essential aspects of domestic incident management. The communications and information management are very important whatever may be the size of the event or whether it is planned event or unplanned incident. Whenever a large number of people are collected, clear information and instructions need be circulated to the general public and the responders as quickly as possible.

The main goals of communications and information management include establishing and maintaining a common operating picture, ensuring accessibility and interoperability. A common operating systems interoperability offer the framework which is necessary to:

(a) Formulate and disseminate indications and warnings

(b) Formulate, perform and communicate operational decisions to the incident site, as well as between incident management bodies across jurisdictions and functional agencies

(c) Prepare for expected requirements and request supporting incident management activities

(d) Develop and ensure complete awareness and understanding of an incident within and across jurisdictions

Earlier to an incident, the entities responsible for taking appropriate pre-incident preventive actions need communications and information management steps and systems to inform and guide various critical activities and agencies involved. These processes include mobilisation of resources, as well as strategic planning by preparedness organisations, multi-agency coordination bodies, agency executives, jurisdictional authorities, and emergency operation centre (EOC) personnel.

During an incident, the individuals responsible for incident management are supposed to follow communications and information processes and systems to inform the formulation, coordination, and performing of operational decisions and request for assistance.

Transportation's intelligent transportation systems offer great opportunity for state and local transportation agencies to provide real-time traveller information to their counterparts situated at the emergency operations centres (EOCs). Effective emergency management and incident response activities depend on effectiveness and flexibility of communications and information systems that provide a common operating picture to emergency management individual and their affiliated organisations.

9.7 SUMMARY

The advanced vehicle control and safety service in vehicles is essentially required to improve the safety of vehicle users as well as road users. The recent developments in sensor technology at an affordable cost has paved the way for mass scale use of advanced vehicle control and safety devices in road vehicles. The advanced systems like lateral, longitudinal and intersection collision avoidance, safety readiness, pre-crash restrain and automated vehicle operations can be very useful in ensuring safety of road vehicles.

The timely response by the emergency vehicle is fundamental to minimise the damage in case of any road accident. Both manual and automatic means of communication can be used to notify the emergency management about any incident. Further, information management can play a vital role in providing real-time information to passengers, analysing the past data to plan for future, predict the travel patter, etc. The regional integrated transportation system (RITIS) in US is a successful example of an integrated information system in transport sector.

Questions

9.1 What do you understand by advanced vehicle control and safety (AVCS) system?

9.2 Describe different user services for AVCS.

9.3 Write short notes on the following:
 (a) Intersection collision avoidance
 (b) Safety readiness
 (c) Pre-crash restraint deployment

9.4 How ITS can be used for emergency management?

9.5 Why ITS information management is important for integrated application of ITS?

9.6 What are the benefits of ITS integration?

9.7 What do you know about regional integrated transportation information system?

9.8 Describe importance of communications and information management for effective incident management.

Bibliography

Advanced Vehicle Control Systems (AVCS) for Maintenance Vehicle Applications, Raytheon E-Systems, Department of Transportation Federal Highway Administration, December 20, 1996.

Branscomb, Lewis M. and James Keller, *Converging Infrastructures: Intelligent Transport and the National Information Infrastructure,* MIT Press, London, 1996.

CATT Lab, A User-Focused R&D Laboratory At The University Of Maryland, United States. (http://www.cattlab.umd.edu/?portfolio=ritis)

European Standard, Intelligent Transport Systems (ITS), Communications Architecture, ETSI EN 302 665 V1.1.1, 2010–2009. (http://www.etsi.org/deliver/etsi_en/302600_302699/302665/01.01.01_60/en_302665v010101p.pdf)

Ganeshkumar, B. and D. Ramesh, "Emergency Response Management and Information System (ERMIS): A GIS Based Software to Resolve the Emergency Recovery Challenges in Madurai City", Tamil Nadu, *Int. Journal of Geomatics and Geosciences*, Vol. 1(1), pp. 1–13, 2010.

Japan Vehicle Information and Communication System Centre, "VICS Evolution, 1990–2008", 2008.

Mathew, Tom V., *Transportation Systems Engineering*, IIT Bombay, August 5, 2014.

Purushothaman, B.M., S. Arunachalam, R. Srinivasan, and S. Suresh Babu, "Emergency Response Management System for Mysore City", *International Journal of Earth Sciences and Engineering*, Vol. 4(1), pp. 48–54, 2011.

Ramachandran, N. and G. Devi, Accident Emergency Response and Routing Software (AERARS) Using Genetic Algorithm, 2011.

Thapar, M.S., Emergency Response Management System for Hyderabad city, 1999. (Accessed from: www.gisdevelopment.net)

Transit Vehicle-to-Infrastructure (V2I) Applications: Near Term Research and Development, Transit Vehicle and Centre Data Exchange: Operational Concept, Final Report, March 2015. (www.its.dot.gov/index.htm)

Washoe County Regional Transportation Plan, "Transportation Management/Intelligent Transportation Systems", Chapter 7, November 21, 2008.

10 ITS Technologies for Vulnerable Road Users

10.1 INTRODUCTION

Vulnerable road user (VRU) is a term used to describe those road users who face the highest risk of being involved in a traffic accident on roads. Thus, vulnerable road users are those who are unprotected by an outside shield, namely pedestrians and two wheelers, as they bear a greater risk of injury in any collision against a vehicle and therefore they need higher protection. Among VRUs, pedestrians and cyclists are most unlikely to inflict injury on any other road user and prone to road accidents, while motorised two wheelers, with heavier machines and higher speeds, may pose a danger to others. As per WHO report on road safety 2013, 50% of all road traffic deaths are among pedestrians, cyclists and motorcyclists as shown in Figure 10.1.

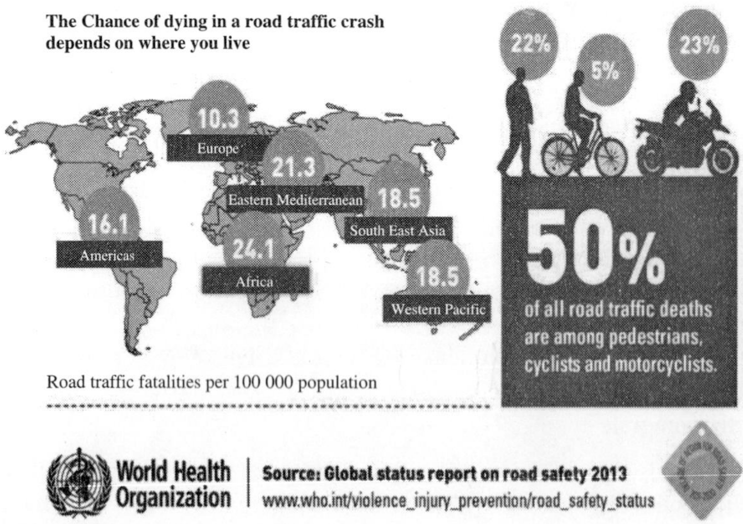

FIGURE 10.1 Fifty percent road traffic deaths are pedestrian (reproduced with permission).

In many countries, mopeds (or light motorcycles) are widely used as preferred and affordable mode of transport. The increasing use of two wheelers enhances the risk of road accidents and deaths. The review of current status in protective measures for VRUs and the recommendations for prospective safety policies required to mitigate the risk will therefore have to be analysed in relation to the use of two wheelers in each country. Further, even among vulnerable road users, some are more vulnerable than others, in particular the elderly, the disabled, and children (Figure 10.2).

FIGURE 10.2 Most vulnerable road users in Indian cities.

Elderly people: They show a gradual reduction of their abilities to cope with complex traffic situations and therefore bear a greater risk of being involved in an accident. They also become physically more fragile, which results into sever consequences in case of their lesions sustained injury in accident as compared to members of the younger age groups. Therefore, many elderly people being aware of their own difficulties try to stay away from traffic, thus reducing their mobility and the scope of their social life. As such improving traffic situations and infrastructural amenities to facilitate the movements of elderly is not only a safety requirement, but also a course of action needed to enable the elderly persons to keep their place in society longer.

Differently abled also called Divyang: They include any individuals with a physical, sensory or mental impairment or some combination of these affecting their movements. They are mostly pedestrians, either walking, with or without artificial aids, or using a wheelchair; or some may be three-wheeler cycle. As the elderly, the disabled face more risk of an accident in difficult traffic situations or on parts of the infrastructure which are not adapted to their abilities, and in some cases, they may also present a lower faculty of recovery from injuries. If Divyang people suffer due to some traffic accident, this results in further aggravating their disability. Creating a conducive physical environment to enable the Divyang to move about with a level of ease and protection is not only a safety measure, it is a basic need to treat the disabled with equity and to fully integrate them in the society.

Children: They are also particularly one of the VRUs as their abilities to cope with traffic evolve over the time with age and remain severely limited in the initial nine or ten years of their life. They, therefore, face high risk in conditions of heavy or fast motorised traffic. Children, being shorter in height may skip the sight of the vehicle drivers who focus their attention on other vehicles and tend to forget about pedestrians or cyclists. Although children may recover more easily and quickly from injuries than most adults, the trauma resulting from a traffic accident is never to be underestimated. The policies for protection of children shall not rely on better adaptation to traffic rules and improved behaviour on their part while crossing the road. They shall be allowed freedom of movement in appropriate surroundings. The policies to protect children shall promote more disciplined and improved behaviour of the vehicle drivers.

10.2 NEED FOR VRU'S SAFETY

Nearly half of those who die on the world's roads are VRUs—pedestrians, cyclists, and riders of motorised two-wheeler and these passengers account for around 50% of global road traffic deaths. This proportion is even greater in low-income countries than in high-income countries. For example, in low-income countries of South East Asia more than 80% of those killed are vulnerable road users, whereas in the high-income countries like America, the corresponding figure is 22%. In some countries, users of unsafe public transport may also be considered to be VRUs. For example, passengers hanging off the sides of buses, hanging on train doors, or overcrowded minibuses. If data on deaths among users of unsafe modes of public transport is included in the global figure of vulnerable road users then it is likely that the proportion of deaths would be even higher than the 50% found in WHO survey is shown in Figure 10.3.

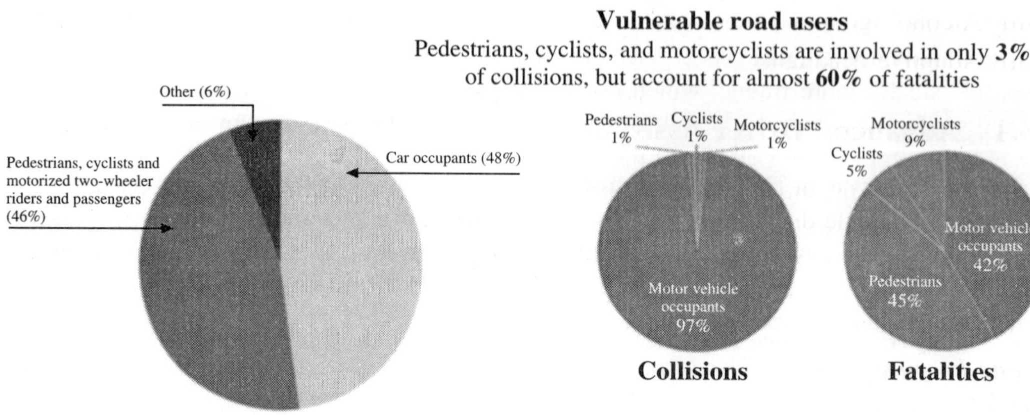

FIGURE 10.3 Vulnerable road users fatalities (*Source:* http://www.who.int/violence_injury_prevention/road_safety_status/key_data/kd_2.pdf?ua=1, reproduced with permission).

10.3 ITS TECHNOLOGIES FOR VRU'S

As discussed the extent of risk faced by VRUs, it is important to find out the ways to mitigate the risk. The modern intelligent transport systems (ITS) tools, if deployed appropriately, can have great potential to protect the VRUs from road accidents. ITS tools are increasingly adopted as a policy by government not only to protect VRUs but also to improve their quality of life. Some of these tools have already been discussed elsewhere in the book as part of other strategies like traffic management, vehicle control, etc. However, even at the cost of repletion, some of the relevant ITS tools are discussed again here in context of VRUs. The following ITS technologies are mostly used for the benefit of VRUs:

(a) Advanced driver assistance system (ADAS)
(b) Intelligent speed adaptation (ISA)
(c) Pedestrian detection systems
(d) Blind spot detection systems
(e) VRU beacon system
(f) Pedestrian traffic signals
(g) Pre-green for bicycles
(h) Average speed cameras
(i) Red light violation detection system
(j) Solely-infrastructure based applications
  (i) Intelligent pedestrian traffic signal
  (ii) Roadside pedestrian presence warning
  (iii) Crossings adaptive lighting (CAL)
(k) e-Call
(l) Cooperative systems, in particular intersection safety (INS)
(m) Alcohol ignition interlock
(n) Adaptive headlights

10.3.1 Advanced Driver Assistance System (ADAS)

Driver assistance system (DAS) in the first place aims at helping the driver task (comfort) or preventing the vehicle driver from getting involved in an accident and/or reducing the severity of the impact. These systems can assist, inform, alert or warn drivers, or sometimes intervene in the actual driving process to enhance safety. The development of ADAS is led by the automotive industry as evident from examples:

(a) ***Lane keeping system:*** Fiat launched the first haptic lane keeping system (LKS) in 2008 which alerts the driver if lane discipline is violated.

(b) ***Night vision and warning (NVW):*** It was introduced by BMW in 2008. The aim of NVW is to extend the visible range for a driver in darkness, including obstacle detection and warning, as well as warning for vulnerable road users.

(c) ***Pedestrian detection systems (PDS):*** It has been launched by Volvo in several 2010 and 2011 models. The PDS detects the presence of VRU and alerts the driver.

(d) ***Driver drowsiness monitoring:*** Mercedes introduced driver drowsiness monitoring (DDM) in 2010. The DDM system monitors the condition of the driver and warns the driver when symptoms of drowsiness are observed.

The pedestrians are the main category that is targeted by currently available applications, but they will be beneficial to other categories of VRU as well.

10.3.2 Intelligent Speed Adaptation

Intelligent speed adaptation (ISA) ensures that speed restriction, if any, in defined areas such as residential streets and school areas is enforced by the system through an alert to the driver or intervention by the system in case of over speeding. The working of ISA is explained in Figure 10.4.

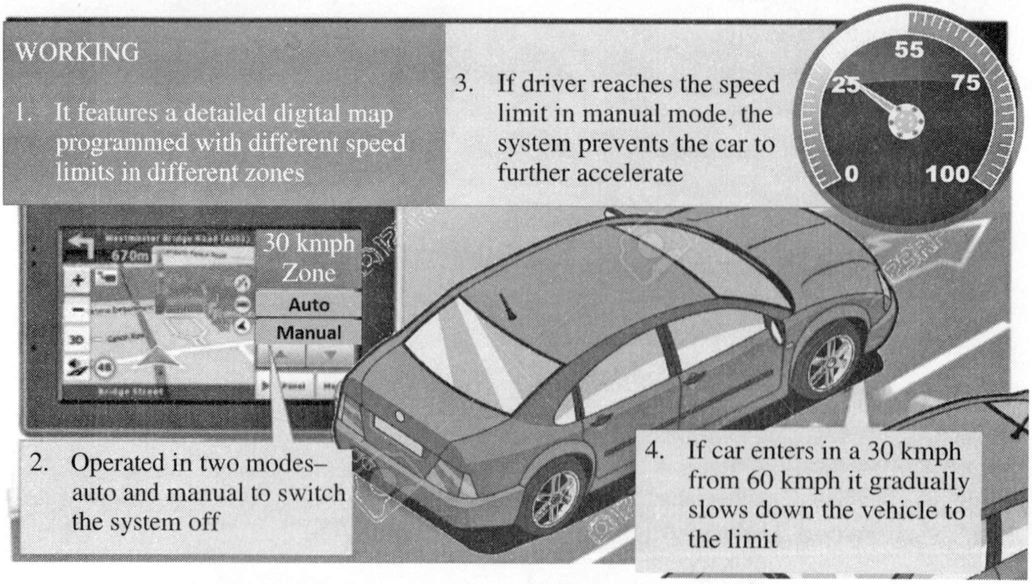

FIGURE 10.4 Intelligent speed adaptation system.

To avoid accidents related to over speeding, Indian Government has recently (*Times of India*, September 10, 2016) announced a policy to make it mandatory to install audio warning system in vehicles which will give a brief beep alarm at a speed of 80 kmph and once speed of 90 kmph is reached, there will be continuous beep. All the existing models of cars from October 2018 will be required to have this feature.

10.3.3 Pedestrian Detection Systems

Pedestrian detection systems use LIDAR, RADAR, or camera technology to detect a pedestrian. The PDS warns the vehicle driver and may apply brakes automatically if an impending impact is anticipated to either avoid the collision or reduce the collision speed.

With regard to deployment potential, the first versions of these systems are already available on high-end vehicles as shown in Figure 10.5.

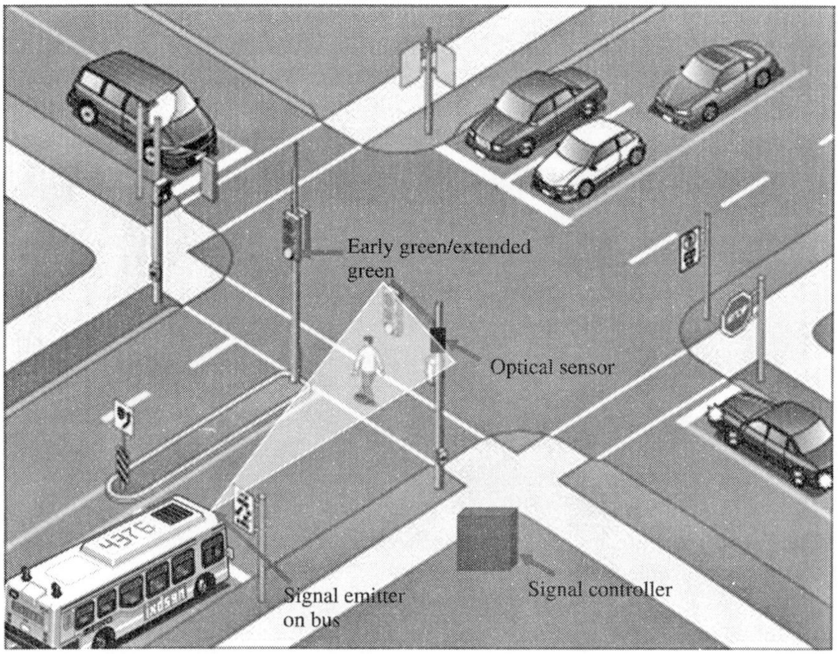

FIGURE 10.5 Pedestrian detection systems.

However, there is a risk of delegation of responsibility to the system and that the driver may become more reckless. In addition to detection of pedestrians and cyclists, experts see a good deployment potential in VRU detection systems for powered two wheelers (PTWs), especially in order to alert the driver of the presence of motorcyclists when entering an intersection.

10.3.4 Blind Spot Detection

Blind spots are areas around a vehicle that are invisible to the driver through the windows or standard mirrors. The driver may fail to notice hidden road users or hazards while manoeuvring, with potentially fatal consequences. Blind spots can exist to the front, rear and sides of vehicles. While reversing the vehicle, the low height objects like children comes under blind spot as they may not be visible to the driver. Vehicle reverse gear sensors also called parking sensors in road vehicles are designed to alert the driver to obstacles while reversing the vehicle. These systems use electromagnetic or ultrasonic sensors. The sensors emit acoustic pulses, with a control unit measuring the return interval of each reflected signal and calculating object distances. The system in turns warns the driver with acoustic tones, the frequency indicating object distance, with faster tones indicating closer proximity and a continuous tone indicating a minimal pre-defined distance. Systems may also include visual aids, such as LED or LCD readouts to indicate object distance.

Blind spot detection (Figure 10.6) is considered by lots of experts as a promising ITS for the safety of VRUs, especially in supporting truck drivers when turning right and avoiding cyclists crashes. Besides, VRU acceptance of blind spot detection system is rated as high, but acceptance by drivers and authorities is lower. They suggest that drivers may worry about the loss of control, and there may also be the problem of drivers relying on this technology rather than checking mirrors.

FIGURE 10.6 Blind spot detector (*Source:* http://www.commercialmotor.com/sites/default/files/news-article-images/2016/03/cyclist%20and%20lorry-9.jpg).

To avoid fatalities of VRUs like children behind the vehicles, Indian Government has recently (*Times of India*, September 10, 2016) announced a policy to make it mandatory to install vehicle reverse gear sensor system in all the existing models of cars from October 2018.

10.3.5 VRU Beacon Systems

This system envisages the use of tags for kids through infrastructure, such as bus stops, combined with warnings both for the bus driver, to inhibit the start at boarding and de-boarding, and for other vehicle drivers to warn of children presence (Figure 10.7).

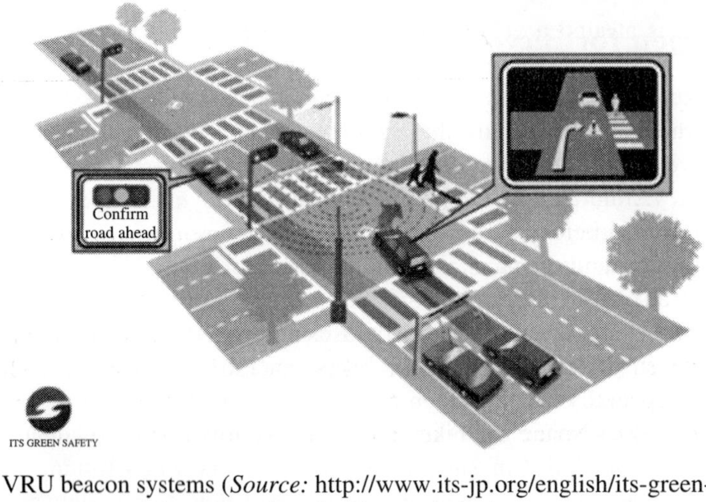

FIGURE 10.7 VRU beacon systems (*Source:* http://www.its-jp.org/english/its-green-safety-showcase/, reproduced with permission).

According to the experts, beacons should indicate type of user, but that new icons may be difficult to comprehend. Regarding negative side effects, it may be difficult for the system to discriminate between "on-board walk"/safe and unsafe conditions. In addition, traffic complexity is increased since traffic communication increases.

10.3.6 Pedestrian Traffic Signals

Smart pedestrian traffic signals allow safe crossing of road by pedestrians by displaying or extending the green aspect of pedestrian signal as per the need of the pedestrians. The system can automatically detect the presence of the pedestrians or may be activated by them. The PTSs are seen as one of the most promising ITS for pedestrians and cyclists. In this sense, it not only enhances VRU mobility, but also VRUs are less likely to cross during the red light phase. On central urban areas, one of the big challenges for traffic management is to deal with the time shared in pedestrian crossings at junctions, since green light for pedestrians is red light for cars.

Acceptance of this system by VRUs is considered high; since it responds to the need of pedestrians, especially elderly pedestrians and persons with reduced mobility, to be able to cross the road at their own speed.

A negative aspect could be that drivers may perceive more frequent stops but these systems may skip cycles when there is no pedestrian crossing the road (anymore), which is likely to increase acceptance and compliance by the drivers. Countdown traffic signals and pre-green for bikes are also included in this category. According to the experts, countdown traffic signals are requested by older road users to remove uncertainty over crossing times, claiming that it avoids red crossing. However, it should be adapted for different walking speeds.

This information benefits all. Having a countdown signal makes pedestrians feel safer at managing the crossing, it calms down drivers because they know how long they have to wait, and it helps the traffic flow because slower pedestrians will avoid starting to cross at the middle of the green light.

10.3.7 Pre-green for Bicycles

At major road intersections separate traffic signals may be implemented for bicycles, cars and pedestrians. Separated traffic signals for bikes and cars have the advantage of displaying a pre-green for bicycle users. An ideal pre-green of 4–5 s gives bicycle riders a head start, making them more visible to drivers as they continue through the intersection. Bicycle traffic signals will also turn red before automobile signals, allowing time for automobiles to continue through the intersection and/or make safer right turns.

The system is also seen as promising ITS tool in order to enhance safety and mobility of cyclists since they are clearly seen. Notwithstanding, drivers acceptance might be low and they might be surprised by green for bikes and still red for cars. One expert claims that this system is especially helpful when there is a significant amount of traffic on that road, but car drivers may block space for bikes and start driving when the signal for cyclists turns green.

10.3.8 Red Light/Speed Cameras

Red light/speed cameras detect the red light violation or speed violation and help the enforcement agencies to penalise the violators. This system is simple and cheap and is the only method of ensuring that traffic lights are obeyed. This system should be considered not only as enforcement aid but also a preventive action. To promote a change in attitudes and behaviour, the installation of red speed cameras is highly recommended. Average speed cameras have been used successfully in UK to reduce drivers' average speed. Expert suggests that its use should be extended, and it should be combined with in-car information on speed limits.

10.3.9 Red Light Violation Detection System

It may be a standalone system which works on fewer resources. Red light violation is one of the major factors for road accidents at intersections because drivers become impatient due to number of reasons like less time assigned to period of green time, waiting time at the signal, impatience of driver, etc. The manual system fails to perform the job due to number of reason like

(a) Faulty registration number plates in shape, fonts, colour, languages

(b) Very high degree of heterogeneous traffic and lack of lane discipline

(c) Inadequate number of enforcement personnel such as traffic police

In case of more violators in manual method at a time, only one or two can be booked and others run away. In that situation, the ITS based red light violation detection system is considered to be one of the best solution.

10.3.10 Solely-infrastructure based Applications

These applications are embedded in the road infrastructure such as signals, warning systems, etc. Some of the infrastructure-based applications are explained as follows:

Intelligent pedestrian traffic (IPT) signal control

State-of-the-art traffic light management systems use advanced sensors to monitor crossing behaviour of pedestrians. In this method, sensors detect the presence of pedestrians and take pre-defined actions for traffic signal control. The pedestrians need not to worry about pressing of any button or wait for their turn (as per fixed signal plan) to cross the road. The system takes care of signal timing automatically. At the same time if pedestrian takes more time to cross the road (may be due to age or any other reason), the pedestrian green light is adjusted accordingly to facilitate the movement of those pedestrians. Figure 10.8 illustrates the intelligent pedestrian detection system.

This system provides a clear benefit to children, people with disabilities and older adult pedestrians. It provides them with the time they effectively need to cross a road. Furthermore, such systems add for sure to the comfort of the older adult pedestrians. Research on impacts would provide better insight into the effectiveness of IPT in increasing road safety in general, and for the VRU in particular.

FIGURE 10.8 Intelligent pedestrian detection system (*Source:* https://www.fhwa.dot.gov/publications/research/safety/pedbike/05085/images/fig1015.gif, reproduced with permission).

Roadside pedestrian presence warning

It warns upcoming traffic for the presence of pedestrians close to the crossing. The application provides information to the connected vehicle from the infrastructure that indicates the possible presence of pedestrians in a crosswalk at a signalised intersection. Such systems have been deployed in Sweden and are reported to result in an average speed reduction of 2.2 kmph on 50 kmph roads. This reduction, as well as the increased attention of the motorist when passing the crossing, is likely to increase crossing safety.

Crossings adaptive lighting (CAL)

Improving the visibility of the pedestrian in darkness is important to provide safe environment to the pedestrians. This requires (better) lighting on pedestrian crossing facilities and situations where pedestrians share the road with other modes of transport. Crossings adaptive lighting (CAL) can significantly contribute to pedestrian visibility and so do improvements in the lighting system of vehicles, e.g., lights that turn with the curvature of the road, will also have a positive influence. The lighting on a (zebra) crossing is automatically switched on when a pedestrian is approaching. This reduces the chance that the pedestrian is not or late observed by the driver, and saves energy as well.

10.3.11 e-Call

e-Call is a vehicle-based system which automatically establishes a voice connection with emergency help system, and sends digital location and vehicle information to emergency services in case of a crash. Twenty-five European countries have signed the e-Call Memorandum of Understanding, committing themselves to the establishment of national public service answering point (PSAP). e-Call (Figure 10.9) does not prevent accidents but mitigates the effects of severe accidents.

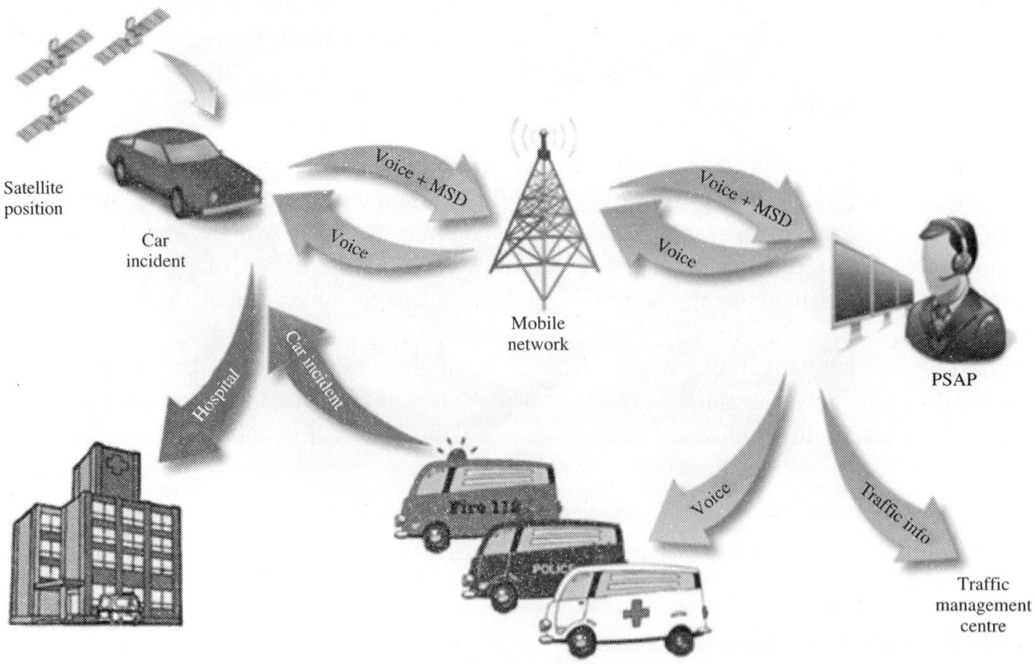

FIGURE 10.9 e-Call system (*Source:* https://ec.europa.eu/digital-single-market/ecall-time-saved-lives-saved, reproduced with permission).

Of the VRU, the younger and older drivers are likely to be benefited the most from e-Call. Other VRU would benefit indirectly when involved in a severe accident with a car or truck, as emergency services will arrive faster and better informed at the scene. However, impact of e-Call will be limited, if it is not installed in the majority of vehicles.

10.3.12 Intelligent Cooperative Intersection Safety System

The Intelligent Cooperative Intersection Safety System (IRIS), is a cooperative roadside application that aims at decreasing the number of accidents at controlled intersections. Vehicle-to-infrastructure communication and data fusion techniques make the core of the system. The system is based in a roadside unit and connected to both physical sensors (to sense the presence of static and/or moving objects on the road) and communication devices to receive data from the vehicles. This data is used to identify conflicting vehicle paths as possible accident sites. It reacts upon this identification by warning one or multiple vehicles about this threat.

10.3.13 Alcohol Ignition Interlock

An alcohol ignition interlock is a device which prevents a vehicle from starting if the driver has been drinking. The device works like a breathalyzer, measuring the alcohol in a person's system. If the amount of alcohol on the driver's breath exceeds a pre-programmed level, then the interlock temporarily locks the vehicle's ignition.

To take the test, the driver simply blows a breath sample into the device. If the sample is good, then the vehicle will start as usual. If alcohol is detected, the driver needs to wait before testing again. The first time alcohol is detected, the wait time is a few minutes, but if subsequent tests are failed the driver is locked out for increasingly longer periods.

10.3.14 Adaptive Headlights

Adaptive headlights are an active safety feature which is designed to make driving at night or in low-light conditions safer by improving visibility around curves and over hills. When driving around a bend in the road, standard headlights continue to shine straight ahead, illuminating the side of the road and leaving the road ahead of you in the dark. Adaptive headlights, on the other hand, turn their beams according to your steering input so that the vehicle's actual path is lit up.

Similarly, when a vehicle with standard headlights crests a hill, the headlight beams temporarily point upwards towards the sky. This makes it difficult for drivers to see the road ahead and for oncoming motorists to see the driver approaching. In contrast, adaptive headlights use a self-levelling system that points the light beam up or down, according to the position of the vehicle.

Adaptive headlights are helpful when driving on winding roads at night, during twilight, or in other low-light conditions. They can address many potentially dangerous situations, including:

- An animal is standing on the road just around a poorly lit curve.
- An oncoming vehicle negotiating a turn accidentally drifts into driver's lane.
- Cresting a hill on a narrow road, driver is unable to see whether another motorist is coming.
- As driver rounds a curve, the headlights temporarily blind oncoming traffic.

Recently Google has announced a concept of soft cars to protect the VRUs from Self Driving car related accidents. The details of Soft car are presented in Box 10.1 as below.

Box 10.1 Soft Car for safety of VRUs

Google's self-driving car unit Waymo has patented a way to make vehicles softer to reduce collision-related injuries. If the rigidity of the car could be taken away from the hard shells of cars, the intensity of the accidents and injuries can be reduced. Recently, the Google-owned company has been issued a patent by the United States Patent and Trademark Office (USPTO) which describes ways to make our cars softer without becoming detrimental to drivers and passengers. The patent describes how "tension members" can be arranged to alter the rigidity of a vehicle's surface. These members, which could be cables or rods, for example, will respond by changing tension should sensors pick up and predict a potential object, hazard, or pedestrian. The vehicle may identify and respond to a potential collision by altering the tension that is applied to one or more tension members, thereby altering the rigidity of the vehicle's surface. Processors and sensors embedded in the vehicle will detect potential hazards and determine whether or not the object and car are on a collision course. If an accident is expected, one or more sections of the car—such as the hood, bumper, or panels—will become softer through cutting, releasing, or loosening the tension members. In turn, this would reduce the solidity of surfaces and the potential harm of impact. If the object is a car, the system will maintain tension to protect the driver and passengers in the car. However, should the potential hazard be a bicyclist, pedestrian, or animal, tension will loosen.

The sensors embedded in the car may also learn from hazard scans over time. For example, should a child impact the vehicle rather than an adult, tension may be loosened more in certain areas to create the softest surface possible based on their shape, size, and height.

Source: http://www.zdnet.com/article/googles-waymo-patents-soft-self-driving-cars-to-keep-you-safe/

10.4 SUMMARY

The deployment of ITS technologies has the potential to enhance safety and mobility of VRUs. However, it is essential for their success, that these systems are tailored to the specific needs of road users.

VRUs are likely to be indirectly benefited from some ITS applications which have been developed for safety of passenger cars. From a review of accident causes, it can be concluded that there are a number of factors that have a large impact on the number and severity of injuries on the side of all or most types of VRU such as:

(a) Speed
(b) Alcohol
(c) Non-observance of the VRU by the vehicle driver
(d) Late and insufficient braking by the vehicle driver in cases of emergency

It is emphasised that many ITS applications that have been investigated for years are still not ready for real-life deployment. Of many others, the impact on traffic safety is not clear or not ready for quantitative assessment. In general, systems that assist the car driver in the driving task will influence the behaviour of the driver.

Questions

10.1 Why ITS is important for vulnerable road users (VRUs)?
10.2 Prepare a list of ITS technologies/applications for VRUs.
10.3 Describe advanced driver assistance system.
10.4 What do you understand by intelligent speed adaptation?
10.5 How pedestrian detection system works?
10.6 How blind spot detection helps in safety of VRUs?
10.7 Explain VRU beacon systems.
10.8 How pedestrian traffic signals is helpful for VRUs?
10.9 Write short notes on:
  (a) Pre-green for bicycles
  (b) Pedestrian traffic signal control
  (c) Roadside pedestrian presence warning
  (d) Crossings adaptive lighting (CAL)
  (e) e-Call

Bibliography

Allsop, Richard, "The Effectiveness of Speed Cameras: A review of evidence", RAC Foundation, November 2010 (http://www.racfoundation.org/assets/rac_foundation/content/downloadables/efficacy_of_speed_cameras_allsop_181110.pdf)

Andriluka, M., S. Roth, and B. Schiele, "People-tracking-by-detection and people-detection-by-tracking", In CVPR, 2008.

Carsten, O. and F. Tate, "Intelligent Speed Adaptation: Accident Savings and Cost-Benefit Analysis", *Accident Analysis and Prevention,* Vol. 37, pp. 407–416, 2005.

Forkenbrock, G., R.L. Hoover, E., Gerdus, T.R. Van Buskirk, and M. Heitz, "Blind spot monitoring in light vehicles—System performance", Report No. DOT HS 812 045, Washington, DC: National Highway Traffic Safety Administration, July 2014.

Global Safety Partnership on Vulnerable Road Users, International Federation of Red Cross and Red Crescent Societies.

Huth, H. and G. Helau, "Predicting the acceptance of advanced rider assistance systems", *Accident Analysis and prevention*, Vol. 50, pp. 51–58, 2013.

Källhamer J., "Pedestrian Injury Mitigation by autonomous braking", Autoliv Research, 2009.

Kulmala, R., "Ex-ante assessment of the safety effects of intelligent transport systems", *Accident Analysis and Prevention*, Vol. 42, pp. 1359–1369, 2010.

Leden, L., P. Gårder, and U. Pulkkinen, "An expert judgement model applied to estimating the safety effect of a bicycle facility", *Accident Analysis and Prevention*, Vol. 32, pp. 589–599, 2000.

Mueller, Friedrich, "Adaptive Lighting System Today and Tomorrow", 2010. (http://www.smvic.com.cn/smvic/res_base/smvic_com_www/upload/article/file/2010_4/12_16/46ehghqzzcx7.pdf)

Nordqvist, M. and N.P. Gregersen, "Study on motorcyclists behaviour and attitude towards road safety 2010", Borlänge, Sweden: SMC, 2011.

Paine, Michael and David Paine, "In-Vehicle Intelligent Speed Advisory Systems", Australia Paper Number 07–0247, Vehicle Design and Research. (http://www-nrd.nhtsa.dot.gov/pdf/esv/esv20/07-0247-W.pdf)

Regan, M., J. Oxley, S. Godley, and C. Tingvall, "Intelligent Transport Systems: Safety and Human Factors Issues", Victoria: Royal Automobile Club of Victoria, 2001.

Riches, Ian, "Automotive Advanced Driver Assistance Systems, Challenges and Opportunities", 2013. (http://on-demand.gputechconf.com/gtc/2013/presentations/S3413-Advanced-Driver-Assistance-Systems-ADAS.pdf)

Safety of Vulnerable Road Users, Scientific Expert Group on the Safety of Vulnerable Road Users, OECD, www.VRUITS.eu.

Scholliersa, Johan, Daniel Bellb, Andrew Morrisc Alejandra, and B. García, "Improving safety and mobility of Vulnerable Road Users through ITS applications", Transport Research Arena, Paris, 2014.

Violence and Injury Prevention, Key Facts, World Health Organisation, 2014. (http://www.who.int/violence_injury_prevention/road_safety_status/key_data/en/)

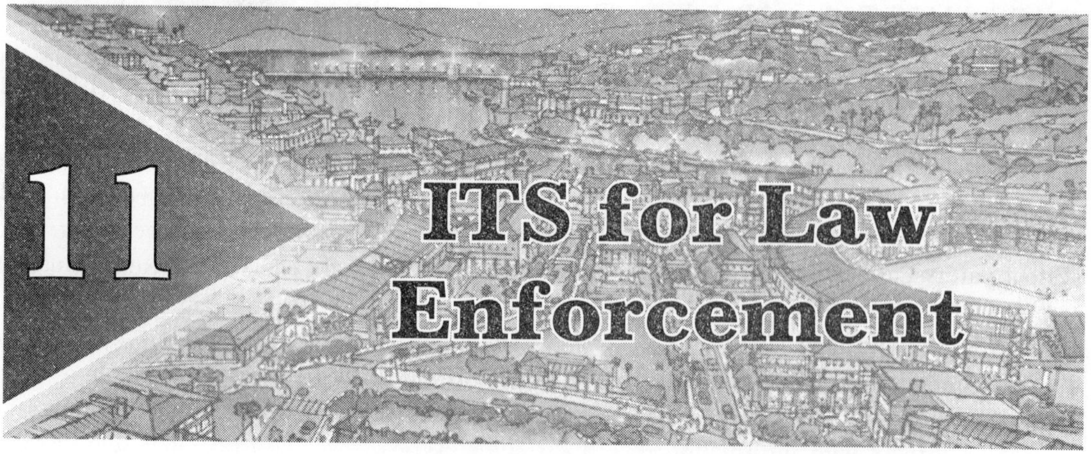

11 ITS for Law Enforcement

11.1 INTRODUCTION

A wide range of ITS technologies either exists or are under development that could be used by law enforcement agencies to further support and enhance their enforcement of safety-critical road rules. The Traffic police is already involved in enforcement of various traffic rules and regulations to ensure safety and security of road users. However, the sheer volume of traffic on roads undermine the capability of traffic police to enforce the rules. Jumping of traffic lights when traffic police is not around, is not very uncommon in India. The ITS technologies could be used to enhance capabilities of traffic police for the enforcement of those road rules that they already enforce and to support enforcement activities that they currently have difficulty with. The ITS technologies including in-vehicle, infrastructure-based and cooperative systems; systems that would both prevent drivers from engaging in illegal driving behaviours and support the enforcement of those road rules that have been violated.

11.2 ITS TECHNOLOGIES TO ENHANCE AND SUPPORT THE ENFORCEMENT TRAFFIC RULES AND REGULATIONS

11.2.1 Speeding

Over speeding is one of the major cause of road accidents. Use of speed cameras to detect over speeding and fine the violators are increasingly being used by traffic police. A number of in-vehicle and infrastructure-based ITS and telematics technologies can further assist police in the enforcement of speed-related road rules. These tools include intelligent speed adaptation (alerting and limiting), in-vehicle dynamic displays, flashing variable speed signs, school zone information built into existing route guidance or other systems, and brief school zone approach messages/alerts transmitted via a DSRC system. These technologies are discussed as follows:

Intelligent speed adaptation

Intelligent speed adaptation (ISA) refers to a system that automatically warns the driver and/or limit vehicle speed when the driver, intentionally or inadvertently, exceeds the stipulated speed limit by a predetermined amount. ISA is one of the in-vehicle ITS tools which continuously monitors the speed of the vehicle, aimed at encouraging or preventing drivers from exceeding the speed limit. The position of the vehicle is established via GPS or roadside beacons. The current speed and position of the vehicle is compared with the local stipulated speed limit. ISA system can issue warnings or limit vehicle speed if the vehicle exceeds this stipulated limit.

In-vehicle information displays

The in-vehicle information displays present drivers with information that would normally appear on static or variable signs outside of the vehicle. The in-vehicle displays can be used to provide drivers with information regarding temporary changes in speed limits due to road works, road or weather conditions or an accident. The roadside beacons or Internet or satellite can transmit this information to on-board unit of in-vehicle information displays. The visual display used could be a standalone display or integrated as part of a multi-functional display which is available in the vehicle to support enforcement of road safety laws or the route navigation display. Further, it could alert drivers, using either visual or auditory alerts, to approaching traffic control devices such as stop, give way and no entry signs, and even inform them if they are travelling in a restricted use lane such as an emergency vehicle lane. Flashing variable speed limit signs are installed at a number of school zones to display the temporarily reduced speed limit during school crossing hours.

11.2.2 Driving Under the Influence of Alcohol and Other Drugs

Prevention of drunk driving has remained a major challenge for law enforcement agencies. Conventionally, breath and salvia testing technologies have been used for a number of years to test consumption of alcohol and other drugs. The modern ITS technologies could be used to assist traffic police to enforce drink and drug driving laws. These include alcohol detection systems, performance tests, electronic licences and keys, closed circuit television (CCTV) to target drug and alcohol testing, and telematics systems that immobilise the vehicles of recidivist offenders during high alcohol and drug use times.

Alcohol sniffer systems

Alcohol 'sniffer' systems can detect the presence of alcohol on the driver's breath using sensors from a remote location. If alcohol is detected, driver is required to blow into a breathalyser unit. If the breathalyser unit confirms the presence of alcohol, then the system either immobilises the vehicle or limits the speed of the vehicle to reduce crash risk and severity.

Performance tests

The technologies are being developed which will be capable of detecting and assessing impairment based on reaction times and coordination. Such performance tests would require the driver to complete and pass a psychomotor driving task (i.e., tracking or reaction time task) when they first enter the vehicle and, depending on the results, will either allow the ignition to be started or will immobilise the vehicle.

Electronic licences

The drunk driving can also be prevented by use of electronic licences and smartcard licence readers. These devices could be used to help enforce alcohol restrictions and alcohol interlock requirements. Electronic licences store information about the driver, such as their age, licence status and any restrictions or conditions on their licence, and use this information to determine if the person is allowed to drive. The electronic licence could be used in conjunction with alcohol interlocks. For example, the system would disable the vehicle if the driver has alcohol in excess of the amount allowed.

CCTV to better target alcohol and drug testing

CCTV or video analytics may be deployed to detect potentially impaired drivers approaching alcohol testing sites. The system helps in better targeting alcohol/drug testing by identifying and targeting those drivers who appear under influence of alcohol based on their observed driving performance. Police monitoring the traffic using this technology could inform to police testers the details of vehicles they should target.

11.2.3 Vehicle Immobilisation Technology

A telematics system installed on vehicles can immobilise them during high alcohol and drug use times (e.g., at night or on the weekends). This system would prevent recidivist drink or drug drivers from driving their vehicle during these higher risk times.

11.2.4 Seatbelt Wearing

In an attempt to increase seatbelt wearing rates, a range of seatbelt reminder systems are being developed and introduced by several vehicle. Most of these seatbelt warning systems are linked to the driver's seat only. The new seatbelt reminder and interlock systems are designed to detect unrestrained occupants in all seating positions. The seatbelt interlock systems are connected to one or more seats that contain sensors both within the seats and in the belt assembly. The vehicle ignition is disabled if these sensors determine that any occupant in the vehicle is not using seat belt.

The frontal image speed cameras have also being developed to take a frontal image of vehicles detected speeding. These cameras are capable of identifying the driver of the vehicle and the front licence plate. The image is detailed enough to determine if the two front occupants of the vehicle are wearing seatbelts.

11.2.5 Giving Way (Emergency Vehicles)

The emergency vehicles, ambulances, fire tenders, etc. need to be given priority in traffic. The ITS technologies can provide drivers with an advanced warning of an approaching emergency vehicle in emergency mode (e.g., lights and sirens active) so they have more opportunity to move out of the emergency vehicle's path. In-vehicle emergency vehicle proximity warning devices have been developed that issue warnings to drivers when an emergency vehicle is approaching. These systems are designed to alert drivers about an approaching emergency vehicle before they are likely to be able to hear the siren in order to provide drivers with more opportunity to clear a path for the emergency vehicle and thus, avoid collisions.

11.2.6 Railway Level Crossing Warnings and Camera

The level crossing gates of railways especially unmanned ones have remained point of conflict and accidents. At unmanned level crossing gates, there is no body to close the gate to road traffic when train passes through the gate location. A range of prototype technologies have been developed and evaluated that detect approaching trains and provide in-vehicle or infrastructure-based warnings to drivers at unmanned railway level crossings. These systems use a range of sensors and transmitters to detect approaching trains, such as sensors to detect the vibrations of approaching trains or transmitters located on trains that are detected by receivers in vehicles approaching the crossing. The approaching train warnings can be provided to drivers as visual and/or auditory warnings located within the vehicle or in the road environment around the crossing.

11.2.7 Keeping a Safe Following Distance

The rear-end collisions constitute a major proportion of all police-reported road crashes, particularly in urban areas. In majority of cases the rear end collision is attributed to one or both of two principal factors: driver inattention or distraction and following a lead vehicle too closely. Rear collision warning devices have been designed to monitor the time headway between vehicles and either alert the driver about an impending collision (Forward Collision Warning) or assist the driver to maintain an appropriate time headway from the lead vehicle by issuing visual and/or auditory alerts when drivers reach an unsafe following distance (Following Distance Warning). Time headway is defined as the distance in metre from a vehicle ahead divided by speed in metre per second.

A number of "close following detection devices" have been, or are currently being developed that can be used to detect drivers who are tailgating. Such systems provide police with an objective measurement of vehicle's following distance. Cameras have been developed that are capable of detecting the distance between vehicles and taking a visual image of those vehicles that are detected travelling too close to the vehicle in front. Laser technology has also developed a hand-held laser speed detection system with a built-in following distance detection device. This "Distance Between Cars" device is capable of determining the distance between two moving vehicles. The device can be used in handheld mode or mounted on a stationary police vehicle. A camera unit can also be attached to the device so that an image of vehicle deemed to be following too closely can be taken (www.lasertechnology.com.au)

11.2.8 Prevention of Use of Mobile Phones in Vehicles

The use of mobile phone while driving can be dangerous as reaction time of the user increases while using the mobile. The use of mobile phone jammers fitted to vehicles has been identified as a possible solution to prevent drivers from using hand-held mobile phones while driving. Mobile phone jammers prevent mobile phone use by transmitting a signal on the same frequency as the mobile phone signal and at a high enough level so that the two signals cancel each other out. The mobile phone detector/alerter fitted in the vehicle that can determine if a mobile phone is being used. This system would then issue a warning signal on the exterior of the vehicle that alerts police to the fact that a mobile phone is being used in a vehicle so that they can investigate further. Mobile phone alerters are currently used in hospitals, where mobile phone signals can interfere with medical equipment.

11.2.9 Licencing and Registration

The various technologies available for law enforcement using licencing and registrations of vehicles are as follows:

Automatic licence plate recognition systems

Automatic licence plate recognition (ALPR) system is capable of automatically detecting and reading the registration/licence plates of vehicles that pass the system's cameras. The system can check the owner and vehicle details from the vehicle registration database. ALPR system can be used by traffic police to detect a number of offences such as:

(a) Theft of vehicle
(b) Vehicles with expired registration
(c) Owners who have been disqualified from driving

ALPR systems can also be used to enforce other types of crime (e.g., notify police that a vehicle owner is wanted for other, non-traffic offences). ALPR systems fitted to police vehicles would prevent police from having to manually enter vehicle licence plate information into their on-board system, decreasing their chance of being involved in a distraction-related incident. ALPR systems are increasingly being used by police world-wide to assist traffic enforcement.

Electronic licences

Electronic licences or keys can be used to prevent drivers from driving outside the conditions of their licence (e.g., preventing drivers from driving during certain hours or from driving certain vehicle types). Electronic licences could be linked to their owners in order to prevent people from using another driver's licence to start the vehicle, by requiring drivers to input a PIN number or scan their fingerprint when they enter the vehicle.

11.2.10 Heavy Vehicles

The detection of over loading in case of commercial vehicle has remained a challenge before the law enforcing agencies. The overloading not only damages road surface but also cause a loss of revenue (road tax, excise tax, etc.) to the authorities. Weigh-in-motion stations and vehicle length measuring devices are installed at a number of locations to detect violations of heavy vehicle mass and dimension limits. Cameras or ALPR systems could also be linked to these measuring stations to identify vehicles that exceed the maximum weight and dimension limits and issue infringement notices to the vehicle owner, driver or other responsible party.

11.3 DETECTION OF VIOLATIONS

A range of technologies exists that could automatically detect traffic violations and identify offending vehicles without the police having to be present. Some of the technologies are discussed as follows:

11.3.1 Closed Circuit Television

Closed circuit television (CCTV) could be fixed at various locations on the road network to monitor driver behaviour and detect the traffic violations, such as travelling in restricted lanes, illegal overtaking, failing to give way to other road users, using hand-held phones and not wearing a seatbelt or motorcycle helmet. CCTV could be linked to imaging technology or automatic vehicle identification systems in order to automatically to send notices and impose penalties to traffic rules violators.

11.3.2 In-vehicle Event Data Recorders

The critical information about the vehicle like speed, location, acceleration, deceleration, etc. with time can be recorded and logged using in-vehicle event data recorders. Event data recorders could be linked to a GPS receiver on the vehicle to establish vehicle location and programmed to record certain driving data. The data from the system can be used by the traffic police to check violations such as excessive speed, etc. The system could thus record violations such as excessive speeds in certain locations, which police can then access to verify that an offence occurred. The use of event data recorders could also deter drivers from violating road rules in the first place because they know their driving behaviours are being recorded.

11.3.3 ALPR Systems and Electronic Vehicle Identification

ALPR systems and electronic vehicle identification (EVI) systems can automatically identify vehicles once they have been detected violating traffic laws. ALPR systems automatically 'read' the licence plate of vehicles to identify the vehicle. Electronic vehicle identification (EVI) devices can identify vehicles through an electronic tag located on the vehicle that contains unique vehicle information using a wireless communication link to read. This electronic tag can be located anywhere on the vehicle and can transmit vehicle information such as registration number, vehicle identification number (VIN), vehicle make, model and dimensions, and vehicle classification category. EVI can be used in a range of applications including traffic law enforcement, by automatically identifying vehicles that are over speeding, violating red lights, or moving in restricted lanes. Electronic tag readers linked to speed and red light cameras can be activated when a vehicle is detected over speeding or violating a red light. System designers and researchers claim that the use of EVI has a number of advantages over automatic licence plate recognition systems, namely being able to detect every vehicle passing the reader unit unlike licence plate recognition systems which cannot read all licence plates under certain lighting conditions or if the plate is dirty. But EVI systems require tags to be placed on the vehicle, either at the time of manufacture or as a retrofitted device, which increases the cost of these systems.

11.3.4 Remote Engine Immobilisation

Remote engine immobilisation works by preventing the vehicle from being restarted once the system has been activated and the engine has been switched off. Engine immobilisation could also be coupled with a vehicle tracking device so that police can track the vehicle at a safe distance.

11.3.5 Remote Speed Governing Device

A remote speed governing device that limits the vehicle to the local speed limit requires the system to be linked with a GPS system and digital road map containing the speed limits, as well as a vehicle tracker. This technology works by reducing vehicle speed to a point where police would be able to more safely intercept the vehicle without the danger of a high speed pursuit.

11.3.6 Vehicle Tagging and Tracking Device

An alternative to remote vehicle immobilisation technology is vehicle tagging and tracking technology. A real-time vehicle tagging and tracking device called *Star Chase* has been developed during 2006 in the US that is capable of tracking offending vehicles in real-time. The system consists of an adhesive vehicle tracking dart containing a GPS receiver, radio transmitter and power supply that is fired by police from a compressed-air launcher onto an offending vehicle. The launcher can be a hand-held device or mounted on the front of a police vehicle. The offending vehicle's position is determined via GPS and wirelessly transmitted to the police where they can track the vehicle in real-time.

11.4 SUMMARY

There exists a range of new and existing ITS and telematics technologies that could be used to assist police for enforcement of traffic laws and help overcome some of the difficulties. Police currently experience a number of problems when enforcing certain road rules. The technologies identified can be used to assist police at all points along the traffic enforcement chain, from detecting traffic violations to processing infringement notices and providing photographic or video evidence of the traffic offence. The technologies identified also have the ability to prevent or deter drivers from engaging in illegal and dangerous driving behaviour in the first place, thereby increasing safety and reducing the need for police intervention.

Questions

11.1 How ITS can be used for law enforcement?

11.2 Describe different ITS and telematics technologies that can assist police in the enforcement of speed-related road rules.

11.3 Describe use of ITS for controlling 'driving under the influence of alcohol'.

11.4 Write short notes on the following:
- (a) Vehicle immobilisation technology
- (b) Seatbelt wearing
- (c) Frontal image speed cameras
- (d) Emergency vehicle in-vehicle warning system
- (e) Railway level crossing warnings and camera

11.5 How automatic licence plate recognition system works and what are its applications?

11.6 Describe various technologies available for detection of violations?

11.7 What do you understand by intercepting vehicles? How are these used for law enforcement?

Bibliography

Al Marzooqi, A., "Road safety systems for monitoring fleet drivers", *Proceedings of the 68th Road Safety Congress: Safer Driving, Reducing Risks, Crashes and Casualties*, March 2003.

Carsten, O. and F. Tate, "External vehicle speed control", Final report: Integration, Institute for Transport Studies, University of Leeds, UK, 2000.

Chevreuil, M. and A. Canal, "Development of automated traffic enforcement systems in France", France, 2003.

Department of Justice, Speed Cameras: The cost of not having speed cameras, Department of Justice, Victorian Government, 2004 (http://www.camerassavelives.vic.gov.au/)

Diamantopoulou, K. and B. Corben, "The impact of speed camera technology on speed limit compliance in multi-lane tunnels", *Proceedings Road Safety Research, Policing and Education Conference*, 4–5, Adelaide, Australia, CD-ROM, pp. 282–287, 2002.

Guha, S., K. Plarre, D. Lissner, S. Mitra, B. Krishna, P. Dutta, and S. Kumar, "Autowitness: locating and tracking stolen property while tolerating GPS and radio outages", In Sensys, November 2010.

Jäger, F., C. Lagauterie, S. Lewis, P. Kok, and U. Grottker, "Speed and red light enforcement using digital cameras with international acceptance", *Proceedings of the 12th World Congress on Intelligent Transport Systems*, San Francisco, California, 2005.

Mathur, S., T. Jin, N. Kasturirangan, J. Chandrasekaran , W. Xue, M. Gruteser, and W. Trappe, "Parknet: drive-by sensing of road-side parking statistics", In Mobisys, 2010.

Nakanishi, Y.J. and W.C. Fleming, TCRP Synthesis 93—Practices to Protect Bus Operators from Passenger Assault, Washington, DC: Transit Cooperative Research Program, Transportation Research Board, 2011.

Peltola, H. and R. Kulmala, "Weather related intelligent speed adaptation—experience from a simulator", *Proceedings of the 7th World Congress on Intelligent Transport Systems*, Turin, Italy, 2000.

Regan, M.A., "A Sign of the Future-1: Intelligent Transport Systems", Chapter 14, In C. Castro and T. Horberry, (Ed), *The Human Factors of Transport Signs*, CRC Press, USA, pp. 213–224, 2004.

Regan, M.A., J.A. Oxley, S.T. Godley, and C. Tingvall, "Intelligent transport systems: Safety and human factors issues", Report No. 01/01, Royal Automobile Club of Victoria (RACV), Melbourne, Australia, 2001.

Regan, M.A., K.L. Young, and N. Haworth, "A review of literature and trials of intelligent speed adaptation devices for light and heavy vehicles", AP-R237, Austroads, Sydney, New South Wales, 2003.

Van Boxtel, A., "Early implementation of intelligent speed adaptation (ISA) in the Netherlands", *Proceedings of the 32nd ISTA Conference*, Vienna, Austria, 1999.

Young, Kristie L. and Michael A. Regan, "Intelligent Transport Systems to Support Police Enforcement of Road Safety Laws", ATSB Research and Analysis Report Road Safety Research Grant Report, 2007.

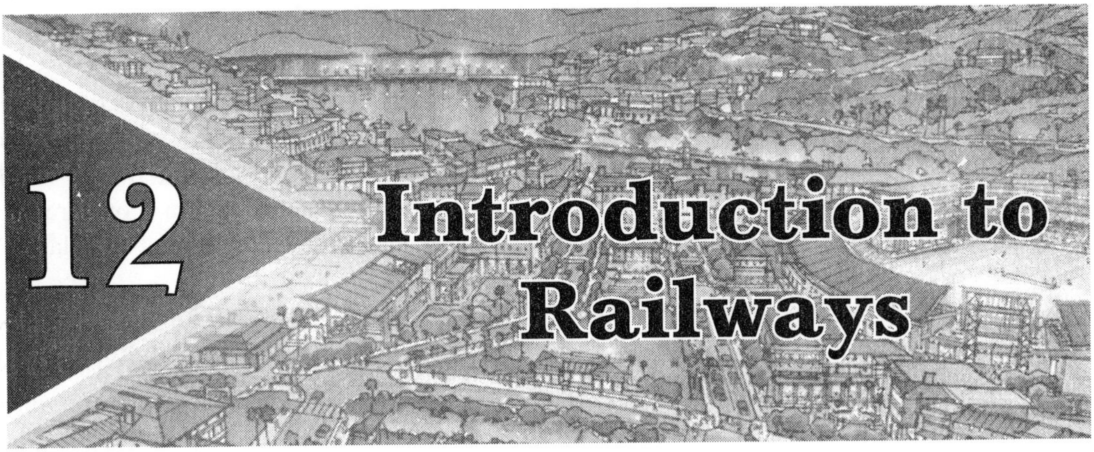

12 Introduction to Railways

12.1 INTRODUCTION

Unlike roadways where knowledge about road, signals, vehicles, etc. are widely available in public domain, the intricacies of railway system mostly remains confined to the railway operators. Before deliberating upon the use of intelligent transport systems in railways, a brief introduction of railway system would be helpful for the readers.

Transportation of passenger and goods by wheeled vehicles running on rails is termed as *rail transport* or *train transport*. The rail vehicle also called train is guided by the tracks on which it run. The track usually consists of two parallel steel rails tied together through sleepers which are laid on the ballast. The movement of trains on track is controlled by signals to ensure safe separation between two trains. The hauling power is provided by locomotives which either draw electric power from a railway electrification system or produce their own power, usually by diesel engines.

12.2 MERITS AND DEMERITS OF RAILWAYS

12.2.1 Merits of Railways over other Modes of Transport

When compared with roadways, the railways offer the following advantages:

(a) Rolling stock in railway transport systems generally has lower frictional resistance due low friction between steel wheels and steel rails when compared with road vehicles. As per studies, the railway system is considered to be six times more energy efficient than roadways.

(b) Being energy efficient, railways is cheaper mode of transport as compared to other modes except waterways.

(c) Its speed over long distances is more than any other mode of transport, except airways. Thus, it is the best choice for long-distance traffic.

(d) Railway transport is economical, quicker and best suited for carrying heavy and bulky goods over long distances.

(e) Railway is the safest form of transport. The chances of accidents and breakdowns of railways are least as compared to other modes of transport.

(f) The rail transport is better organised than any other form of transport. It has fixed routes and schedules. Its service is more certain, uniform and regular as compared to other modes of transport.

12.2.2 Demerits of Railways over other Modes of Transport

(a) The railway projects are capital intensive (requiring huge investment). The cost of construction, maintenance and overhead expenses are quite high as compared to other modes of transport.

(b) Railway system is not flexible. Its routes and timings cannot be adjusted to individual requirements.

(c) It cannot provide door to door service.

(d) Being capital intensive, railways are mostly natural monopolies which is by and large against public interest at large. Even if controlled and managed by the government, lack of competition may breed inefficiency and high costs.

(e) Railway system is not suitable and economical for short distance and small traffic of goods.

(f) The railway must have full load for its ideal and economic operation. As it has a very large carrying capacity, under-utilisation of its capacity, in most of the regions, is a great financial problem and loss to the economy.

12.3 COMPONENTS OF RAILWAY SYSTEM

The railway system consists of the following four components:

(a) Track

(b) Rolling stock

  (i) Locomotive

  (ii) Passenger coaches

  (iii) Wagons

  (iv) Self-propelled vehicles

(c) Train control system

(d) Traction system

(e) Station

12.3.1 Track

The track or permanent way is the set of rails on which trains run. Generally, the track consists of two parallel steel rails having a specified distance in between (gauge) and tied to sleepers which are laid perpendicular to the rails on the ballast of specified thickness spread over the formation as shown in Figure 12.1. The rails are joined with each other by fishplates and bolts or welded together. The rails are fastened to sleepers by various fittings like keys, spikes, clips, etc. The sleepers are spaced at a fixed distance as per the gauge and held in position by embedding in ballast or concrete in case of ballast less track being used by metro rail. Railway track is shown in Figure 12.2.

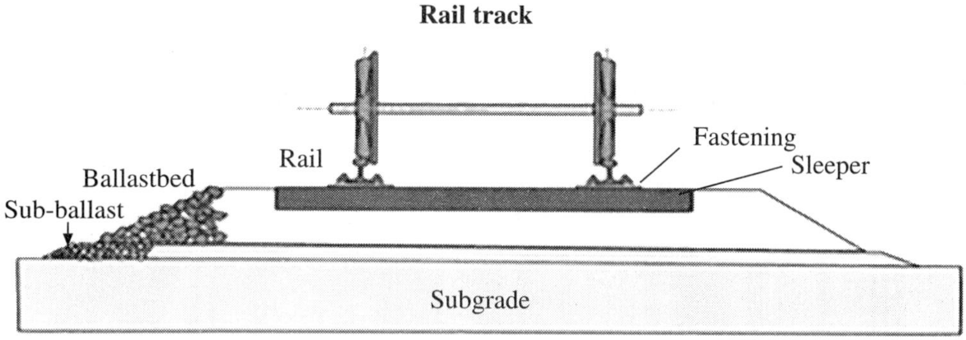

FIGURE 12.1 Components of railway track (*Source:* http://image.slidesharecdn.com/tn-1-07b-b-railwayengineering-141016055340-conversion-gate02/95/railway-engineeringb-4-638.jpg?cb=1413441659).

FIGURE 12.2 Railway track.

12.3.2 Rolling Stock

The railway vehicle which runs on the track is termed as *rolling stock*. It usually includes both powered and unpowered vehicles, for example locomotives, coaches, and wagons. A train is a set of rolling stock comprising of locomotive and coaches/wagons or self-propelled vehicles. The following types of rolling stock are generally used in railways:

Locomotive or Engine

A locomotive or an engine in a railway system is the rolling stock which provides motive power to the train. It is the invention of steam locomotive in England way back in the 19th century which led to evolution of rail road industry world over. Locomotives may generate their power from fuel (coal, petroleum or natural gas), or they may take power from an outside source of electricity. The locomotives are commonly classified on the basis of their source of energy. The most common types of locomotives are:

(a) Steam engines—not used now in India except on some heritage lines (Figure 12.3)
(b) Diesel engines (Figure 12.4)
(c) Electric engines (Figure 12.5)

FIGURE 12.3 Steam locomotive.

FIGURE 12.4 Diesel locomotive.

FIGURE 12.5 Electric locomotive, Indian Railways.

Passenger coach

A passenger coach or car is a rolling stock meant to carry passengers (Figure 12.6). The coach has seats or berths to provide sitting or sleeping facilities to the passengers. Depending upon the facility, the passenger coaches are generally of following types (Table 12.1):

TABLE 12.1 Types of passenger coach

S.No.	*Type*	*Facilities*
1	Ordinary chair car	Non-air conditioned, chair type, only sitting
2	AC chair car	Air conditioned, chair type, only sitting
3	Executive chair car	Air conditioned, only sitting, wider seats
4	General class	Non-air conditioned, bench type, sitting
5	Sleeper	Non-air conditioned, berths, sleeping and sitting, three tiers across the width of the coach
6	3rd AC	Air conditioned, berths, sleeping and sitting, three tiers across the width of the coach
7	2nd AC	Air conditioned, berths, sleeping and sitting, two tiers across the width of the coach
8	1st AC	Air conditioned, berths, sleeping and sitting, cabin type accommodation

There are many other variants of the coaches like Rajdhani, Shatabdi, Garibrath, LHB, Hum Safar, Mahamana, UDAY, etc. Description of each type of the coach is beyond the scope of this book.

FIGURE 12.6 Passenger coach, Indian Railways.

Wagons

The wagons are type of rolling stock which are designed to carry goods, livestock, containers, oil products, etc. A set of wagons attached with a locomotive forms the goods train. A variety of wagon types are in use to handle different types of goods as shown in Figures 12.7–12.9. The following types of wagons are mostly used in railways (Table 12.2):

TABLE 12.2 Types of wagons

S.No.	*Type*	*Commodity*
1	Open box type (open from top)	Coal, minerals, ores, iron coils
2	Covered	Food grain, fertiliser, cement
3	Flat	Containers, iron coils, military equipment
4	Tank	Oil, gas, milk, water
5	Hopper (bottom discharges)	Coal, ballast

FIGURE 12.7 Open wagon, Indian Railways.

FIGURE 12.8 Covered wagon, Indian Railways.

FIGURE 12.9 Containers loaded on flat wagons, Indian Railways.

Self-propelled multiple units

The self-propelled vehicles do not need any locomotive to haul them. The self-propelled multiple units (MU) are rail coaches coupled with other units of the same type having one motor car, driving car and trailing cars. The motor car is equipped with the engine to provide motive power. One of the part of driving car is driver cab having equipment to control the movement of the train. The passenger accommodation is available in all the cars. A combination of one driving car, one motor car and trailing cars which is capable of moving on its own is called a unit. The train set is formed by multiple of these units. Multiple units are classified by their power source and are of two main types: (a) electric multiple unit (EMU) and (b) diesel multiple unit (DMU). Diesel-powered units may be further classified by their transmission type: diesel-electric (DEMU), diesel-mechanical (DMMU) and diesel-hydraulic (DHMU). The major advantage of multiple units is high acceleration and deceleration which makes them suitable for suburban services, trams, etc. Further, there is no need to reverse locomotive at the terminals as multiple units can run in either direction.

For example, Indian Railways EMU unit consists of one motor car, one driving car and one trailing car, i.e., 3 cars per unit. The EMU train set may be of 3, 6, 9, 12, ... cars. EMU train of Indian Railways and EMU based Delhi Metro train are shown in Figures 12.10 and 12.11 respectively.

FIGURE 12.10 EMU train, Indian Railways.

FIGURE 12.11 EMU based Delhi Metro train.

12.3.3 Train Control System

The train control system consists of signals, interlocking and control panels at stations and at central control.

Signals

Unlike road vehicles where the driver is able to control the speed of vehicle in the face of an obstruction, for railway systems, the signalling system is required to alert the train driver in advance against any obstruction ahead as the stopping distance for rail vehicle is more than human sighting distance which means if driver applies the break after seeing the obstruction visually, he would not be able to stop the train before the obstructions. The basic purpose of signally system is to maintain adequate safe separation between two trains along with maintaining an optimum speed of the train for efficient operations. The primitive signalling system consists of different combinations of flags, signal arms, colour light, etc. Conventional signals are shown in Figure 12.12.

The block section system of signalling control is used most widely in railroad industry. In this system (also called *absolute block system*) only one train is allowed to occupy one section called *block section*. The limits of the block section is defined by two adjoining stations, i.e., only one train is allowed on a track between two stations. With advancement in the technology, the automatic block system, central traffic control (CTC), continuous automatic train control (CATC) and communication based train control (CBTC) have been developed which have been discussed in detail in Chapter 13 on 'Train Control System'.

The signals have also evolved from road signals—flag types, signal arms, colour light to cab signals (indication in driver's cab).

FIGURE 12.12 Conventional signals.

Interlocking

An interlocking is an arrangement of signals that prevents conflicting movements through an arrangement of tracks, points and crossings Figure 12.12(a). The points and crossings facilitate movement of train from one track to other track. An interlocking is designed so that it is impossible to display a signal to proceed unless the route to be used is proven safe. At the same time, it must be impossible to operate the points while the train is moving on it. Further, it must be impossible to take OFF position (proceed) at one and same time for two fixed signals which would lead to conflicting movements. Interlocking can be categorised as mechanical, electrical (electro-mechanical or relay-based), or electronic/computer-based depending upon the technology used. The electronic interlocking is used in the most advanced railway systems.

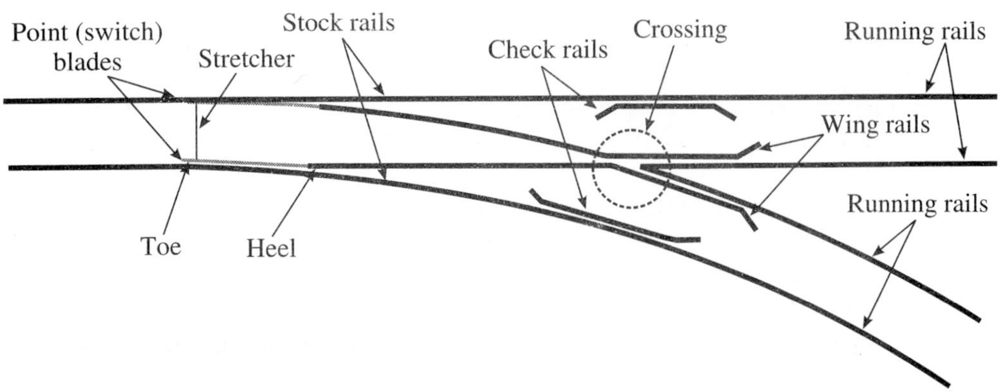

FIGURE 12.12 (a) Simple turnout of railway track showing Points and Crossing and other parts.

Control panels

The third major component of train control system is 'control' from where the signals, points, and crossings are operated. Mostly 'control' is available at two levels—local and central.

Local control: The local control is available at the station level usually called *block station.* The block station is equipped with control panel from where the Station Master can change the points, crossings and signals to receive and dispatch the train at and from the station (Figure 12.13). The safety of operation is ensured by the interlocking system.

FIGURE 12.13 Station control panel at station, Indian Railways.

Central control: The central control is the placed from where the train movement over a section/network of railway is monitored and controlled. The decision to regulate the trains giving precedence to one train over other or crossing at one station (in single line section) is taken from the central control. Delhi Central Control is shown in Figure 12.14.

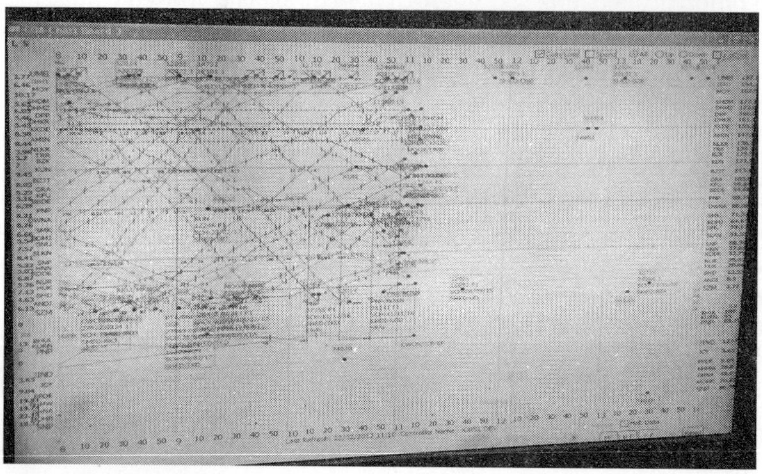

FIGURE 12.14 Delhi Central Control, Indian Railways.

If points and crossings at station level can be controlled from central control, it is called *central traffic control (CTC)*. In CTC system, the local control is only used when the central control has failed or in emergency situations. The CTC system is used in Mumbai Suburban system of Indian Railways (Figure 12.15).

FIGURE 12.15 Mumbai Suburban CTC, Indian Railways (*Source:* Sh. Rahul Himalyan, DyCOM/ Western Railways, reproduced with permission).

12.3.4 Traction System

The traction to railways is provided by locomotives or multiple units which are powered by steam, diesel or electric based engines. The sources of power for steam and diesel based locomotives are part of the locomotive itself. In case of electric-based locomotive or multiple units, the source of power is either overhead line or an additional rail (third rail). These days mostly alternate current (AC) based traction system is used in electric locomotives and multiple units. Indian Railways use 25 kV AC overhead traction system for powering its electric locomotives and electrical multiple units (EMUs)/mainline electrical multiple units (MEMUs) (Figure 12.16).

FIGURE 12.16 Over head traction system, Indian Railways.

12.3.5 Railway Station

A railway station is a facility where trains regularly stop for boarding and alighting by the passengers or loading and unloading of freight or for both the purposes. The stations which defines the boundary of a block section are called *block stations.* The block stations usually have points and crossings to facilitate the train to change the track. The block station controls the movement of the train in adjacent block section with the cooperation of block station at the other end of the block section. The block station is equipped with the local control panel for controlling points, crossings and signals within the station limit and block instrument (Figure 12.17) to allow entry and dispatch of train into the block section in consultation with station at other end. The block stations usually have loop lines to allow crossing of two trains (in single line section), to allow precedence to one train over other (Double line section). Depending upon the location, the block station may be a terminal (end of the line), roadside or wayside (on one line) or junction (connection of two lines) station (Figure 12.18).

FIGURE 12.17 Neil's ball block instrument, IR.

FIGURE 12.18 Jind Railway Station, IR.

The stations with only stopping facility without any signals, points, crossing, block instrument are called halt or flag station. The halt stations do not have any loop line. Both block station and halt station have platforms to facilitate boarding and alighting by the passengers.

12.4 SUMMARY

This chapter has been focused on conventional railway system. The basic concepts of railway systems have been discussed very briefly with the intention to make readers aware of them. The railway system consists of fixed infrastructure and moving stock. The fixed infrastructure consists of track, rolling stock, signalling, traction systems and locomotive, coaches and wagons are the moving infrastructure. The in-depth study of railway systems is beyond the scope of this book.

The wide scale application of advanced IT system has been a game changer in the railway technology. The subsequent chapters discusses various application of ITS in railways.

Questions

12.1 What are the advantages of railways over roadways?

12.2 What are different components of a railway system?

12.3 What types of locomotives are used in Indian Railways?

12.4 What do you understand by interlocking in the context of railway signalling system?

12.5 What do you understand by overhead traction system and what is its purpose in a railway system?

12.6 What is the purpose of a Railway Station from operational point of view?

Bibliography

Aura, Y., Y. Mochinaga, and H. Nagasawa, *Railway Technology Today 3*, Railway Electric Power Feeding Systems, in Japan Railway and Transport Review, 16, 1998. (http://www.ejrcf.or.jp/jrtr/jrtr16/pdf/f48_technology.pdf)

Banerjee, Somnath, "Basics of Railway Signaling". (http://indianbusinesspages.com/Downloads/BSRSSample.pdf)

Chandra, Satish and M.M. Agarwal, *Railway Engineering*, 2nd ed., Oxford, Delhi, 2013.

General Rules of Indian Railways.

Handbook on Basic Concepts of Railway Signalling, "Principles of Train working and need for signalling", Chapter 1 (http://static.scbist.com/scb/uploaded/331_1354714435.pdf).

Lautala, Pasi and William Sproule, *Tuning Transatlantic Cooperation In Rail Higher Education*, Handbook for Rail Higher Education, 2011. (http://www.ncl.ac.uk/newrail/assets/docs/tunrailhb.pdf)

Lozano, José A., Jesús Félez, Juan de Dios Sanz, and José M. Mera, "Railway Traction", Universidad Politécnica de Madrid, Spain. (http://cdn.intechopen.com/pdfs/34430/InTech-Railway_traction.pdf)

Operating Manual of Indian Railways.

Silajev, Igor, "CBTC–Communication Based Train Control", Rail Signalling Solutions. (http://www.rpknis.rs/ictforum2010/PPTprezentacije/English/CBTC-Communication%20Based%20Train%20Control%20-%20Igor%20Silajev.pdf).

The Railways and Other Guided Transport Systems (Safety) Regulations 2006 (as amended), Office of Rail Regulation, UK, 2014. (http://orr.gov.uk/__data/assets/pdf_file/0020/2567/rogs-guidance.pdf)

"Train Control and Signaling", Chapter 3, TCRP Report 13: Rail Transit Capacity. (http://www.tcrponline.org/PDFDocuments/TCRP%20RPT%2013-B.PDF)

13 Intelligent Train Control System (ITCS)

13.1 INTRODUCTION

The ITCS is used to monitor and control the movement of the trains to ensure safe and efficient train operations. The ITCS is the brain of the train operations, which controls all functions related with train running from a central control. Conventionally, the train control system envisages a signalling system (Figure 13.1), however, extensive use of IT has widened the scope of signalling system from adequate separation of trains while maintaining a reasonable speed to automatic train operations.

For Railway systems, the signals are essentially required to control the railway traffic and to maintain adequate distance between trains to avoid any possibility of collision. The enormous weight and inertia of a train makes it difficult to quickly stop when encountering an obstacle. The braking distance of the trains is more the distance upto which the train driver can see. Without a signal at an adequate distance ahead of the obstruction, the train driver would not be able to stop the train before the obstruction.

The train driver is solely responsible for observing the signals and any mistake on his part will result into serious mishap. The accidents on account of human error has always been a cause of serious concern for safety of railways. The maximum permissible speed of the trains is also limited by the human capacity to observe the roadside signals. The huge demand for high speed rail led the railway engineers to bring innovations in signalling system also. The revolution in IT system at the same time proved to be boon for railway signalling system. Today the modern train control system is a perfect example of close integration of IT with hardware of signalling systems of Railways. This chapter discusses the evolution of train control system from basic signalling to most advanced communication based train control (CBTC) system being used in driverless operation of metro trains.

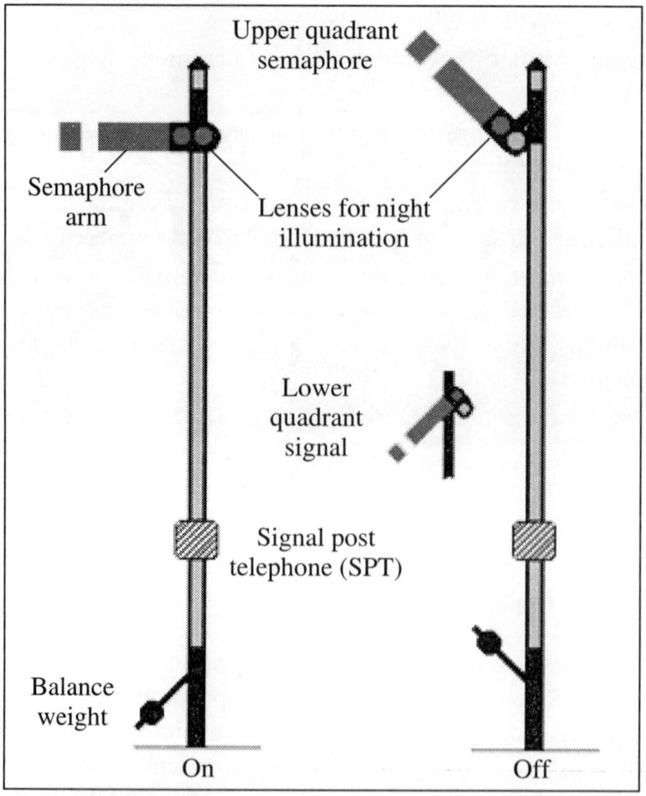

FIGURE 13.1 Early age signals being used in railways (*Source:* http://www.railway-technical.com/Semaphtype.gif, reproduced with permission).

13.2 TRAIN CONTROL SYSTEMS USED IN RAILWAYS

Train control is the process by which the movement of train is regulated for the purposes of safety and efficiency. The process is carried out by a combination of elements, some men, and some machines—located on the train, along the track, in stations, and at remote central facilities. These elements interact to form a command and control system with four major functions:

(a) *Train protection:* Prevention of collisions and derailments
(b) *Train operation:* Control of train movement and stopping at stations
(c) *Train supervision:* Direction of train movement in relation to schedule
(d) *Communication:* Exchange of information among the elements of the system

13.2.1 Train Protection

Train protection entails functions whose purpose is to assure the safety of train movement by preventing collisions and derailments. Train protection function includes detection of location of train on track, safe separation between trains, prevention of conflicting movements, speed enforcement, train and track surveillance.

The functions that make up train protection are:

(a) *Train detection:* Monitoring of the track to determine the presence and location of trains

(b) *Train separation:* Assuring that trains on the same track maintain a safe following distance to prevent collisions

(c) *Route interlocking:* Preventing trains on crossing, merging, or branching routes from making conflicting (unsafe) moves that would cause a collision or derailment

(d) *Overspeed protection:* Assuring that train speed remains at or below the commanded or stipulated civil speed limit as to prevent collisions resulting from going too fast to stop within the available distance and to prevent derailments due to excessive speed on curves or through switches

(e) *Train and track surveillance:* Observing conditions on and in the vicinity of the track ahead of the train and monitoring safety-related conditions on-board the train

The train protection functions are critical for safety of the trains so in case of any conflict with any other control function, the train protection function overrides the other function. For example, if a train needs to be stopped for safety reasons (such as maintaining adequate distance from next train), there would be a conflict between train protection function and train schedule maintenance function. To ensure safety, the train protection function will override the schedule maintenance function of train supervision.

13.2.2 Train Operation

Train operation consists of functions which are necessary to move the train and to stop it at stations. Train movement, as controlled by train operation functions, is under the direction of train supervisory functions for efficient operations and always within the constraints of train protection functions for safe operations.

Train operation function includes controlling the speed as required by train protection function (speed enforcement, maintaining the speed to adhere to the schedule, stopping the train at right location at the platform, opening and closing of the doors of the train when train has stopped at the right location.

13.2.3 Train Supervision

Train supervision functions monitors the movement of trains continuously in order to maintain intended traffic patterns and minimise the effect of train delays on the operating schedule. Train supervision includes certain information processing and recording activities not directly concerned with train safety and movement but necessary to the general scheme of operations.

The train supervision functions include schedule design and implementation, selecting and assigning routes to be followed by trains, controlling train departures from terminals or intermediate stations in accordance with the schedule, monitor trains movement against the schedule, adjusting movement commands and revising the schedule in response to train, traffic, and environmental conditions. The advanced train supervision system also generates alarms to alert malfunctions, breakdowns, or problems, and record their time, location, and nature.

13.2.4 Communication

The functioning of train protection, operation and supervision depends on exchange of information in the system. The communication in itself facilitates other systems to work. For working of train protection system, information regarding location of individual trains, their speed, status of points, crossings, and signals is required. Similarly for efficient train operation, in addition to above information, the location of stations, platforms, and stopping time at the station are required to be known by the system. Train supervision requires the information regarding current running of the train, schedule, reversal/end station, condition of train and track to generate alerts. This exchange of information may take any of several forms—voice, visual, auditory, and digital or analog electrical signals.

In earlier days, the train control functions discussed above have been performed manually. As the technology has evolved, progressively these functions are being transferred to machine for sake of improvement in safety and efficiency of the system. It does not necessarily suggest a completely automated system. It can be applied to a system where certain functions or groups of functions are performed automatically while others are performed manually. Automatic train protection (ATP), automatic train operation (ATO), and automatic train supervision (ATS) are used to designate major groups of functions that may be automated. For example, if a system is said to have ATP, it means that train protection is accomplished (either completely or mostly) by automatic devices without direct human involvement. If a system is described as having ATC consisting of ATP and some ATS, this indicates that train protection is assured by automatic devices and that train supervision is a mixture of manual and automatic elements. By implication, train operation in such a system would be manual. An advanced automatic train control system is fundamental to modern rapid transit systems which require to run trains at very high frequency.

13.3 SYSTEM OF WORKING

System of working for a railway means how the vital functions of train control system—protection, operation, supervision and communication are performed. There are following four main systems of train working in railways:

13.3.1 Absolute Block System

The absolute block system of train control is designed to ensure the safe operation of a railway by allowing only one train to occupy a defined section of track (block) at a time. The railway line is divided in block sections defined by the boundaries of two adjacent stations. At a time only one train is allowed to run in a block section. The entry into a block section is controlled by the station master of the dispatching station with the help of an instrument called *block instrument*. The dispatching station allows the train to enter the block section ahead only when, i.e., it is clear of any obstruction which is ensured by the block instrument. So, in this system of working except some functions of train protection viz. route interlocking, block protection, the other train control functions are performed manually. Figure 13.2 explains the working of absolute block system. Only one train is permitted in a block section. Once a block section is occupied, the signal protecting the block shall show 'red' to prevent entry of another train in the block section. If a block section is not occupied, the signal protecting the block section shall show 'green' aspect to allow entry of a train in the block section.

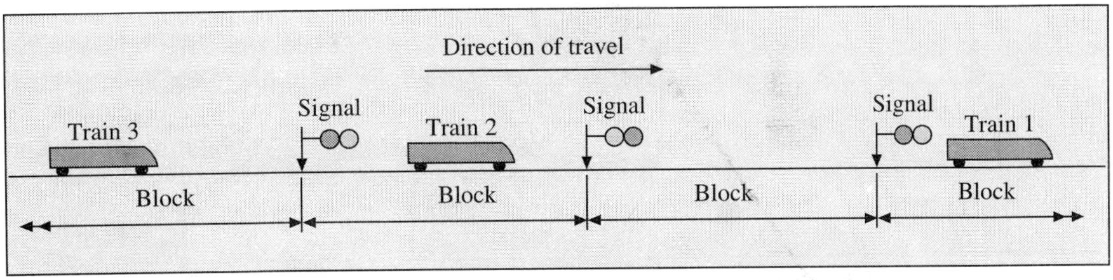

FIGURE 13.2 Working of Absolute Block System (*Source:* http://www.railway-technical.com/sig201.gif, reproduced with permission).

The absolute block system is the most prevalent system of working being used by Indian Railways as this system is based on mechanical gears which does not require any IT application. Over the time with use of electronic relays and intervention of IT, the efficiency of absolute block system has reasonably improved.

13.3.2 Automatic Block System

Like absolute block system, the automatic block system of train control is also based on space interval of the trains, however, in this system the section between two stations (block section) is further divided into smaller subsections called *automatic signalling sections*. Each signalling section is protected by a stop signal at its entry location. The system is able to sense the occupation of a signalling section by the train or any other obstruction. An automatic block system uses the track circuit or axle counter to automatically detect trains in blocks and to control the signals for each block. The system automatically allows the movement of a train into a signalling section only when there is no other train or any obstruction in this signalling section. There is no manual intervention required to control the movement of the trains in 'automatic signalling territory.' This system may be called *intelligent block system* as it is automatic in nature based on application of electronic circuits. However, this system still requires intervention of station master at each station to decide routing, precedence, stopping, passing and starting the trains from the station. In this system, the train protection is automatically ensured by the system. The train operation and train supervision are still carried out manually. The communication is based on system to send information automatically about track occupation, signal aspects, etc. as well as manual (telephonic). Figure 13.3 explains the working of automatic block system being used in Indian Railways.

13.3.3 Central Traffic Control System

Traffic control or operation control centre in a railway system controls and commands the whole system including the stations and trains. All vital decisions like which train will depart first from a station, which train will get precedence, which train needs to be stopped for crossing in a single line section are taken from the traffic control/OCC.

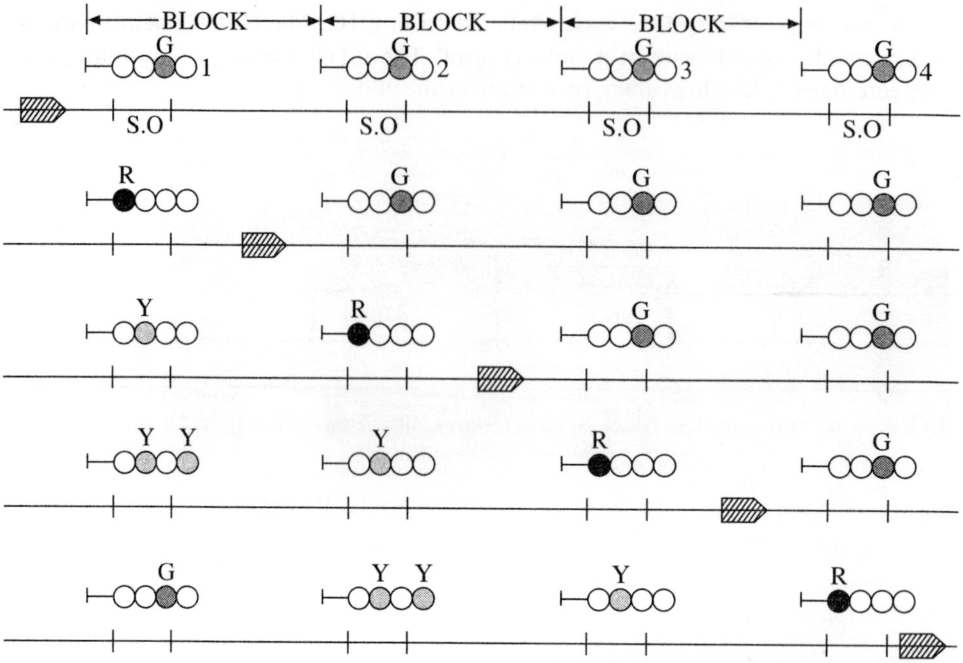

FOR SIGNAL.1. To ASSUME YELLOW-LINE MUST BE CLEAR FOR ONE BLOCK AND ONE OVERLAP
FOR SIGNAL.1. To ASSUME DOUBLE YELLOW-LINE MUST BE CLEAR FOR TWO BLOCK AND ONE OVERLAP
FOR SIGNAL.1. To ASSUME GREEN-LINE MUST BE CLEAR FOR THREE BLOCK AND ONE OVERLAP

FIGURE 13.3 Working of automatic block system.

Train traffic control requires full and continuous knowledge of the train operations, i.e., train location with time on round the clock basis. In a conventional traffic control system, telephonic communication is used for exchanging information between traffic control and station, but this process is too slow and inefficient for modern rail traffic volumes.

Centralised traffic control (CTC) provides the traffic control centre with information on the situation of all trains in all track sections and permits the centre to control train routes directly. Centralised traffic control (CTC) consolidates train routing decisions that were previously carried out by station master at the station to a control centre called central traffic control (CTC). The system consists of a centralised train controller's office that controls railway signalling system and traffic flows in the section of the rail system designated as CTC territory (Figure 13.4).

The heart of the centre is a number of centralised display and control panels, connected to stations and trains by various types of equipment such as radio equipment, command telephones, train schedule recorders, train number display units, etc. The control panel is equipped with a graphical representation of the railway network and the real-time status of the trains running over different sections. On this panel, the controller can keep track of trains' locations across the territory controlled by him. Like automatic block system, the CTC uses track circuits or axle counter to detect the occupancy of the track. In CTC system, the station masters are not required at stations for the purpose of train operations. The command to set any route (track) for train movement as per requirement can be executed from CTC. The CTC system involves use of advanced computer based system of train control. The CTC system is truly intelligent transport system in context of railways.

FIGURE 13.4 Mumbai Suburban CTC, Indian Railways (*Source:* Rahul Himalyan, DyCOM/Western Railways, reproduced with permission).

In the context of train control systems, the CTC system is a semi-automatic system where train protection functions are performed by the system and remaining ones are still manual or semiautomatic.

13.3.4 Continuous Automatic Train Control (CATC) System

The above three systems are essentially different types signalling systems which indicate the availability of the routes through roadside static signals. The train is being controlled by the train driver (manual train operation) depending upon the aspect (Green/Red/Yellow) of the signal ahead. The capacity and safety of the system is limited by the human capacity to site the signal and the reaction time to take appropriate action. Any error by the train driver to observe and comply the signal aspect may result into very serious accident.

However, the growing demand for high speed trains for intercity travel and high frequency services for intra-city travel cannot be met by a system limited by the human capabilities. The system to run high speed or high frequency services needs to be automated so that even if train driver fails to act timely, the system should offer safety protection to the train operations. The continuous automatic train control system entails use of modern IT systems to automate the train working in order to ensure safety along with high speed and frequency of trains. The indication of the signal is available in the driver's cab in the form of target speed and target distance (also called *cab signalling*) (Figure 13.5).

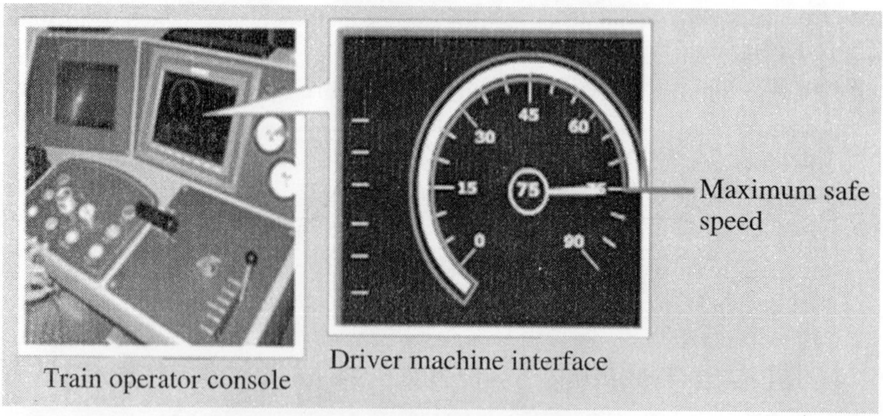

FIGURE 13.5 Cab signal inside driver's cab (*Source:* DMRC, reproduced with permission).

In CATC system, most of the train control functions are performed by the system depending upon the system configuration adopted by the railway operator. Here, a typical CATC has been discussed to make the readers understand the fundamental features of CATC. However, the CATC system may change from one operator to other operator depending upon the system specifications and technology used. But guiding principles of train working remain more or less same irrespective of the technology used.

The continuous automatic train control system works on the principle of target speed and target distance with cab signalling by means of continuous transmission from track to train through track circuits, ensuring safe movement of all trains under all operating conditions by continuously generating a safe operating envelope defined by the limit of movement authority and the maximum safe speed. The salient features of CATC system are:

(a) There is continuous transfer of information between train and track duly monitored and controlled from a central control.

(b) The train driver gets the signals in the cab in the form of 'target speed' and 'target distance'. The target speed is the maximum speed that he is allowed to drive the train and target distance is the safe distance ahead upto which he can drive the train. He does not have to rely on track side signals.

(c) If the train driver fails to comply with the signal, the system is able to control the speed of the train to avoid any untoward incident.

(d) The system is responsible for supervision, control and safe operations of the trains.

Components of CATC System

The CATC system consists of two major components—automatic train supervision (ATS), automatic train control (ATC). Depending upon degree of automation, ATC can be an Automatic Train protection (ATP) subsystem or an Automatic Train operation (ATO) subsytem. The ATS functions are performed by Central Area Train Supervision (CATS) from Operation Control Centre and by Local Area Train Supervision (LATS) from station level control center (Figure 13.6).

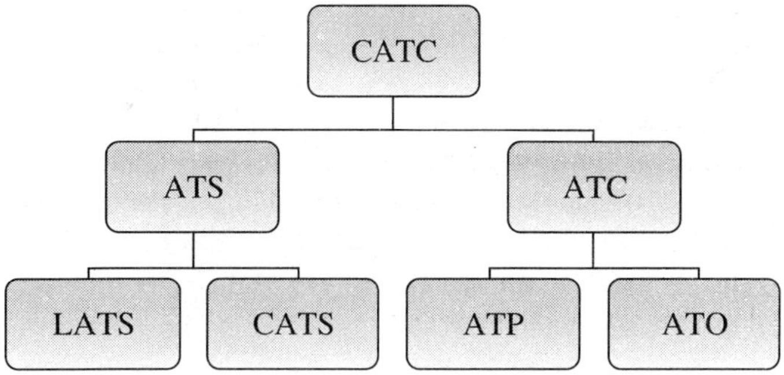

FIGURE 13.6 Components of CTC system.

Automatic train supervision (ATS)

ATS is at the heart and brain of various systems operational in CATC based train control system. It acts as an intelligent interface where one can see all trains running in real-time in entire network from a remote central location called operation control centre (OCC). ATS is fully automatic and needs no manual intervention in all the operational tasks besides ensuring safe and efficient train operations. ATS subsystem monitors the system status continuously and provides the appropriate controls to direct the operation of trains in order to maintain intended traffic patterns and minimise the effect of train delays on the operating schedule. The system also has the capability to organise the train movements in an optimum manner in case of abnormalities. Using networked computers and automated functions, ATS ensures efficient train operation, automatic monitoring and control of interlocking through automatic train tracking to automatic route setting. ATS assigns identification numbers to trains and provides all necessary real-time train information to the operator. ATS also caters for tools for time table generation and modification as also for the management of signaling equipment.

A typical ATS provides interface with following subsystems of train operations (Figure 13.7):

(a) *Automatic train control (ATC):* Responsible for safe and efficient train operations

(b) *Radio system:* Responsible for radio based communication system between train and OCC

(c) *Master clock:* GPS based clock system to ensure that same time is picked by all systems and station

(d) *Mimic panel display/Video wall:* It provides train running indication in OCC

(e) *Passenger information display system (PIDS):* It provides real-time information about expected arrival of the trains at LED/LCD panels installed at Platforms, concourse and other locations of the station

(f) *Supervisory control and data Acquisition (SCADA):* It is a system for remote monitoring and control used in controlling the tunnel ventilation system

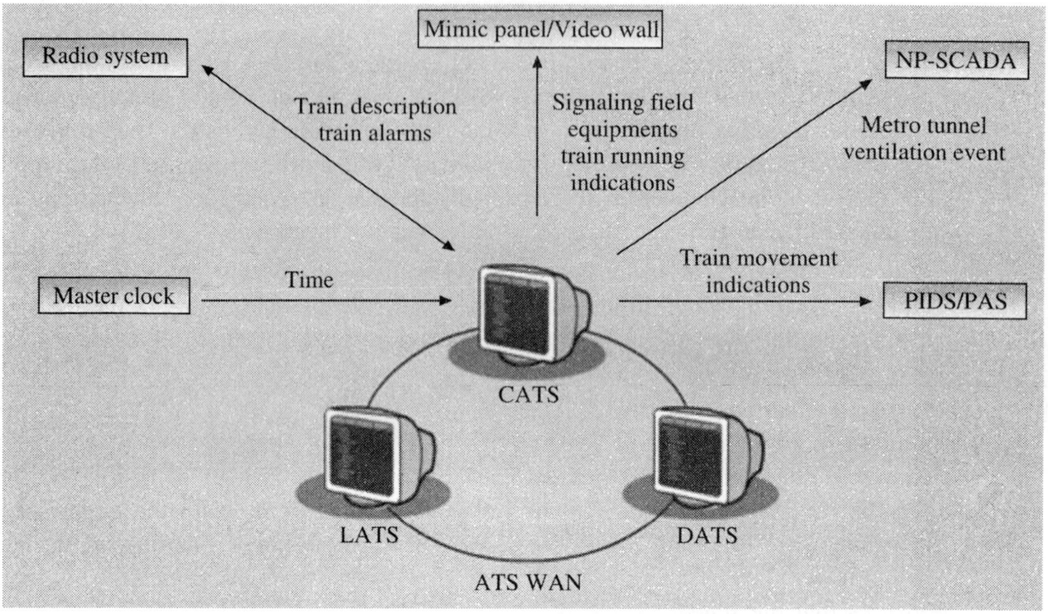

FIGURE 13.7 A typical ATS interface with various subsystems.

ATS performs following functions:

(a) To monitor and control the traffic through MMI
(b) To perform automatic route setting and automatic train regulation
(c) To monitor and regulate train movement continuously
(d) To display information on passenger information display system (PIDS)
(e) To generate schedules
(f) To generate alarms and events

Automatic train protection (ATP) system: All the functions under ATP are vital functions and govern the overall functioning of CATC system. This vital system maintains the safety of train operation including separation of trains, enforcement of speed restriction, opening of platform side doors only and safe operations through interlocking. The main function of ATP system is to ensure safe train separation and safe train movement which is performed by the on-board ATP equipment. It has in its store the needed rolling stock (train) characteristics. The safety information in respect to field entities shall be provided by the computer based interlocking for the track occupancy and by the wayside ATP equipment for the track profile like curves, gradients, location/position of signals and points, location of stopping point, and permanent and temporary speed restrictions, etc. It receives information from the track, calculates the speed versus location profile and deduces orders such as 'stop at this point' or 'reduce the speed to this level at this point' etc. The communication link between the track and train is used to transmit the train ID and alarms and messages to the track side ATP equipment.

Functions of on-board ATP equipment

(a) *Tracking speed profiles:* The on-board equipment continuously computes the necessary track related speed profiles (normal working profile, full service braking profile and emergency braking profile) from the line data and train data as the train moves along the track. Based on the dynamic track related speed profiles, the ATP system monitors and enforces the changes in the target distance/target speed, thus maintaining a safe distance between trains.

(b) *Maintaining safety in train operations:* For this, the system automatically initiates braking when the safety limits are exceeded. The automatic braking of the train is also initiated in the event of missing or incorrect information from the wayside equipment or a fault in on-board equipment.

(c) *Stopping location monitoring:* The stopping location of the train at the platform is predefined into the system as stopping window. The system ensures that the train comes to stop within the stopping window (usually –0.7 m to +1.0 m). The system regards the stopping window as a 'fail-safe stopping point' with the target speed of 0 kmph. This function ensures that a train stops does not run beyond designated stopping window. The monitoring of train stop in the target area of a station also ensures that the scheduled trains stop at stations.

(d) *Opening and closing of doors:*

  (i) Releasing doors on the correct side at stations when the train has come to a stop within the stopping window (usually –0.7 m to +1.0 m).

  (ii) The system ensures that no movement of train is possible until all doors of the train are detected closed and locked. The loss of door closed signal causes the ATP system to initiate a brake application.

(e) *Monitoring of speed restriction sections:* Sometimes the speed restrictions are imposed on the track due to some temporary maintenance works or permanent geometrical constraints. The system ensures that the train observes the permitted speed within the restricted section until the tail end of the train has passed the end point.

(f) *Monitoring direction of travel and backward rolling:* The unauthorised reverse movement of train is prevented by the application of emergency brakes, after a pre-determined distance.

Functions of ATP wayside equipment: The signalling equipment are installed along the track to perform following functions:

(a) Storage of line parameters like line gradient, curve, track section length, civil speed restriction sections, temporary speed restriction sections, and station stopping points, etc.

(b) Generation and transmission of necessary data in fail-safe manner to the on-board equipment for the safe train movement.

(c) In execution of its functions, ATP interfaces with external subsystems such as CBI, various train-borne equipment, ATS and ATO subsystems and driver's MMI, etc.

Automatic train operation (ATO): Under automatic train operation (ATO), the train is driven by the system, i.e., acceleration, regulation, and stopping of the train are performed by the system. The train driver is required to close the doors and push start button to start the train. The automatic train operation (ATO) system ensures automatic control of train running from station to station, while remaining within the safety envelope calculated by the ATP. Train doors at the stations are also opened automatically by the system. The closing of the doors, in the system is activated by the train driver, on receipt of a command from the ATS system to leave the station. When the doors are closed, the train driver starts the train. The ATO system then operates the train automatically to the next station.

Technology used for CATC system

There are a number of variants of CATC available as developed by different firms across the world. The technology of ATC has evolved over the time and still changing at a fast pace with innovations taking place every day. There are following two major technologies being adopted in recent times for ATC:

Track circuit based train control system: For safety and efficient operation of a transit system, it is imperative to know the locations of trains at all times. The sensing device providing this information is the track circuit. The *track circuit* is an electrical circuit consisting of a power source, the running rails, and a signal receiver (relay). The track is divided into electrically isolated segments (called blocks) by insulated joints placed at intervals in the running rails. Thus, the basic principle of the track circuit is the shunting phenomenon produced by the train wheels passing along the electrically energised running rails. The presence of the train is detected in the track circuit as a reduction of electrical current, by means of the relay—is used to control the secondary signal circuit and operate various types of track occupancy indicators.

In track circuit based automatic train control system, the whole track is circuited and the presence of the train is known to the system from the track circuit. The system being aware of the location of the occupied track circuit, sends the signal to the following train in form of target speed/distance to ensure adequate separation between the two trains. There is continuous flow of information among track, train and the control centre. The system is fail safe in the sense that if track circuit fails, it gives an indication of occupation of the track.

In conventional track circuit based system, the system knows the location of the track circuit being occupied by the train. So if the length of the track is 400 m, the train can be anywhere on this track circuit, the system will only know that this track is occupied. In this system, the minimum distance between the two trains is limited by the length of track circuit in addition to braking distance of the train. This system is not suitable for running the trains at very close headway like 60s or 90s.

Communication based train control (CBTC) system: The rising demand for high frequency services in urban areas cannot be met by conventional track circuit based system. For high frequency services, it is important to know the exact location of the train so that safe distance between the two trains can minimised to the level of braking distance of the train. In CBTC system, the telecommunications between the train and track equipment is used for the traffic management and infrastructure control. In this, the exact position of a train is known more accurately than with the traditional signalling systems (Figure 13.8). This results in a more efficient and safe way to manage the railway traffic. Metros (and other railway systems) are able to improve headways while maintaining or even improving safety.

A CBTC system is a "continuous and automatic train control system utilising high-resolution train location determination, independent of track circuits; continuous, high-capacity and bidirectional train-to-wayside data communications; and train borne and wayside processors capable of implementing automatic train protection (ATP) functions, as well as optional automatic train operation (ATO) and automatic train supervision (ATS) functions", as defined in the IEEE 1474 standard.

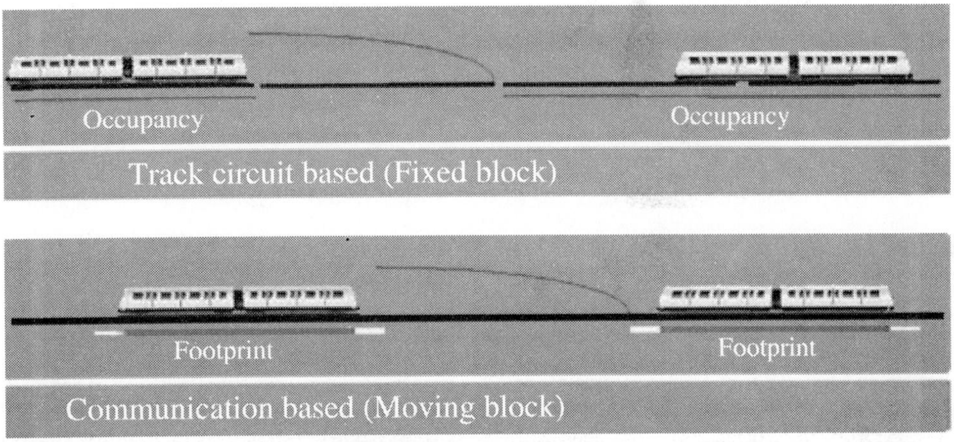

FIGURE 13.8 Fixed block vs moving block (*Source:* https://upload.wikimedia.org/wikipedia/commons/5/53/FB_vs_MB.jpg, Fee to use).

Driverless train operations (DTO)

The latest development in the IT and associated reliability of the system has paved the way for driverless operations of the trains. In DTO, the entire train operations (start, control, regulate, stop, maintain adequate distance, opening and closing of the door, etc.) are managed by the system only without any intervention of the train driver. Every train is having a train attendant who is responsible for commercial functions like ticket checking, attending passengers etc. He/she is trained to drive the train in case of any failure or emergency. The advanced version of DTO system is unmanned train operation (UTO). In UTO system, there is no attendant in the train even for emergency duties. Multiple level of redundancies in the system ensures high level of reliability of the system. The DTO and UTO systems are examples of high level of intelligent transport system being deployed in Urban Railways.

The level of application of ITS in Urban Railways can be defined in terms of grades of automation (GoA) (Figure 13.9). According to the international association of public transport (UITP), there are five grades of automation (GoA) of trains:

(a) **GoA 0:** It is on-sight train operation, similar to a tram running in street traffic.

(b) **GoA 1:** It is manual train operation where a train driver controls starting and stopping, operation of doors and handling of emergencies or sudden diversions.

(c) **GoA 2:** It is semi-automatic train operation (STO) where stopping is automated but a driver in the cab starts the train, operates the doors, drives the train if needed and handles emergencies. Many ATO systems are GoA 2.

(d) **GoA 3:** It is driverless train operation (DTO) where starting and stopping are automated but a train attendant operates the doors and drives the train in case of emergencies.

(e) **GoA 4:** It is unattended train operation (UTO) where starting and stopping, operation of doors and handling of emergencies are fully automated without any on-train staff.

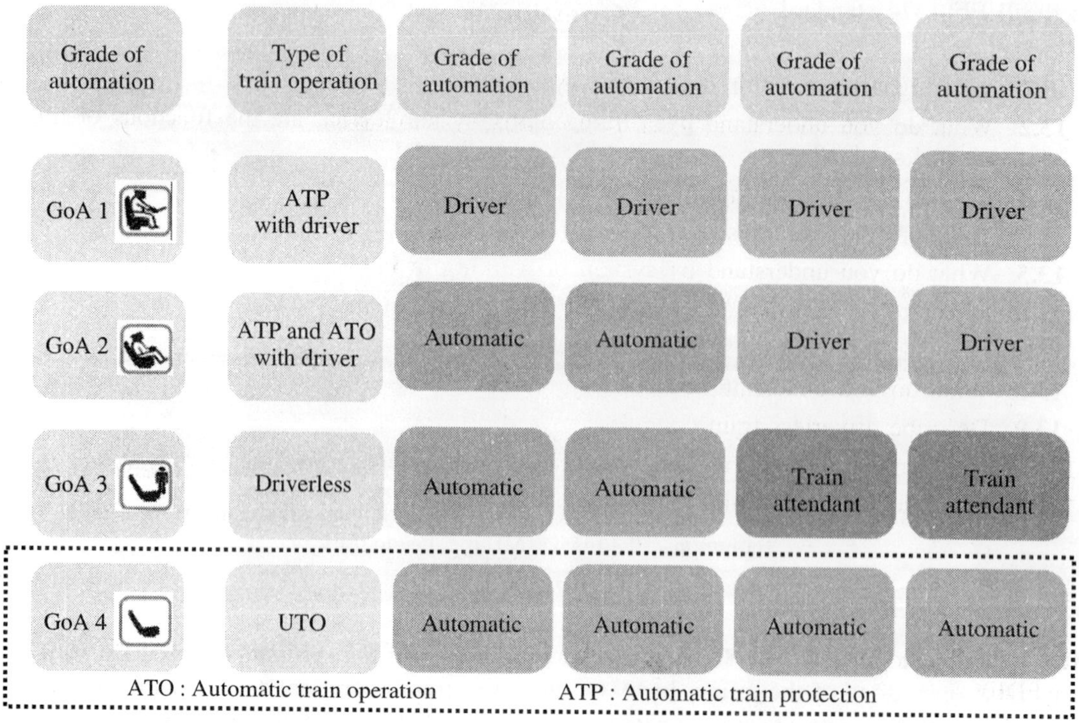

FIGURE 13.9 Grades of automation in train control system (*Source:* http://metroautomation.org/automation-essentials/).

13.4 SUMMARY

The train control system is analogous to the sensory organs and central nervous system of the human body. It senses and processes information, makes decisions, and transmits commands. Train protection, operation, supervision and communication are vital functions of a train control system. The train protection system is required for prevention of collisions and derailments. The train operation entails control of train movement and stopping at stations. The train supervision is required to monitor and direct the train movement as per the schedule and communication is required to exchange of information among the elements of the system. Depending upon how these vital functions are performed, there are four major system of train working—absolute block system, automatic block system, central traffic control (CTC) and continuous automatic train control (CATC) system. Over the time, the IT application in train operation has revolutionised the train control system which has not only improved the safety of train operations but also enhanced the efficiency and capacity of train operation. The communication based train control (CBTC) is the most advanced form of CATC which can allow movement of driverless trains at a close headway of 60 s. With CBTC system, it is possible to run the trains without any train driver at a headway of even 60 s. The automatic train control is perhaps the best application of ITS in railways.

Questions

13.1 Why signals are required in a railway system?

13.2 What do you understand by a train control system? What are the functions of train control system?

13.3 Explain the need of train protection system.

13.4 What are the functions of train supervision system?

13.5 What do you understand by system of working in railways?

13.6 Differentiate between absolute block system and automatic block system.

13.7 Describe the working of continuous automatic train control (CATC) system.

13.8 What do you understand by communication based train control system?

13.9 Describe driverless train operations.

13.10 Differentiate between moving and fixed block system.

13.11 What do you understand by grades of automation (GoA)?

Bibliography

Automatic Train Control, Chapter 3. (https://www.princeton.edu/~ota/disk3/1976/7614/761405.PDF)

Bajpai, Abhishek, et al., "An Introduction to Train Management System of Western Railway", 2007. (http://www.irfca.org/docs/wr-tms.pdf)

Banerjee, Somnath, "Basics of Railway Signaling". (http://indianbusinesspages.com/Downloads/BSRSSample.pdf)

General Rules of Indian Railways.

Goverde, Rob M.P., Francesco Corman, and Andrea D'Ariano, "Railway line capacity consumption of different railway signalling systems under scheduled and disturbed conditions", 2012.

Handbook on Basic Concepts of Railway Signalling, "Principles of Train working and need for signalling", Chapter 1. (http://static.scbist.com/scb/uploaded/331_1354714435.pdf)

Ning, Bin, *Advanced Train Control System*, WIT Press, Boston, 2010. (http://static.scbist.com/scb/uploaded/331__ning_b_ed_advanced_train_control_systems.pdf)

Operating Manual of Indian Railways.

Press Kit Metro Automation Facts, Figures and Trends, "A global bid for automation: UITP Observatory of Automated Metros confirms sustained growth rates for the coming years", UITP. (http://www.uitp.org/sites/default/files/Metro%20automation%20-%20facts%20and%20figures.pdf)

Silajev, Igor, "CBTC–Communication Based Train Control", Rail Signalling Solutions. (http://www.rpknis.rs/ictforum2010/PPTprezentacije/English/CBTC-Communication%20Based%20Train%20Control%20-%20Igor%20Silajev.pdf)

"Train Control and Signaling", Chapter 3, TCRP Report 13: Rail Transit Capacity. (http://www.tcrponline.org/PDFDocuments/TCRP%20RPT%2013-B.PDF)

14 Intelligent Trains

14.1 INTRODUCTION

The definition of the train is "a series of connected railroad cars pulled or pushed by one or more locomotives". The conventional trains consist of coaches/wagons coupled together by mechanical means and pneumatically attached for continuity of the brake system. There is no electronic communication among different coaches of the train. The subsystems of the trains like braking system, traction system, air-conditioning, door system are only checked in the depot as part of the preventive maintenance. The train driver does not get any alarm or message about the suboptimal functioning of any subsystem during the run of the train. He is not able to pre-empt any failure as no advanced information is available with him regarding functioning of different subsystems. The failure of any subsystem leads to break down of the train on line causing great inconvenience to the passengers and other operational difficulties. The use of information technology in the on-board management of subsystems of the trains has completely changed the control, monitoring and maintenance philosophy of the trains. The coaches of modern rolling stock are integrated through electronic communication system controlled by a central computer called *train integrated management system (TIMS)*.

The TIMS is a train information system that performs continuous monitoring of control information in the driving cab with instruments for monitoring as well as checks on train status to ease the burden on operator (train driver) and conductors, properly provide emergency information to the train crew and to streamline the checking procedure. With integrated control of train-mounted devices using software logic and serial transmission, the system makes the train car system simpler and more reliable, establishing an optimal train car system. The modern trains are intelligent enough to diagnose the faults, take corrective actions, generate alarms, create log of events, generate maintenance requirement, etc. All the coaches in an intelligent train are electronically integrated by a system of train lines and work as a unit. The subsystems of a typical trains are shown in Figure 14.1. The subsystems of the train are interfaced with TIMS.

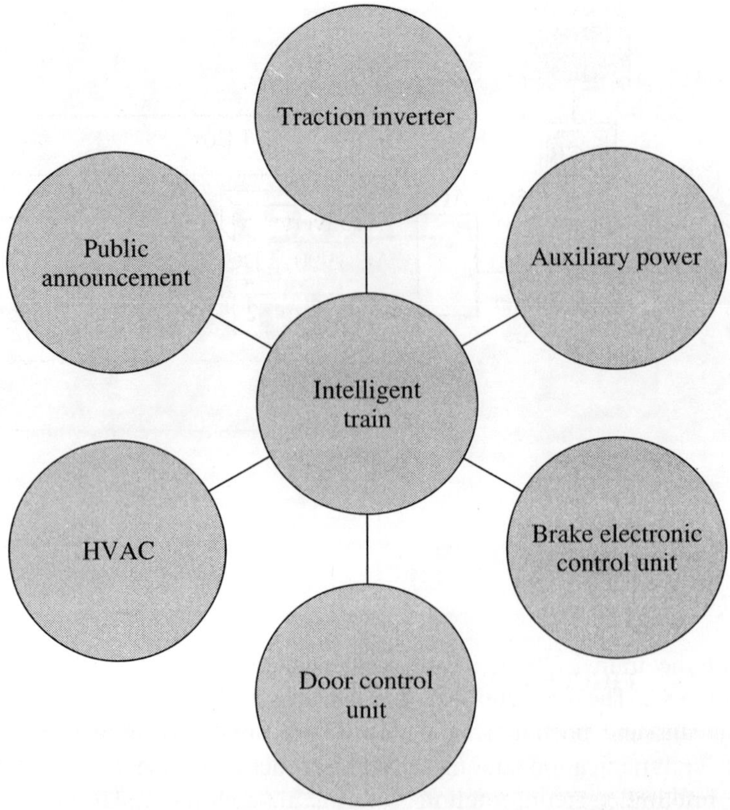

FIGURE 14.1 Subsystems of a typical intelligent metro train.

The subsystems of a typical metro train are described in the following sections.

14.2 TRACTION INVERTER—CONVERTER/INVERTER (CI)

The traction power is required to run the traction motors in the train. The traction power is mostly sourced either from overhead line or 3rd line along the track. The traction power is supplied from pantograph (a device for collecting an electrical current to power an electric locomotive/ EMU) in case of overhead traction system through main transformer to the converter-inverter (CI). The CI generates a variable voltage and variable frequency (VVVF) output through converter and inverter to feed the traction motors for providing the desired motoring/braking demand generated by the train driver. It provides required voltage at suitable frequency to feed the traction motor. The inverter works on VVVF principle to control the number of rotation of the traction motor, i.e., the speed of train.

In the typical subsystem of a metro train (Figure 14.2), the electricity (25 kV AC) is supplied to main transformer through vacuum circuit breaker (VCB). The same is stepped down to 1058 V (AC) and supplied to the converter which converts AC supply into DC for running the traction motors. The main transformer also feeds to static inverter (SIV) for auxiliary supply in the train.

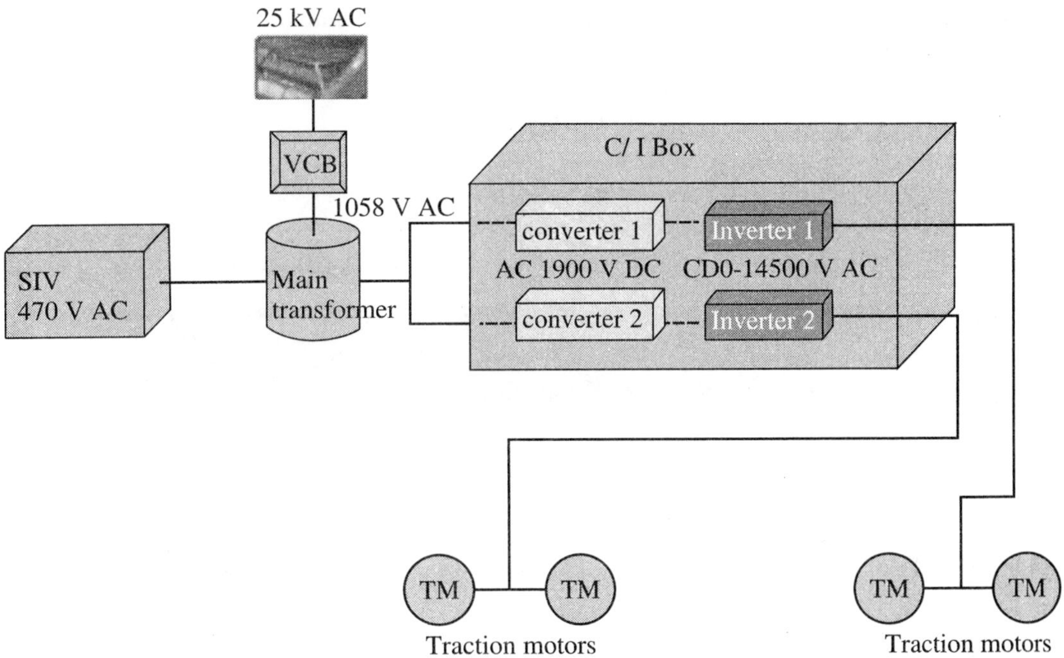

FIGURE 14.2 Illustrative diagram of traction inverter.

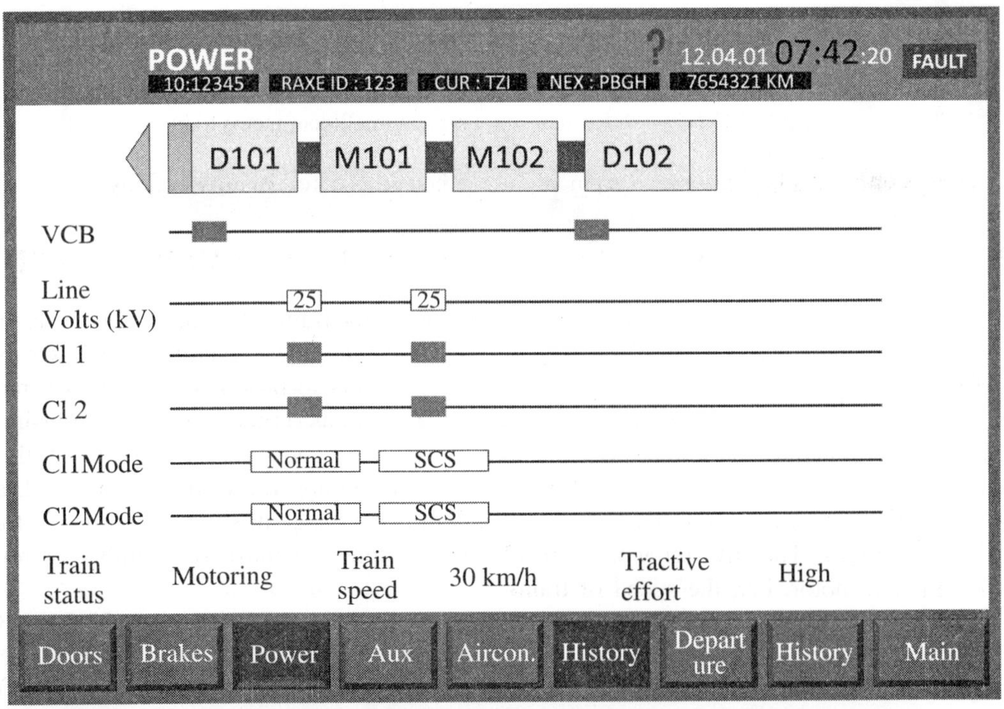

FIGURE 14.3 TIMS page on CI system.

The power page of TIMS indicates the status of overhead line voltage, converter-inverter, VCBs, train running, speed, etc. (Figure 14.3).

14.3 AUXILIARY POWER SUPPLY

The auxiliary power in the trains is required for lighting, air conditioning, compressors, blower fan motors, oil pump motors, etc. The static inverter (SIV) converts the power, supplied from main transformer to the auxiliary power (Figure 14.4).

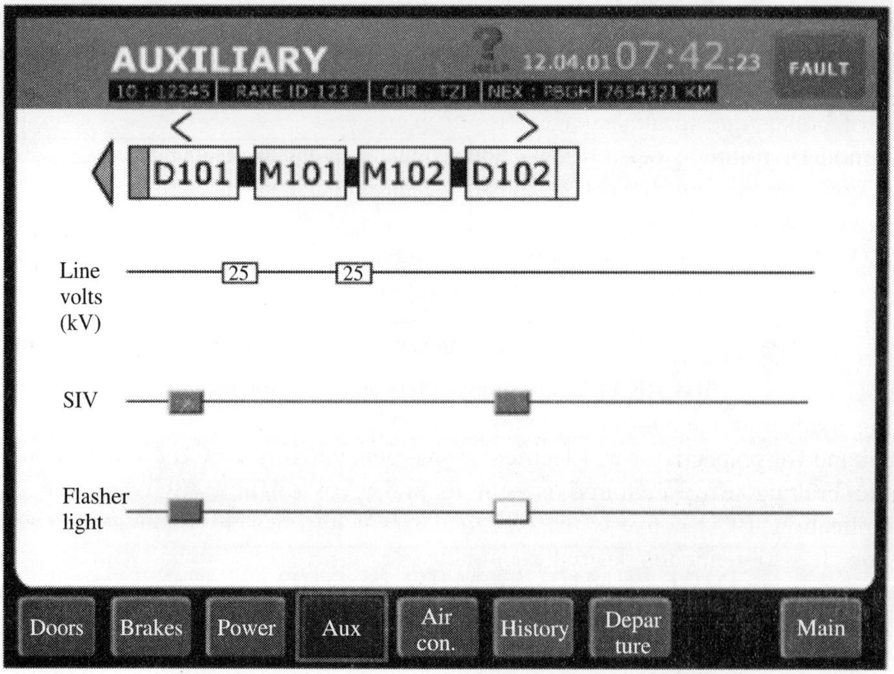

FIGURE 14.4 TIMS page on auxiliary power supply.

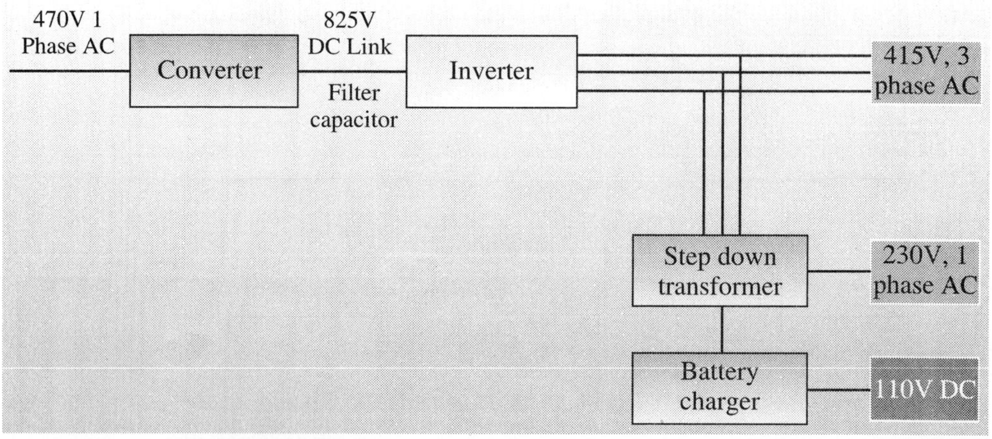

FIGURE 14.5 Illustrative diagram of auxiliary power supply.

In the above typical subsystem of a metro train (Figure 14.5), the main transformer steps down 25 kV (AC) to 470 kV (AC) for auxiliary power supply. SIV converts 470 kV AC into 415 V (3-phase AC), 230 V (AC) and 110 V (DC) for various auxiliary systems in the train.

14.4 BRAKE ELECTRONIC CONTROL UNIT (BRAKE SYSTEM)

Conventionally pneumatic valves are used to control and generate brake applications on the cars along the length of the train. Generally, the conventional system consists of a brake pipe that runs along the length of the train which supplies air to reservoirs mounted on each of the cars. The brakes are released when the brake pipe and car components are charged with air. The air pressure is dropped by the train driver to apply the brakes. The conventional braking system is inefficient due to high reaction time of the system, uneven braking effort, frequent jamming of brakes due to air leakage, loss of energy etc.

The modern rolling stock employs electronically controlled pneumatic (ECP) brakes in place of conventional air brakes system. ECP braking uses electronic controls to activate air-powered brakes on the cars. The ECP brakes are controlled brake electronic control unit (BECU) (Figure 14.6). It is a microprocessor based electronic brake control unit. The train driver generates demand for brake using traction brake controller (TBC). This demand is converted into the form of pulse width modulation (PWM) signal by PWM generator. PWM signal carrying the information of amount of braking effort demanded goes to BECU's of each car. BECU process is the signal, considering load factor, regenerative braking effort possible depending upon train speed, brake blending principal and it also calculates electro pneumatic brake demand for respective car. Electrical signal generated by BECU corresponding to electro pneumatic braking effort required is sent to brake control unit (BCU) of the coaches for brake application. BCU converts the electrical signal into desired pneumatic signal for brake application. It also generates desired pneumatic pressure in brake cylinder pipeline. Pneumatic pressure generated by BCU goes to all TBUs of that car for brake application on all the wheels. The TIMS page of BECU system for a typical metro train is shown in Figure 14.7.

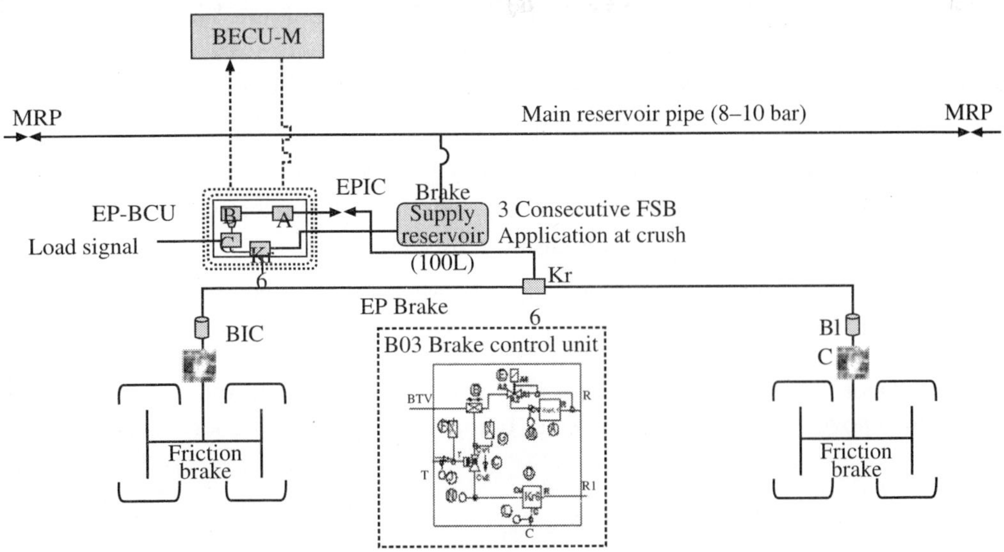

FIGURE 14.6 Illustrative diagram of BECU.

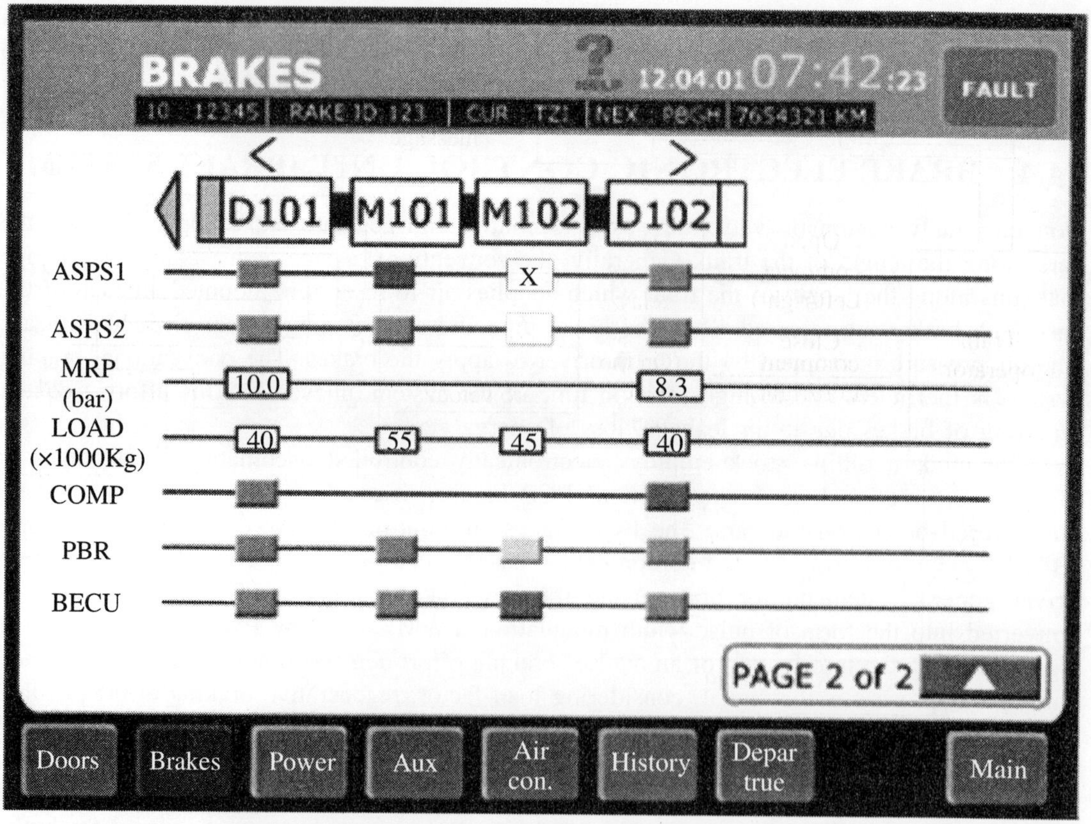

FIGURE 14.7 TIMS page of BECU.

14.5 DOOR CONTROL UNITS (DCU)

A door control unit is the nerve centre of a door system of the train, the door is controlled and monitored to ensure smooth operations while ensuring safety of passengers. The train doors need to be opened and closed at each stopping station. The DCU ensures that:

(a) train doors are opened only when train stops at right location at the platform.
(b) train doors are closed properly without any obstruction before start of the train.
(c) any defect in the train door is reported to the train driver through TIMS.
(d) doors remain closed and locked during train movement.

The train doors are operated electronically, via electronic door control unit (DCU), except when the manual emergency release mechanism is used. The DCU activates door opening and closing on receipt of signals from the train driver's panel. Door status signals, such as "Door Open" and "Door Closed and Locked" are returned from the DCU to the train driver's panel. DCU provides power required by each door components and reports door status to TIMS.

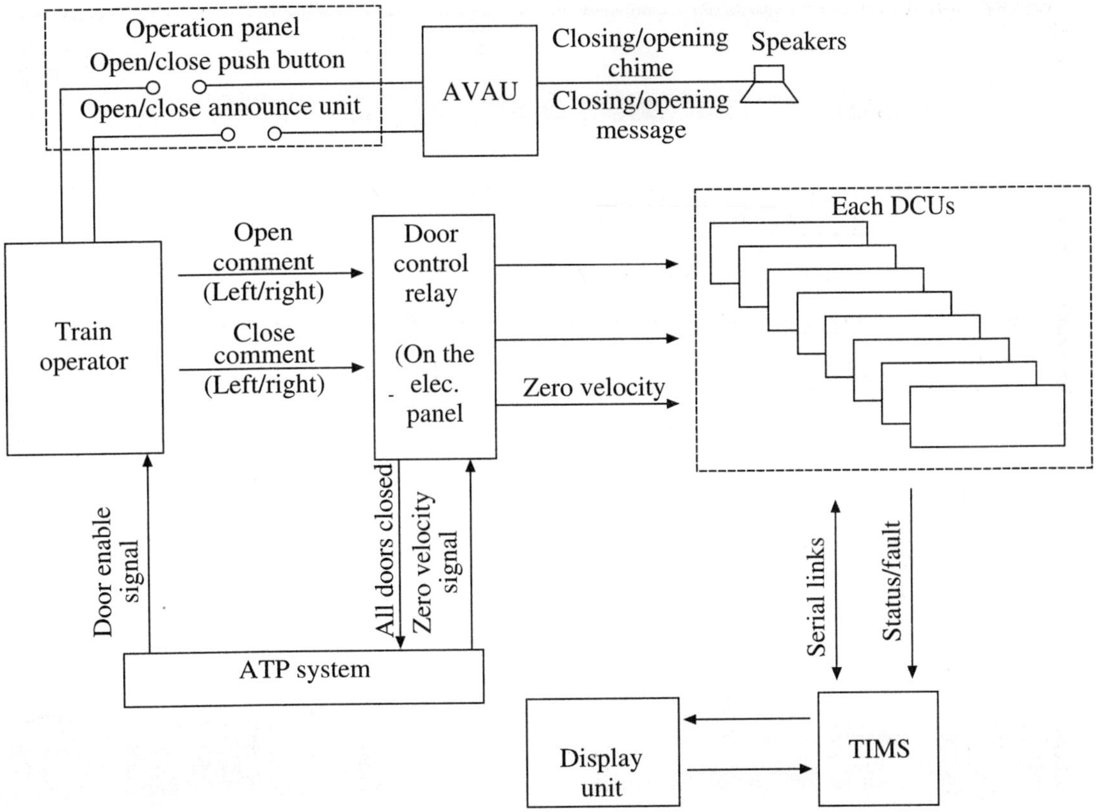

FIGURE 14.8 Illustrative diagram of door control unit.

In the above typical subsystem of a metro train (Figure 14.8), the door open/close command as executed by the train operator is sent to each door control unit (DCU) through door control relay. Simultaneously, the information is sent to automatic voice announcement unit (AVAU) to make announcement regarding closing of the train doors. System also ensures that the door operation is only possible when train is stand still, i.e., zero velocity.

In the above typical subsystem of a metro train, the status of doors is indicated in TIMS. In Figure 14.9, the D101 and D102 are driving cars and M101 and M102 are motor cars. The box ( ) indicates the status of door in each car. The closed door is indicated by yellow colour, open door by pink colour and defective door by red colour. All door close relay (ADCR) and door proving relay (DPR) when energised indicate that all the doors are properly closed and locked. If any of these relays are de-energised, the doors are either not closed or are/is defective. Train cannot start unless all doors are closed and proved by the system. If all doors are closed but relay still remain de-energised due to some fault, the train can be started after operating door proving loop cut out switch (DLPCOS) which disables the door proving loop.

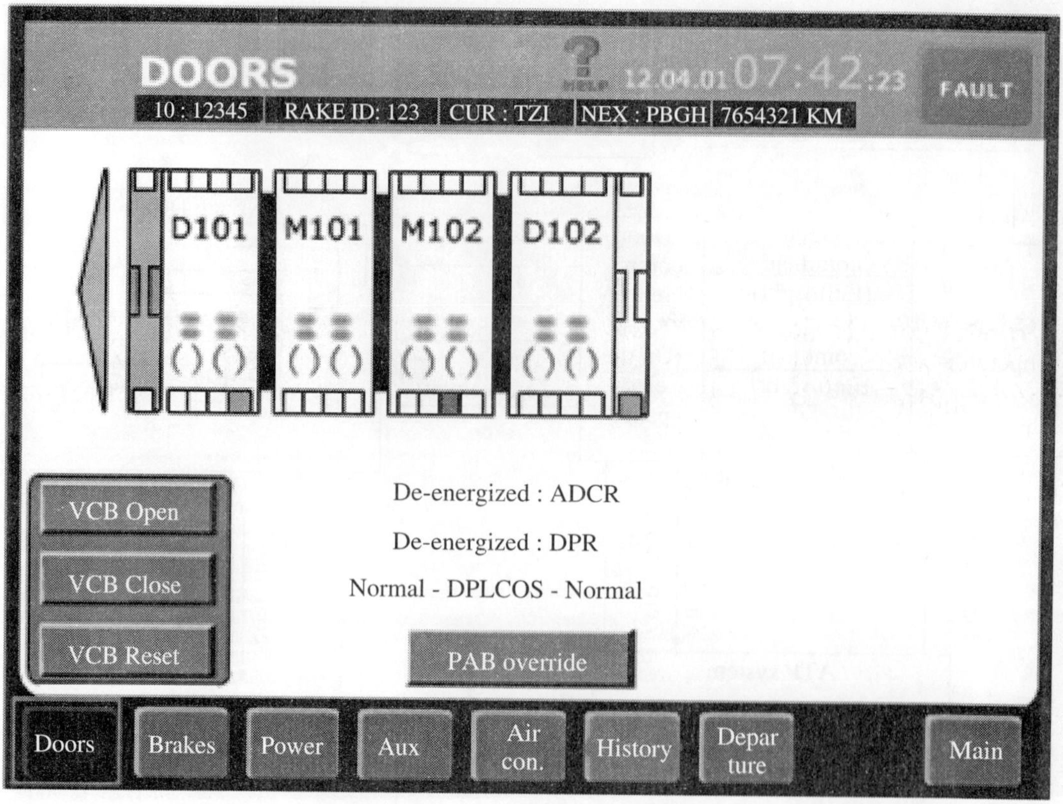

FIGURE 14.9 TIMS page of DCU.

14.6 HVAC

The heating, ventilating, and air conditioning (HVAC) units and distribution systems are essentially required for climate control in the passenger cars as well as the driver's cab. The right amount of conditioned air at desirable temperature and humidity levels should be delivered to the target locations to ensure the comfort of the occupants, i.e., drivers and the passengers. At the same time, the energy required for the flow delivery should be minimised for better fuel economy, while the adverse effects of the associated acoustic noises should be limited. The temperature and humidity in the trains are maintained automatically depending upon the ambient temperature, humidity, and passenger load inside the train. The system is designed to automatically adjust the intake of fresh air with the recirculated air to optimise the comfort level and energy use. In case of smoke outside the train, the intake of fresh air is immediately stopped. In modern rolling stock, in case of non-functioning of HVAC in one car, the air conditioning is supplied from the adjacent car to ensure comfort the passengers.

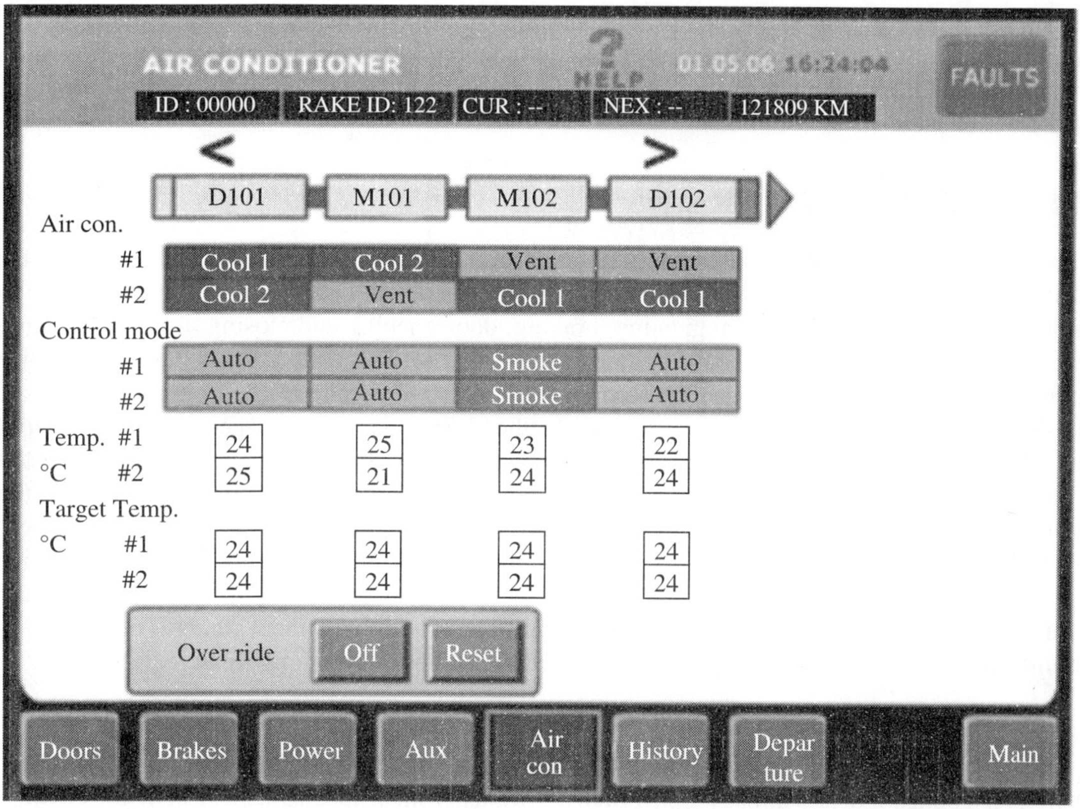

FIGURE 14.10 HVAC system inside trains.

In the above typical subsystem of a metro train (Figure 14.10), the air conditioning page of TIMS indicates the status of ACs of each car, mode of operation (cool/ventilation), inside temperature in each car. The functioning of ACs can be controlled from TIMS.

14.7 AUTOMATIC VOICE ANNOUNCEMENT UNIT (AVAU) AND PUBLIC ANNOUNCEMENT (PA) SYSTEM

Timely information to passengers about next station, side on which trains doors would open, interchange station, delay in train and any emergency situation is required for convenience of the passengers and to handle the emergency situations. In the urban transport system, the stopping stations are at very close distance usually 1–2 km. So train stops on average after every 2 min. It is not practically possible to do manual announcements every 2 min by the train driver who is busy in other train operations duties also. The modern rolling stock is equipped with automatic voice announcement unit (AVAU). The AVAU system programmed to make pre-fed scheduled announcements automatically as per the location of the train. The PA system is also available with the train driver to make unscheduled announcements whenever required. The trains are also equipped with passenger emergency alarms (PEA) using which a passenger inside the train can contact the train driver in case of any emergency. The PEA is one of the vital safety equipment inside the train. Train announcement system is intelligent enough to adjust volume of announcements accordingly to the noise level in the coach so that it remains audible and pleasant to commuters.

14.8 DRIVERLESS TRAINS

The advent of communication based train control (CBTC) system has paved the way for driverless operations of the trains. In driverless train operations, all the functions of the driver from starting the train, driving, regulating, braking, door opening and closing, trouble shooting, etc. are performed by the on-board computer system. The driverless train is equipped with a number of self-diagnostic tools which monitors and relays the health of the system on real-time basis to the central control. The train remains in the sleep mode when placed in the train depot or siding. For inducting the train into service, a 'wake up' command is required to be executed from the control centre which may be remotely located. Before entering into the main line, the train performs self-checks of its vital safety systems such as braking, doors, signaling system interface and if any system is not working satisfactorily, the train is not inducted into the service by the system. While working on main line also there are multiple levels of redundancies in the train which improve its reliability manifold as compared to conventional trains. The driverless trains are truly intelligent having capabilities to run without any human intervention.

14.9 SUMMARY

The ITS application has already converted the trains from a mechanical system to software based electronic system. The modern trains are capable of performing functions as programmed in the software controlling of the train. The various subsystems of the train like traction, auxiliary power supply, braking system, door control unit, HVAC, and PA system are controlled by the train information management system (TIMS) which is the control centre of the train. The driverless trains are fully controlled by the system without any manual intervention. The future is likely to witness further consolidation of information and communication technology (ICT) to make the trains akin to robots which are intelligent, having learning capabilities from experience and able to take certain decisions depending upon its experience.

Questions

14.1 How intelligent trains are different from the conventional trains?

14.2 What are different subsystems of a train?

14.3 What do you understand by train control management system (TCMS)? What are the functions of TCMS?

14.4 Write short note on the followings:
- (a) Traction inverter—Converter/Inverter (CI)
- (b) HVAC
- (c) Brake electronic control unit

(d) Door control unit
(e) Auxiliary power supply

14.5 What do you know about driverless trains?

Bibliography

Baba, Yuichi, Atsushi Hiratsuka, Eiji Sasaki, Osamu Yamamoto, and Masakazu Miyamoto, "Radio-based Train Control System", *Hitachi Review*, Vol. 61 (2012), No. 7, 2012. (http://www.hitachi.com/rev/pdf/2012/r2012_07_109.pdf)

Liu, Changyuan, Xiaoming Li, and Panpan Yang, "Train Control Management System Safety Assessment", *Proceedings of the 2013 International Conference on Electrical and Information Technologies for Rail Transportation (EITRT 2013)*, Vol. II, Lecture Notes in Electrical Engineering, Vol. 288, pp. 583–591, 2014.

Łukasik, Zbigniew, Waldemar Nowakowski, and Tomasz Ciszewski, "Train Control And Monitoring System Simulator", in *Indian Journal of Applied Research*, Vol. 4, Issue 12, December 2014.

Mitsubishi Electric, Ethernet-based TCMS (Train Control and Monitoring System). (http://www.mitsubishielectric.com/bu/transportation/download/pdf/01_TCMS.pdf)

Mohammed, Thabit Sultan, Wisam Fahmi Al-Azzo, Mohammed Ahmed Akaak, and Mohammed Laheeb Suroor, "Full Automation in Driverless Trains: A Microcontroller-Based Prototype", *International Journal of Advanced Research in Electrical, Electronics and Instrumentation Engineering*, Vol. 3, Issue 7, July 2014.

Neil, G., "On-board Train Control and Monitoring Systems", IET Professional Development Course on Electric Traction Systems, pp. 223–246, 2012.

Richard Tracey, Greater London Assembly Member, "Driverless Trains: Efficient, Reliable, the Future", 2010. (http://glaconservatives.co.uk/wp-content/uploads/2012/02/Driverless-Trains.pdf)

Selectron, "Train Control and Monitoring System TCMS in Rail Vehicles", System Overview. (http://www.selectron.ch/selectron-wAssets/docs/flyer/en/Flyer-System-overview-general_EN.pdf)

Train Design Documents, DMRC.

15 Intelligent Maintenance System

15.1 INTRODUCTION

The safe and efficient running of trains largely depends upon the quality of maintenance of assets which include track, train control system, traction system and the rolling stock (Figure 15.1). The maintenance of assets is one of the high cost activity. Around one third of the cost of operations is attributed to the maintenance of the assets. With increased in emphasis on more efficient, cost-effective and performance oriented railway system, the need of optimisation of maintenance of assets has assumed greater importance over the time. While a number of information and communication technologies (ICTs) driven innovation are visible in train operations and customer services, a number of initiatives have also been taken to improve the maintenance practices.

The areas related to asset maintenance and condition monitoring are fit cases for improvements which can be achieved using advanced ICT solutions. The minimisation of critical failures affecting the quality of the train services is the prime objective of any maintenance regime while at the same time keeping maintenance costs as low as possible, in order to reduce life-cycle spending and improve return of investment.

15.2 MAINTENANCE PHILOSOPHY

Railways are made up of complex mechanical, electrical and electronic systems with hundreds of thousands of moving parts. For a reliable railway service, the equipment must be kept in good working condition and regular maintenance is the essential ingredient to achieve this. A railway system cannot function efficiently for a long time if assets start deteriorating and needs frequent adequate maintenance. Without maintenance, the life-span of the assets would reduce and failures would increase. Although maintenance is expensive, it will become more expensive to replace the failing equipment early in its life because maintenance has been neglected. Because of these reasons, the maintenance of assets has been considered an essential

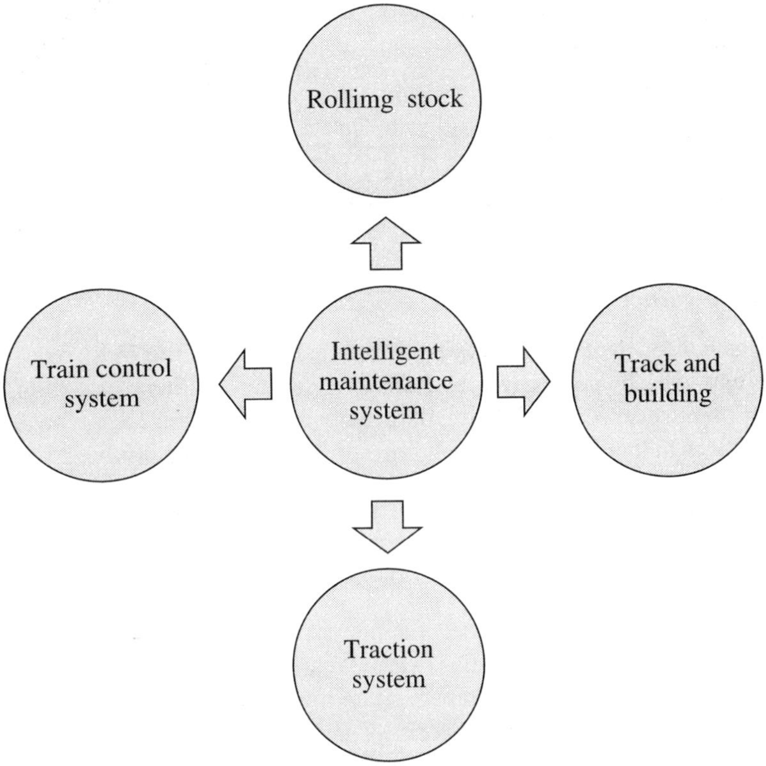

FIGURE 15.1 Intelligent maintenance system.

activity of railway systems. The maintenance system has evolved over the time from reactive maintenance to preventive maintenance and to predictive maintenance.

Railways maintenance practices are traditionally divided between reactive (also called corrective) and preventative, but in the last few decades, many more approaches have been developed:

Reactive (or corrective) maintenance

The reactive maintenance aims to fix assets when they have failed or no longer work well. Maintenance in reaction to unplanned failures should be held as low as possible, but all railway systems have some reactive maintenance that may arise from perfectly valid strategies:

(a) Assets that are not critical to service may be allowed to run to failure in order to reduce the frequency and cost of maintenance interventions. For example, one would not like to replace a tube light unless it has stopped working.

(b) Condition-based maintenance (CBM) is always both planned and reactive, with interventions taking place whenever an asset's condition and criticality requires it, normally before the risk of failure becomes acute.

Preventive or preventative maintenance

Preventive or preventative maintenance involves when assets are maintained before they reach the point of failure, or below acceptable performance. The most basic form of this is planned periodic maintenance, where certain assets are maintained in a certain way at a certain frequency. This idea can be further developed by cutting up and re-grouping the various maintenance activities together to match maintenance shift resources, known as the *balanced exam approach.*

Risk-based maintenance

Risk-based maintenance involves identifying which assets cause the most risk to the reliability of the service and which are the main causes of failure in those assets. It also involves maintaining critical assets to a higher standard than those that are less critical. A key test of railway systems is whether they are applying risk-based or priority-based maintenance and whether they are applying resources in proportion to the breakdown of failure incidents or causes of passenger delay.

Condition-based maintenance (CBM) maintenance

CBM is the next step up from risk-based maintenance. CBM maintains assets based on the whether the asset's condition still remains within required parameters. The aim of condition-based maintenance is to select the most appropriate intervention for the state of the asset at the most appropriate time. It relies on some form of asset condition monitoring (ACM). In most cases, ACM is used as a trigger to fine-tune the timing of corrective or preventive maintenance, as well as correcting or replacing assets at risk before they fail.

Predictive maintenance

Predictive maintenance is the next level of sophistication beyond CBM. This is achieved by automatic and real-time monitoring of the parameters under which equipment operates, in order to determine root causes of failure and predict if equipment including subsystems and components will fail within a given timeframe. Where real-time monitoring is not yet available, several railway systems use a combination of inspections and advanced computer modelling to predict future degradation profiles and inform decision makers about maintenance and asset management strategy.

Reliability-centred maintenance (RCM)

RCM is a process to ensure that assets continue to do what their users require in their present operating context. A fundamental underlying principle is that only those maintenance actions that contribute to reliability should be retained and all tasks should be carried out with the frequency that most contributes to reliability. Also important is the reduction of cost that arises from RCM. Any maintenance action that does not contribute to reliability (and some maintenance interventions may result in reduced reliability) can be reduced in frequency or eliminated altogether. Whereas other forms of maintenance focus mainly on the asset itself, proactive maintenance is a strategy for influencing the failure root-causes that may lie in the environment or operating conditions of the asset. Rather than responding to the active failure symptoms, faults, or machine wear conditions, the maintainer attempts to eliminate the conditions that produce degradation in the asset condition.

Intelligent maintenance system

Today the application of ICT allows real-time monitoring of the equipment and generates alerts whenever required. An intelligent maintenance system is a system that utilises the collected data from the machinery in order to predict and prevent the potential failures in them. The occurrence of failures in the machinery can be costly and even catastrophic. In order to avoid them, there needs to be a system which analyses the behaviour of the machine and provides alarms and instructions for preventive maintenance. Analysing the behaviour of the machines has become possible by means of advanced sensors, data collection systems, data storage/transfer capabilities and data analytic tools developed for such purpose. The aggregation of data collection, storage, transformation, analysis and decision making for smart maintenance is called an *intelligent maintenance system (IMS).*

15.3 MAINTENANCE OF ROLLING STOCK

The conventional maintenance system of rolling stock has been a mix of corrective and preventive maintenance. The preventive maintenance based on mileage (kilometres or miles run by the train) has been most prevalent. The metro trains are mostly being maintained during the night time and inducted for service early in the morning. The maintenance activities are scheduled during the night in such a way that required attention is paid to the equipment as per predefined periodicity. The fitness to run the rolling stock is given for a specific distance or time period. For instance in DMRC, the fitness for rolling stock is given for 72 h, which means that after 72 h, the scheduled maintenance of the rolling stock would be required and a fresh fitness certificate will have to be issued for inducting the train in the service irrespective of the distance run by the train. This is a well-tested and established system of preventive maintenance being adopted in Railway systems. The intelligent trains, today, consist of many complex systems and subsystems which interact among each other. Failure in one system may lead to fault in many other systems/subsystems. At times supposedly small incidents like doors not getting closed may have disproportionate effect on train services, delays, inconvenience to the passengers and cascading effect on other train services. For such reasons monitoring and diagnostic systems on-board trains have become more important, and also more sophisticated. As discussed earlier, the intelligent maintenance system with the help of ICT has been a game changer in the maintenance practices especially for rolling stock.

Typically following functions are performed through ITS tools in modern metro train sets to monitor the heath of the equipment and generate advanced information for any maintenance requirement:

(a) Remote access to train's on-board database system (ODBS)
(b) Train control management system
(c) HVAC maintenance and testing
(d) Passenger doors maintenance and testing

(e) Testing brake control system of the train
(f) Analysis of event and condition data of train system

These tools complement each other for failure investigation and taking decision regarding various maintenance issues on the train. The further details of these tools is available in Appendix A.

15.4 INTELLIGENT TRACK MAINTENANCE SYSTEM

The track is the foundation of the railway system. A strong and well-maintained foundation is fundamental to smooth and efficient running of the system. The track is also the most capital intensive investment for a railway system. Proper care and maintenance of track is essential not only for safety and operational efficiency but also for bringing down the life-cycle cost of the track system. The conventional method of track maintenance involves a schedule of inspections to measure the various parameters of track manually and a regime of periodic maintenance. This manual method of maintenance requires more of maintenance time and is also quite costly. Today, as railways increase their capacity and speeds, it is more important than ever to be completely aware of the state of the infrastructure's condition to ensure the highest quality and safety to those using it, as well as being in the position to maintain at cost as low as possible. The advancement in sensor technology and use of computer software to monitor and analyse the track parameters on real-time basis has changed the way the track maintenance has been undertaken. Some of the intelligent means used for maintenance of track are presented as follows:

15.4.1 Track Geometry

The deficiencies in track geometry are one of the major causes of all maintenance and accident costs for many of the railways world over. There may be so many different factors can adversely affect the correct geometry of the track, even daily use, that railways cannot afford to continuously monitor infrastructure's condition. For effective maintenance of assets, it is essential to assess the health of various parts of the asset on the basis of some objective parameters.

Indian railway system of measuring track geometry

In Indian Railways, the following four kinds of track defects are measured to assess the health of the track:

(a) ***Gauge defect:*** It is defined as deviation from the nominal distance between the inside gauge (running) faces of two rails of track at gauge measuring point (13–14 mm below the rail head) from the design gauge say 1676 mm for broad gauge, 1000 mm for metre gauge or any such given distance for any other gauge.

(b) ***Twist defect:*** Twist is the difference between two cross-level measurements a certain distance apart. Whereas Cross level is the measurement of the difference in elevation (height) between the top surface of the two rails at any point of track. Twist is measured in terms of rate of variation in cross level per meter.

(c) ***Alignment defect:*** It is defined as misalignment of rail gauge face (at 13–14 mm below the rail head) measured as mid chord offset (versine) on stipulated chord or string held between two points on gauge face at specified points say 7.2 m or 9.6 m or such given distance apart. It is separately measured for left as well as right rail.

(d) ***Unevenness defect:*** It is defined as undulation of rail top table (in vertical plane) measured as mid chord offset (versine) on stipulated chord or string held between two points located on rail top table at specified points say 3.6 m or 9.6 m or such given distance apart. It is separately measured for left as well as right rail.

In older days, these defects were measured manually through inspection by foot, trollies, locomotives and rear vehicles by the permanent staff. These inspections, though important, are qualitative and enable assessment based on individual experience. Further, the recording and data management of these parameters have been a voluminous task which affects the interpretation and decision making. To overcome these problems, automatic systems of measurement of geometric parameters of track are increasingly being adopted by railway engineers.

There are three different systems/mechanical means available with Indian Railways to assess the track parameters and behaviour as follows:

(a) Track recording cars (TRC)

(b) Oscillation Monitoring System or portable accelerometers (OMS)

(c) Oscillograph car

Out of the three, only the first one, i.e., track recording cars have the ability to measure the track parameters, whereas the other two measure the accelerations and the ride indices. Incidentally TRC does that too. The functioning of TRC has been explained in greater details here.

Track recording car (TRC): A track recording car is an automated track inspection vehicle used to measure several geometric parameters of the track without obstructing normal railroad operations. The cars use a variety of sensors, measuring systems, and data management systems to create a profile of the track being inspected.

The following parameters are being recorded on continuous basis during the run by TRC used by Indian Railways:

(a) Unevenness—Left rail

(b) Unevenness—Right rail

(c) Gauge

(d) Twist

(e) Curvature/alignment—Left rail

(f) Curvature/alignment—Right rail

(g) Speed

The TRC generates continuous record of these parameters and generates a parameter called *track geometry index* (TGI). TGI is a value computed by comparing the measured standard deviation (SD) of four parameters viz. unevenness, alignment, gauge and twist, respectively, with the SD values for very good track and that for track needing immediate maintenance. Higher values indicate better track.

The quality of the track and the requirement of maintenance can be determined on the basis of TGI value of the track.

Modern track geometry measurement system

There are advanced equipment available which can be fitted on revenue service trains to give maintenance engineers access to frequent track geometry data, without affecting network availability. One such track, geometry measurement system called Opto-Inertial is explained in Box 15.1.

Box 15.1 Modern track geometry measurement system

The new generation of opto-inertial track geometry measurement system (Figure 15.2), branded Laserail™, is a non-contact track measuring system with GPS navigation and dual optical measurement boxes to measure, record and analyse the track geometry. The system, that is fully compliant with the EN-13848 international standard, offers highly accurate measurements at speed up to 400 kmph (250 mph) with real-time report of exceptions. System architecture is completely modular, guarantees small space requirements and it is also configurable to measure full rail profile parameters, implementing two more optical measurement assemblies.

Thanks to improved software functionalities, the possibility of performing fully unattended measurements without occupying any room on-board for PCs or power pack is a proven reality.

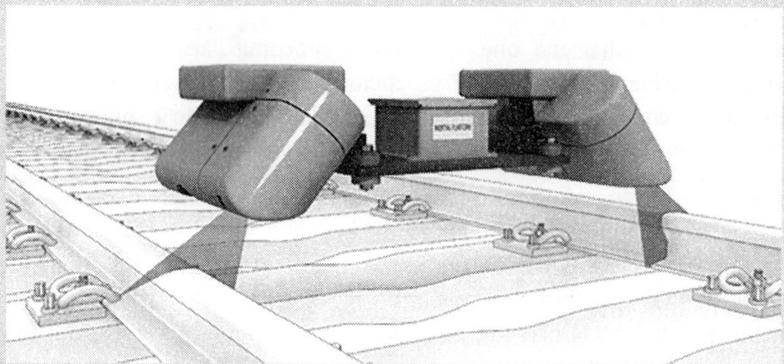

FIGURE 15.2 Track geometry measurement system. (*Source:* http://www.mermecgroup.com/inspection-technology/track-measurement/185/1/track-geometry.php, reproduced with permission).

15.4.2 Rail Profile

The continuous monitoring of rail profile has become more important with increase in speed and volume of the traffic. Knowing the rail wear condition is fundamental because it is strictly related to the key safety aspects of the railway infrastructure.

With rail profile system (RPS) available in market, it is possible to measure the complete cross-section of a rail head and calculate the key data and wear parameters. Along with the essential measurements like head width and crown radius, the wear at the top and the sides can be determined by comparing with the standard profile. One such rail profile system of MERMEC Group is explained in Box 15.2.

Box 15.2 Rail profile system

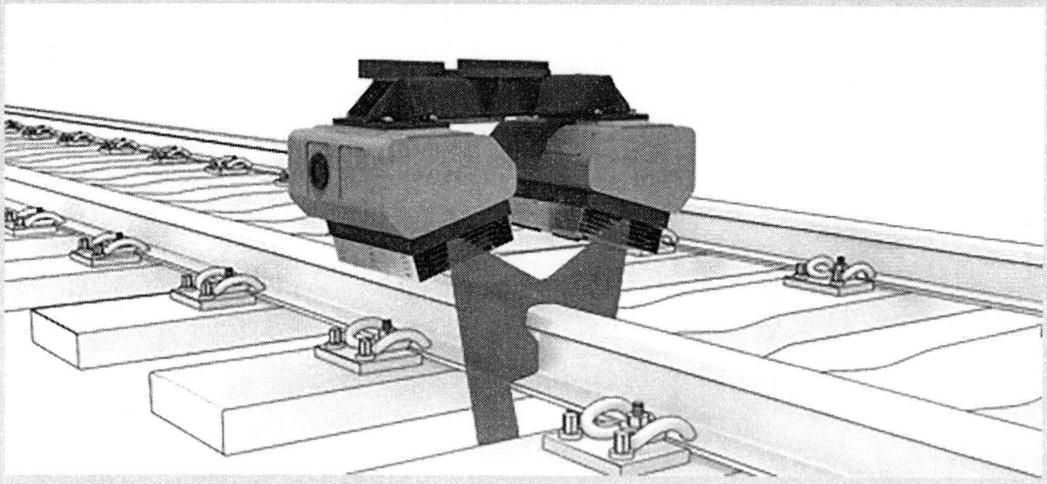

FIGURE 15.3 Rail profile system (*Source:* http://www.mermecgroup.com/inspection-technology/track-measurement/60/1/rail-profile.php, reproduced with permission).

RPS incorporates the latest laser and video camera technology to provide accurate and immediate report on the profile and wear condition of the rail whilst travelling at track speeds (Figure 15.3). The video cameras capture full cross-sectional rail profiles from the base/Web fillet area up to the top-of-rail surface to allow comprehensive and accurate rail measurements.

The system, which can be mounted on every revenue generating vehicles, becomes an important tool in wear trending, which can maximise the effectiveness of the grinding program.

15.4.3 Rail Corrugation

Irregularities on wheels and rails give rise to noise, ground-borne vibration and more general dynamic loading, which increase damage of components of both vehicle and track. Both wheels and rails are prone to develop quasi-sinusoidal irregularities, which are known as *corrugation.* Rail corrugation defects can create a vicious circle, causing problems on wheels, which lead to further rail problems. The increased dynamic interaction forces between wheel and rail caused by corrugation give rise to fastening, sleeper and ballast deterioration reducing the life of track and vehicle components.

The rail corrugation system (RCS) detects and measures rail corrugations using optical-measurements of rail surface profile by recording surface displacements with cameras and lasers through the versine transfer function. The RCS of MERMEC Group is explained in Box 15.3.

Box 15.3 Rail corrugation system

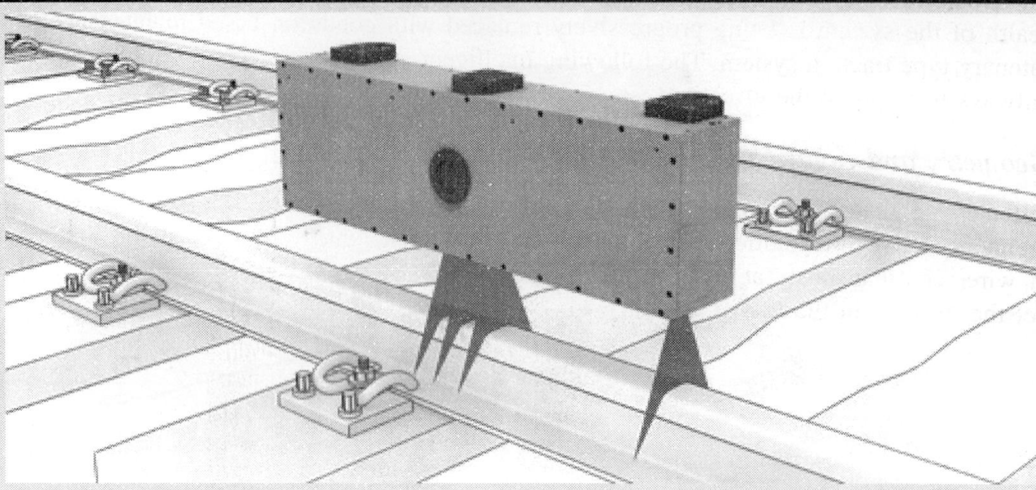

FIGURE 15.4 Rail corrugation system (*Source:* http://www.mermecgroup.com/inspection-technology/track-measurement/61/1/rail-corrugation.php. reproduced with permission).

Fully compliant with the EN-13231-3 international standard the MERMEC Group Rail Corrugation System (RCS) detects and measures rail corrugations ranging from short (30–100 mm) to the long (300–1000 mm) wavelengths. The system uses optical-measurements of rail surface profile by recording surface displacements with cameras and lasers through the versine transfer function (Figure 15.4).

The measurement can be done with the train travelling in both directions and with all weather conditions. Each system offers real-time data analysis and post processing analysis. Real-time functions include data acquisition, storage and diagram visualisation in all the wavelengths. Post processing allows for recorded data visualisation and standard deviation calculations for each wavelength.

15.4.4 Track Video Surveillance

The track video surveillance solution provides a driver-eye view of the leading and trailing track and infrastructure, minimising any potential disruption to the train operator and network. As matter of fact, video recorded from the surveillance system provides assistance in the investigation of operational incidents as well as in evaluation of infrastructure's condition.

DVR cards (also known as digitizer or video capture cards) along with software application enable advanced services and functionalities, such as centralised command and control, and events management. Images are captured direct from the CCD sensor, encoded digitally within the camera, and then transmitted to the DVR via an Ethernet link enabling real-time or recording analysis.

15.5 INTELLIGENT TRACTION MAINTENANCE SYSTEM

The maintenance of overhead catenary system has remained one of the major challenges for railway operators owing number of factors involved like span, stagger, wire height, tension, wave propagation velocity, bending rigidity, vibration endurance, wear, etc. The ever increasing speed and volume of traffic and limitation of availability of maintenance time have further exacerbated the problem. The conventional method of periodic maintenance irrespective of the health of the system is being progressively replaced with condition based maintenance of the catenary type traction system. The following intelligent systems are being employed by various railways to improve the effectiveness of the maintenance of the catenary system:

Geometry and contact wire wear

Catenary geometry and wear system (CGWS) is a non-contact measuring system using laser scanners and high-resolution digital camera to measure height, residual thickness and stagger of wires simultaneously at a high speed (Figure 15.5). The system helps maintenance staff to get the most from the life of contact wires while increasing the safety of the line.

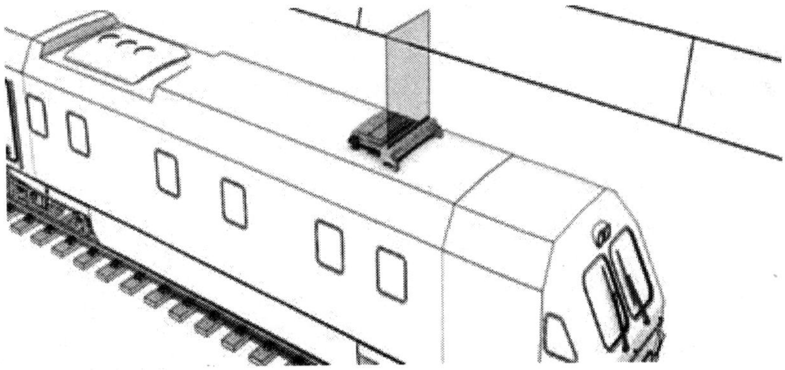

FIGURE 15.5 Geometry and contact wire wear (*Source:* http://www.mermecgroup.com/inspection-technology/catenary-measurement/78/1/geometry-e-wear.php, reproduced with permission).

Electric arc detection (EAD)

The arcing between the pantograph and the contact wire is a common phenomenon. It is more predominant at high speed, increasing load, and in cold icy weather conditions.

Arcing at the pantograph (a device for collecting an electrical current to power an electric locomotive/EMU) distorts the regular waveform of the supply voltage and current and generates transients which propagate along the railway system, including tracks, track circuits, other locomotives on the same track, substation supply transformer, vehicle transformer, etc.

The electric arcs detection system (EADS) detects the presence of electrical arcs and their duration even at a very high speeds. The data can be correlated with the contact line data. Once the locations of arcing are identified, the catenary wire can be suitably maintained as per requirement.

Contact wire thermal scanning

Overhead catenaries can become overheated, expanding to such an extent that they almost touch the wagons. When overheated, contact wire starts to expand, loses tension and starts to sag. In the worst case, it gets wound up around the locomotive's pantograph. The train stops and the line is blocked. Repairing overhead lines takes time and one can only imagine the consequences on busy lines at rush hour. The thermal scan monitoring system is an infrared imaging system designed to carry out the thermal image of the contact wire.

15.6 INTELLIGENT SIGNALLING MAINTENANCE SYSTEM

The operational efficiency of the modern railway system largely depends on proper functioning of the signalling system. The computerised automatic train operations are completely based on electronic systems. The train control system is designed as fail safe system which means that in case of any failure, the systems automatically adopt the safest mode and the safest mode in railway system is suspension of train operations. Minimisation of failures of signalling system has remained a high priority item for railway operators. The maintenance of signalling system depends on the technology adopted by the supplier of signalling systems as such the maintenance systems are also designed by the supplier of the system. The modern signalling systems are capable of generating logs of all events and transmit the same to central server for further analysis and requisite condition based maintenance. The system capabilities varies with system to system as per the design of the supplier.

15.7 SUMMARY

There is a need to progressively shift from conventional preventive maintenance activity to intelligent maintenance system which is based on real-time monitoring of the health of the assets. The intelligent maintenance system apart from being cost effective improves the reliability of the assets. Thanks to convergence of information and communication technology (ICT), the real-time monitoring of assets through online applications has become possible. The modern rolling stock is fitted with number of sensors to monitor the functioning of various subsystems of the train like train control system, brake system, traction system, door system, etc. The real-time generation of events helps in condition based maintenance of rolling stock. The track recording car (TRC) used for monitoring the track parameters has been a significant step in application of ICT for track maintenance. The maintenance of overhead traction system, signalling systems are also going for condition based maintenance system progressively. The day is not very far when all vital assets of the rolling stock will be equipped with sensors to generate real-time diagnosis of the assets and maintenance will be based on the detailed report generated by the system.

Questions

15.1 Why an effective maintenance system is important for health of railway system?

15.2 Which is the most prevalent maintenance practice in Indian Railways?

15.3 What are different approaches to maintenance systems in railways?

15.4 Differentiate between condition based maintenance and predictive maintenance.

15.5 Describe reliability-centred maintenance system.

15.6 Explain intelligent maintenance system for rolling stock.

15.7 What are different track defects which are measured to assess the health of the track?

15.8 What do you understand by track recording car (TRC)? What track parameters are being measured by TRC?

15.9 What do you understand by rail corrugation? How is it measured?

15.10 Describe role of ITS in maintenance of overhead traction system of railways.

Bibliography

Felix A. Marten, Jr., "Reliability Centred Maintenance: A Case Study of Railway Transit Maintenance to Achieve Optimal Performance", a publication of the Mineta Transportation Institute College of Business, San José State University San José, CA 95192-0219 Created by Congress in 1991, 2010.

Fredriksson, Gustav and Hanna Larsson, "An analysis of maintenance strategies and development of a model for strategy formulation—A case study", Department of Product and Production Development Division of Production Systems, Chalmers University of Technology, Göteborg, Sweden, 2012.

Geometry and Contact Wire Wear (CGWS), MERMEC group. (http://www.mermecgroup.com/inspection-technology/catenary-measurement/78/1/geometry-e-wear.php)

Gulati, R., *Maintenance and Reliability Best Practices*, 2nd ed., Industrial Press, New York, 2012.

Indian Railway Permanent Way Manual, "Track Recording, Analysis and Monitoring", Chapter 6. (http://www.indianrailways.gov.in/railwayboard/uploads/codesmanual/IRPWM/PermanentWayManualCh6_data.htm)

O&M Best Practices Guide, "Types of Maintenance Programs", Chapter 5, Release 3.0. (https://www1.eere.energy.gov/femp/pdfs/OM_5.pdf)

Rail Corrugation, System of MERMEC group. (http://www.mermecgroup.com/inspection-technology/track-measurement/61/1/rail-corrugation.php)

Rail Profile, System of MERMEC group. (www.mermecgroup.com/inspection-technology/track-measurement/60/1/rail-profile.php)

Tayal, C.P., "Track Monitoring", Indian Railway Institute of Civil Engineering, Pune, 2013. (http://www.wiki.iricen.gov.in/doku/lib/exe/fetch.php?media=iricen_books:track_monitoring.pdf)

Track Geometry, Measurement System of MERMEC group. (http://www.mermecgroup.com/inspection-technology/track-measurement/185/1/track-geometry.php)

Train Design document of DMRC.

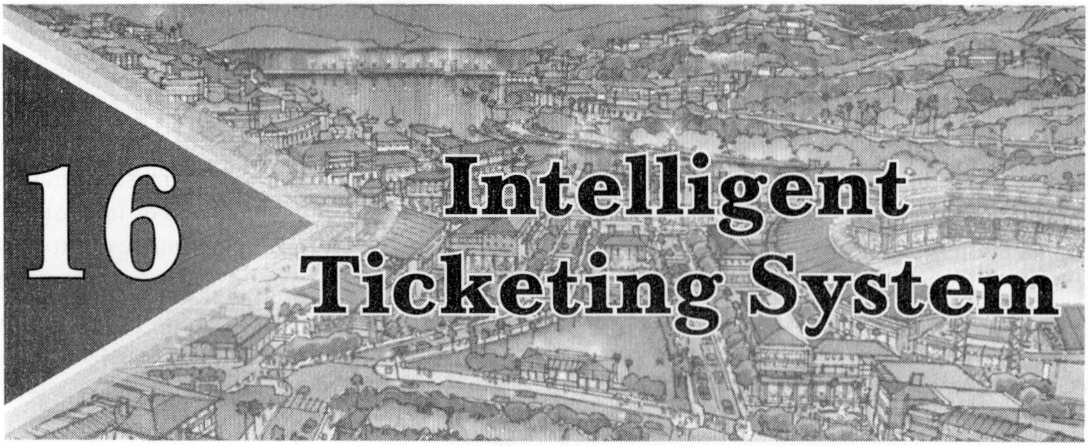

16 Intelligent Ticketing System

16.1 INTRODUCTION

An efficient system of fare collection is fundamental to financial sustainability of any transit system. The early age fare collection system was based on paper tickets for per trip basis or multiple journey monthly passes allowing for any combination of stations. The ticket could have been checked at any stage of journey by ticket checking staff. This system, being manual was inefficient and resulted in large scale fare evasion. The cost of preventing fare evasion in manual system of ticket checking became too high over the time as the cost of manpower increased. In 1960s with the advent of electronic technology for machine reading of magnetically stored information, the automation of the ticketing system became feasible. The automation of ticketing activities gave birth to automated fare collection (AFC) system. Introduced on PACTCO's Lindenwold Line in 1969 and San Francisco BART in 1972, the AFC has become a standard method of fare collection in almost all upcoming metro systems.

Automatic fare collection (AFC) system allows entry to the station only to the passenger having valid fare media, collects exact fare as per the prevailing fare structure from the fare media presented by the passenger and allows exit from the station. So AFC system works as entry/exit control as well as means of fare collection automatically. Only a passenger having valid ticker can enter into the station and make an exit from the station after deduction of the correct fare.

The AFC systems of 1970s are based on magnetic strip technology wherein the ticket consisted of a magnetic strip having stored information about the destination station in case of single journey ticket or balance amount per trips in case of multiple journey tickets. In multiple journey ticket, after every journey, the fare per trip was subtracted from the store value. At the originating station, the passenger has to place the ticket at entry gate, which checks the validity of ticket and allows entry. At exit station again the passenger will have to show the ticket at exit gate which allows the exit if ticket is valid and subtracts the value of fare per trip from the store value in case of multiple journey tickets.

Further, the use of radio frequency identification (RFID) has paved the way of contactless smart card based AFC systems. The contactless technology is being adopted by most of the new metro rail systems and the older ones are also replacing the magnetic system with the contactless technology. The modern automatic fare collection system (AFC) is a contactless smartcard-based end-to-end solution for fare collection and payment. The state-of-the-art solution is uniquely designed with the demand of revenue services for modern transit operation in mind. Furthermore, with the advent of smartcard technology and proliferation of its business applications, AFC also enables transit operators to expand revenue opportunities, exploits the benefits of payment integration with other transit operators as well as non-transit service providers.

16.2 FUNCTIONS AND BENEFITS OF AFC SYSTEM

16.2.1 Functions of AFC System

The AFC system performs following functions:

(a) *Access control:* It only allows entry to valid fare product holder.

(b) *Ticket reading/writing:* The fare product is read to check its validity and entry/exit location is written on the fare product.

(c) *Transaction collection and reports:* The applicable fare is collected from the fare product and reports on revenue collection and ridership, origin-destination, etc. are generated.

(d) *Stock management:* The stock of fare product like token at different stations can be monitored and managed by the system.

(e) *Equipment supervision:* The system keeps an eye on the health of all equipment and any defect affecting operation of equipment is reported as alarm/alert to central server.

16.2.2 Benefits of Modern AFC System

The AFC system offers following benefits over conventional paper ticket system of fare collection:

(a) Improve efficiency of the fare collection system

(b) Reduce operating and maintenance cost of fare collection for transit operators

(c) Ease-of-use for commuters—customer friendly

(d) Allows increased passenger flow in and out of stations and transit vehicles

(e) Allows integration of fares of multiple modes

(f) Allows real-time monitoring of passenger flow and demand assessment

(g) Allows implementation of dynamic fares—different peak and non-peak hour fares

(h) Open up proactive marketing opportunities. Multi-application capability and interoperability enable effective targeted pricing, loyalty programs, cross-marketing with other services, etc.

(i) Controls avenues of fare evasion and revenue loss through open standard and secured transaction technology
(j) Provide reliable and uninterrupted operation for train, bus, road toll, retail applications, etc.

16.3 SYSTEM ARCHITECTURE

AFC is not a mere technology change replacing conventional magnetic tickets/paper tickets with contactless smartcards. It provides a complete, modular, standard-based open solution that meets the operation needs and revenue growth objectives. AFC system consists of central computer system, station computer system and station equipment. For a complex railway AFC system, the central computer system will consist of central clearing house and line central computer system. The system architecture may change as per the design specification and the requirements of the operator. Here, a typical system architecture of AFC system has been presented.

16.3.1 Central Computer (CC) System

Central computer system is responsible for generating reports, sending control command, downloading system parameter, receiving ticketing data from station computer and ticket price list to station computer. A typical CC system is shown in Figure 16.1. The following functions are performed by a typical central computer (CC) system:

(a) It communicates with all station computers (SC) to collect the details of the CSC and CST usage, accounting, operational and auditing data
(b) It stores the transaction and audit data in order to assume the central functions based on collected transactions
(c) It maintains and distributes the equipment operating data (EOD) which includes system parameters, fare table and program to the equipment via the SC
(d) It transmits the keys defined on the CC to the equipment
(e) It performs equipment administration
(f) It performs daily and monthly data backup and housekeeping functions
(g) It provides information to master clock for clock synchronization
(h) It provides history of card or token usage
(i) It also taked backup, archive and restore the vital data.
(j) It provides alarms to signal the operator of the occurrence of any fault

Components of CC

The multifarious tasks assigned to CC are carried out by its different components. As shown in Figure 16.1, a typical central computer system consists of following components:

Data centre (DC): A data centre (or datacentre) is a facility composed of networked computers and storage that businesses or other organisations use to organise, process, store and disseminate large amounts of data. For a typical AFC system, the DC performs following functions:

(a) Store data sent by station's computers
(b) Configure the entire AFC system
(c) Transmit operating parameters, blacklist, and agent list to the AFC equipment via the station's computers
(d) Provides facilities for entire AFC system supervision
(e) Give detailed analysis of passenger traffic, transaction history for any specific CSC

The DC is usually composed of two servers clustered in order to provide backup for each other. The capacity of server depends upon the volume of the data to be handled and specified speed of handling the data.

Middle ware server (MS): The middle ware server (MS) broadcasts the time to the master clock system (MCS) for synchronization via network time protocol (NTP) over the LAN. It also runs booking office machine (BOM) to generate AFC business reports.

Archiving server (AS): This server archives the data using AFC server archive software in order to save on a digital linear tape (DLT) the partial or global data from the central system. DLT is a magnetic tape used for storage of data.

Administration console and certificate authority (AC and CA): The central administration functions of CC are done on AC and CA console via the microsoft management console (MMC). It also manages security access modules for the AFC system.

Network management console (NMC): The NMC manages the central LAN active network equipment and the LAN Station's active network equipment via the WAN. It uses open view network node manager (NNM) for network management.

Locals workstations (LW): The LWs are located within central environment associated with report printer and communicate with DC via LAN. The LWs work as man machine interface for fare configuration, financial and statistic reporting functions that are elaborated from DC data in Data Base Management System (DBMS).

Bulk initialization machines (BIM): The initialization of contactless smart card (CSC) is performed by bulk initialization machine (BIM) at CC and is done to activate CSCs. Once initialized, the CSCs are issued to the users (passengers) for their usage.

Production server (PS): The PS ensures exchange of data between DC and bulk initialization machines (BIM). It also acts as a station computer for a production center and transmits the operating parameters to BIM. It further collates production shift transaction data from BIM for transmission to the DC.

These central subsystems are interconnected on a redundant local area network (LAN). The central computer is able to communicate with the stations computers (SC) via the communication wide area network (WAN).

16.3.2 Station Computer (SC) System

The primary role of the station computer is to provide the usage data collection, downloading of fare related parameters to the AFC equipment; control, monitoring and management of the

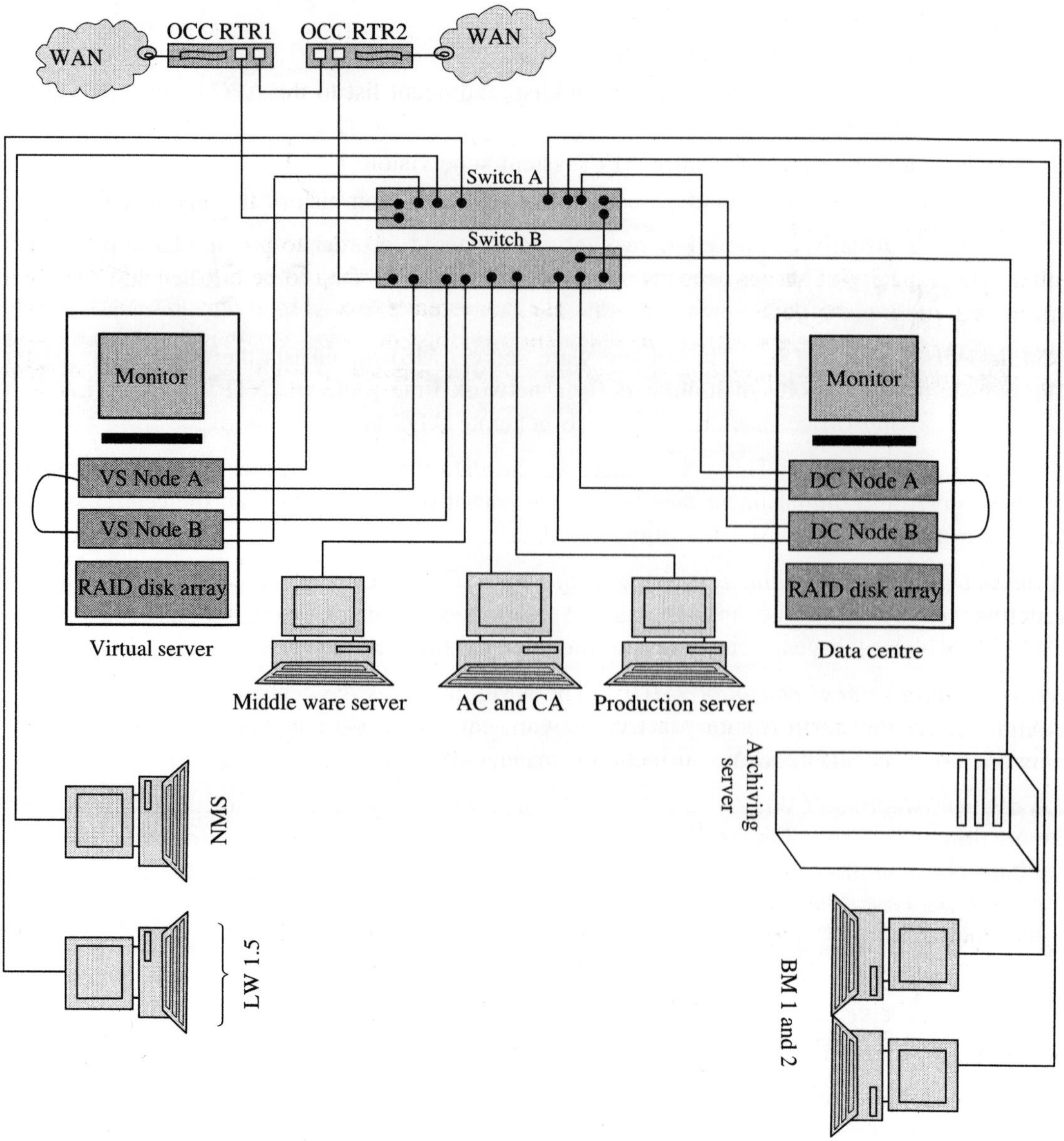

FIGURE 16.1 Central architecture of a typical AFC system. (Reproduced with permission from the source)

AFC equipment. Its secondary role is to provide the station reports (if required) and support the sales office and customer service. It is connected to central computer (CC) through wide area network (WAN). There is continuous exchange of data between CC and SC. It is able to function independently should there be a failure in the communication link with the central computer. However, once connectivity is re-established, the transaction data is transmitted from SC to the CC. Station architecture for a typical SC is shown in Figure 16.2.

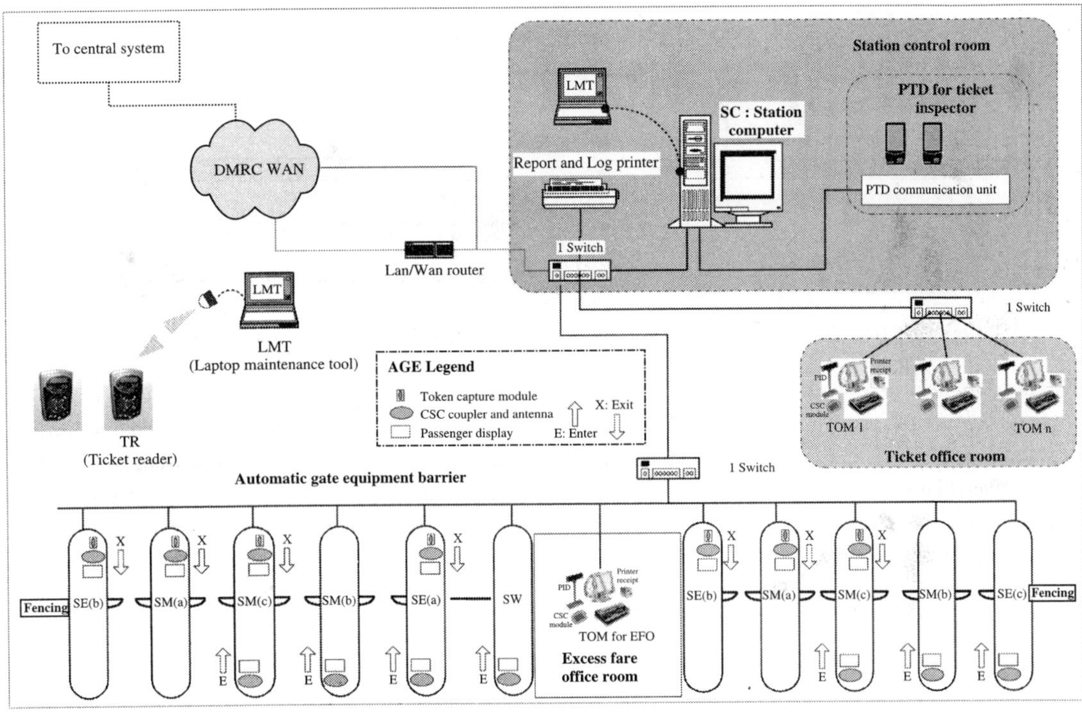

FIGURE 16.2 Station architecture.
(Reproduced with permission from the source)

16.3.3 Station Equipment

These are all the front-end equipment/devices, which are used to serve the commuter. These consist of ticket vending machines, fare gates, booking office machines, mobile and hand-held terminals and ticket recharging machines, etc. Typically the requirement for each implement may vary in terms of the quantity of the equipment/devices.

16.4 FARE MEDIA

The contactless fare media is used in automatic fare collection system. Broadly there are two types of fare media widely used in modern metros: (a) Contactless smart token (CST) (Figure 16.3), and (b) Contactless smart card (CSC) (Figure 16.4). The CST is used for single journey or return journey purpose. The CSC enables customers to load a variety of different journey products, ranging from single trips to monthly tickets, onto the same credit-card sized and to re-use the same card over and over again without having to buy a new ticket for each journey.

The CSC/CST is a contactless means of payment using a smart chip and radio frequency identification (RFID) that enables customers to simply hold their card near to a card reader (located at all entrances and exits to stations) in order for the system to register their journey.

FIGURE 16.3 Contactless smart token.

FIGURE 16.4 Contactless smart card.

The card is preloaded with some credit or trips which is being exhausted at the end of each trip as per the fare structure of the system. The AFC system calculates the exact fare and deduct the same from CSC at the time of exit. In case of CST, the CST is issued for a particular destination. The AFC system of exit station checks the destination for which the CST has been issued. If exit station matches with destination station recorded in CST or exit station is before the destination station along the direction of traffic, the exit AFC will allow the exit from the station. If exit station is after the destination station recorded in CST in the direction of traffic, the exit is not allowed by the AFC exit gates unless fare is adjusted after payment of excess fare. Flow chart for travelling from station A to station B using CST for DMRC passengers is presented in Figure 16.5.

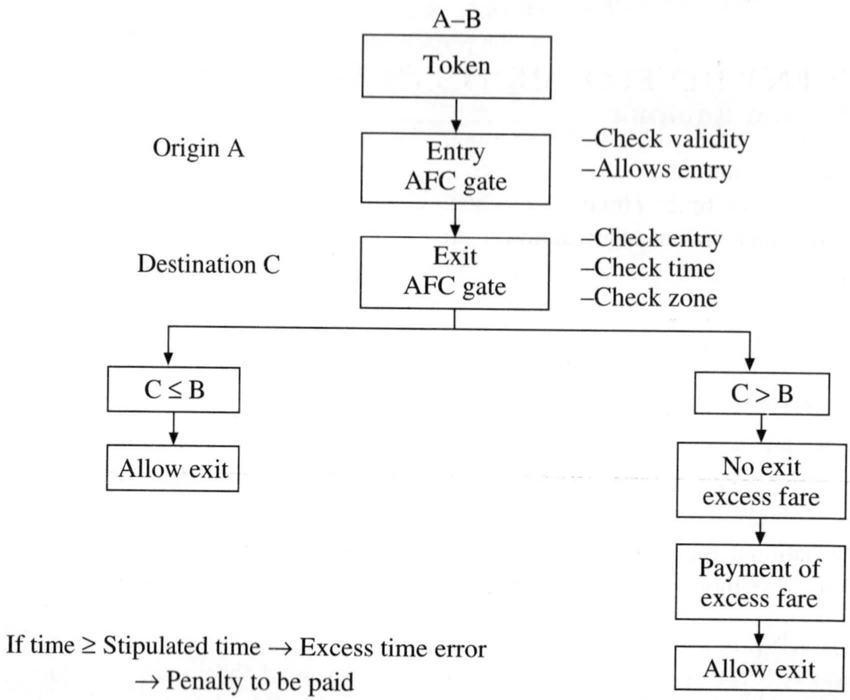

FIGURE 16.5 Typical flow chart for travelling from station A to B using CST.

The flow chart for travelling in DMRC using smart card (CSC) is presented in Figure 16.6.

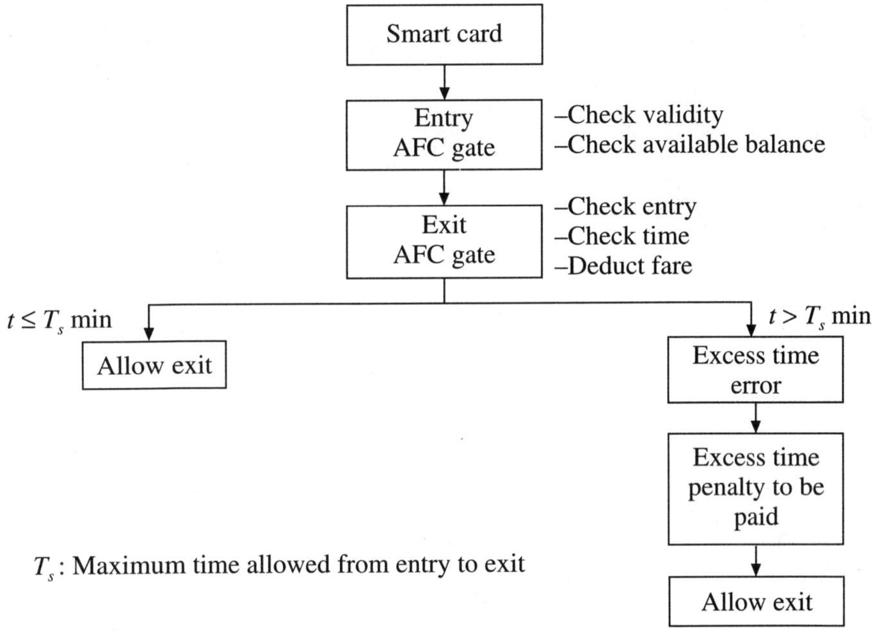

FIGURE 16.6 Typical flow chart for travelling using CSC.

16.5 RECENT DEVELOPMENTS IN AFC SYSTEM

The AFC system is capable of generating lots of data about the travel pattern of the passengers on real-time basis. This data can be used to understand the demand pattern across the lines at different moment of time. The data can be used for various purposes like how many trains to be run during peak and non-peak hours, how many AFC gates are required at entry/exit points, facilities required at the station keeping in mind peak load, etc. The modern AFC system allows the following new features which can be very effectively used to manage passenger flow in the system:

Differential fares

The demand in peak hours remain substantially higher than that in non-peak hours. One of the way to ease the demand in peak hours is to introduce differential fares, i.e., higher fares in peak hours and lower fares on non-peak hour. The differential fare system has been deployed by many international metro systems like London Underground, Washington Metro, Sydney, Melbourne, etc. The modern AFC system allows charging of fares as per the time of the day.

Demand estimation

To offer an optimum level of capacity by train services at any time, it is important to know the hourly demand throughout the day. The AFC system captures the data of entry and exit at the station. The station entry exit data is used to assess the hourly demand (passengers/hour) along any section of the network. This demand data is used to determine the optimum number of trains to be run during different periods of the day—more trains during peak hours, less trains during non-peak hours. The modern AFC systems are capable of generating the hourly demand data along the network of the system. The AFC system can also generate origin destination (OD) data of the passengers which is very useful to understand the pattern of flow of the passengers using the system.

Real-time monitoring of demand

As AFC system captures the entry–exit data on real-time basis, the level of demand on different sections/stations can be assessed on real-time basis (Figure 16.7). The real-time information about the congestion can be used for inducting extra trains in rush hours.

FIGURE 16.7 Station entry/exit data on real-time basis, OCC, MTR Hong Kong (*Source:* Author).

[The entry/exit of major station on real-time basis (refreshed after every 15 min) is displayed in the operation control centre (OCC) of MTR, Hong Kong to assess the level of congestion.]

Delhi Metro assesses the demand on daily basis on the basis of demand by 11:00 AM and decides number of trains to be inducted/withdrawn during evening peak hours. If demand is high in morning peak as compared to average of last 4 weeks, additional trains are inducted in evening peak hours.

Integrated ticketing

Integration of ticketing entails adoption of one fare media for multiple modes of transport across the city or country. Many countries have adopted integrated electronic cards which can be used across all public modes of transport in the city. In most cases, integrated ticketing is made possible by electronic ticketing technologies such as magnetic stripe cards or smart cards. Oyster of London Underground, Octopus of Hong Kong MRTS, ez-Link of Singapore, PASMO of Tokyo Metro are some of the notable example of integrated fare media. Some of the integrated fare cards can also be used for paying for goods and other services. The AFC system make it possible to use one card across multiple modes of transportation.

Concession and promotional schemes

The metro systems offer a number of concessions to children, students, senior citizens, family, etc. The AFC system has capability to offer a number of fare products for different customer segments. The fare is deducted from the fare media as per the applicable fare structure for that customer segment. The personalised travel card with name and photo of the users are being issued to prevent its misuse. The travel cards are tagged for the particular customer segment and the system identifies the card type from the tag and collects fares accordingly. The system is also able to generate audio-visual alarm at entry/exit AFC gates whenever a concessional product is being tapped so as to detect any misuse. If an adult is using concessional product (say child fare product), he/she can be checked at entry/exit point from the alarm system.

Mobile ticketing

Increasing use of mobile phones and innovations in payment mechanism through mobiles have ushered an era of m-ticketing (mobile ticketing). The m-ticketing uses the concept of near field communication (NFC) for electronic ticketing in railway system. NFC is a short-range wireless technology for two-way interactions between electronic devices. NFC technology enables contactless digital transactions between NFC enabled devices when these devices are brought closer within a stipulated range (around 20 cm). For train ticketing system, the AFC system can be equipped with NFC enabled devices at the entry and exit gates. The NFC equipped mobile phone can work as the travel card which will have preloaded money (m-purse). Similar to travel card, when mobile phone is tapped at entry and exit gates, the fare as per prevailing fare structure is deducted from the m-purse (Figure 16.8).

FIGURE 16.8 Mobile ticketing (*Source:* R.K. Dagar, DGM, DMRC taken during visit of Tokyo Metro).

Biometric ticketing

The next generation ticketing system is based on biometric recognition of the user. Everybody has unique biometric features such as finger prints, iris, DNA, face, facial infrared thermogram, palm prints, voice, etc. The biometrics can be used to identify the person by the system. The finger print is the most widely used biometric system as on date. The biometric identification can replace the conventional RFID based travel card as a ticketing solution. In biometric ticketing system, the user will have to register his finger prints (on any other biometric feature) with the metro operator and open an account. He may load his account from online time to time or link his account with his bank for auto-top up. The information about all the registered users is maintained in a centralised database which can be accessed from any of the station at any time. At the time of travel, he will have to scan his fingers at entry AFC gates, the system will scan his finger prints and identify that passenger from the centralised database, if the finger print matches with the pre-stored finger print, the entry gate will open and the information about the starting station of the passenger is tagged in the centralised database. At the exit station, the passenger again scans his fingers at the exit AFC gate, the scanner cross checks the finger prints from the database and if it matches, the fare as per the applicable fare from the origin station to destination station is deducted from the account of the passenger.

Evolution of ticketing system is shown in Figure 16.9.

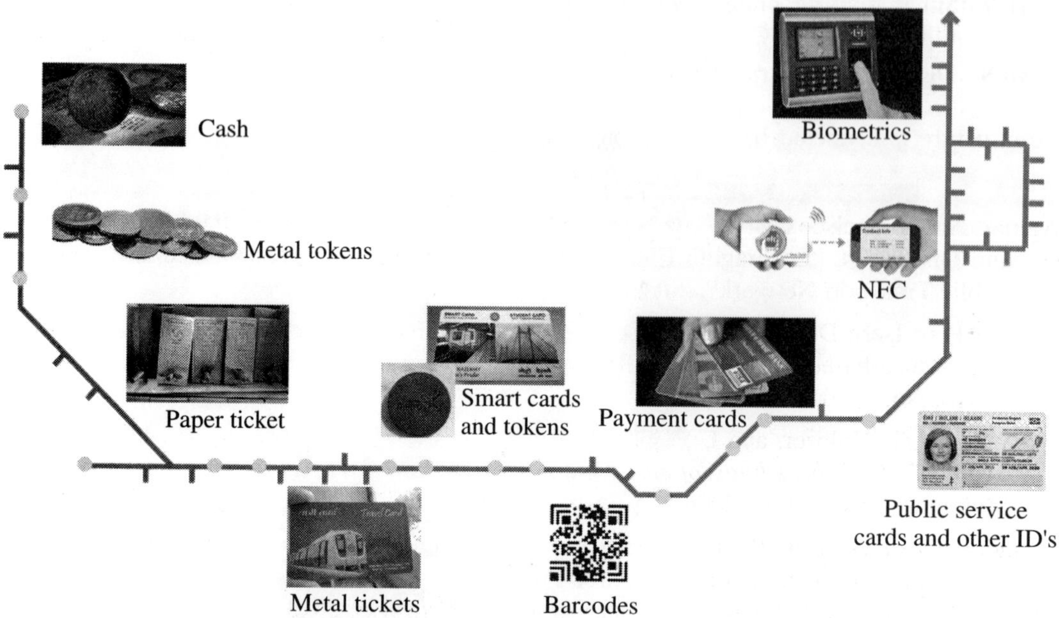

FIGURE 16.9 Evolution of ticketing system.

16.6 SUMMARY

The jounrney from paper ticketing to biometric ticketing has been made possible due to fast changes in technology. The adoption of ICT has not only made ticketing very convenient for the users but also improved the fare collection for the operators. The over all cost of fare collection for the operator has come down significantly as the need for manual checking of tickets has been almost done away with. The users are also benefitted by the convenience of the new system as they do not have to worry about standing in queues for buying the ticket, charging travel card, paying excess fare for travelling beyond destination, etc. In recent past, the pace of change of ticketing system in railways has been overwhelming thanks to fast convergence of IT, communication and biotechnology.

Questions

16.1 What do you understand by automatic fare collection (AFC) System? What are the functions of AFC system?

16.2 What are the benefits of AFC system for a mass transit system?

16.3 Describe typical system architecture on an AFC system.

16.4 Which are different fare media used in Delhi Metro AFC system?

16.5 How integrated multi-modal ticketing system can be developed with the help of AFC?

16.6 How AFC system can be used for travel demand management?

16.7 What do you understand by mobile ticketing? What are the benefits of mobile ticketing system?

16.8 Describe biometric ticketing system.

Bibliography

Aguiar, Ana, Francisco Maria Cruz Nunes, Manuel Joao Fernandes Silva, Paula Alexandra Silva, and Dirk Elias, "Leveraging Electronic Ticketing to Provide Personalised Navigation in a Public Transport Network", 2012.

Andreu Pere Isern-Deya, et al., "A Secure Automatic Fare Collection System for Time-based or Distance-based Services with Revocable Anonymity for Users", *The Computer Journal*, Vol. 0, No. 0, 2011.

Butty'an, L., T. Holczer, and I. Vajda, "Providing location privacy in automated fare collection systems", *In Proceedings of the 15th IST Mobile and Wireless Communication Summit*, Mykonos, Greece, June 2006.

Chandra, Deka Ganesh, Ravi Prakash, and Swati Lamdharia, "Mobile Ticketing System for Automatic Fare Collection Model for Public Transport", *5th International Conference on Computational Intelligence and Communication Networks,* 2013.

DMRC AFC system design documents.

Fleishman, Dan, "Transit Fare Policy, Structure and Technology", MIT—Transit Management Course, March 16, 2010. (http://ocw.mit.edu/courses/civil-and-environmental-engineering/1-258j-public-transportation-systems-spring-2010/lecture-notes/MIT1_258JS10_lec12.pdf)

Foth, M. and R. Schroeter, "Enhancing the Experience of Public Transport Users with Urban Screens and Mobile Applications", *14th Int'l Academic Mind Trek Conf.: Envisioning Future Media Environments (Mind Trek 10), ACM*, pp. 33–40, 2010.

Mala, S. Archana and N. Leela, "Automated Fare Collection System for Public Transport Using GPS", *International Journal of Innovative Research in Science, Engineering and Technology*, Vol. 3, Special Issue 1, January 2014.

Takahashi, Masami, Satoshi Katagata, Jun Ando, Shin-Ya Shiroto, and Masahiro Saito, "Automatic Fare Collection System with Contactless IC (Smart) Card", East Japan Railway Company, Yoyogi Shibuya-ku, Tokyo, Japan, 2001. (http://www.uic.org/cdrom/2001/wcrr2001/pdf/sessions/4_5/238.pdf)

TCRP program report 94, "Fare Policies, Structures and Technologies: Update", Sponsored by the Federal Transit Administration, US. (http://onlinepubs.trb.org/onlinepubs/tcrp/tcrp_rpt_94.pdf)

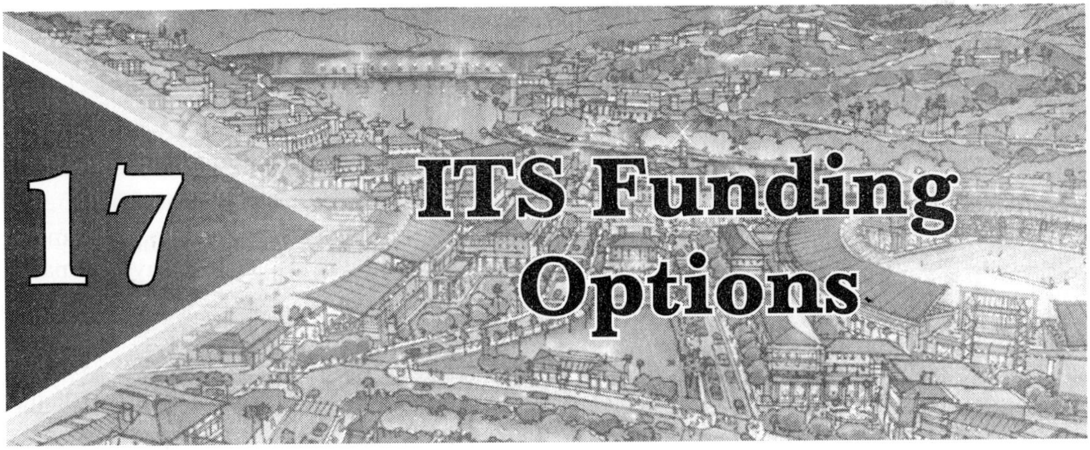

17 ITS Funding Options

17.1 INTRODUCTION

The need and importance of ITS for smooth, effective and efficient transportation services cannot be overemphasised. However, ITS solutions are capital intensive requiring substantial investment. Although the importance of ITS is widely appreciated world over, the limited finances are the biggest challenge being faced not only by developing countries but also by the developed ones. Unlike hard core infrastructure like roads, rails, stops, and ITS solutions are considered as add-on services and become the first casualty in case of shortage of funds. Thus, it is important to understand the various funding options which can be exercised to make ITS a reality. The key objective of selecting a funding option should be the part of revenue generated from the system may go for running of system and part may go in clearing the borrowings. The following options individually or in combination are mostly used to fund in ITS projects:

(a) Financial support from state (Government)

(b) Private investment from corporates

(c) Public-private partnership

(d) User fee

(e) Other measures

  (i) Cross subsidy

  (ii) Tax rebate on import

  (iii) Mandatory use of ITS equipment like CCTV, GPS for security reasons

17.2 FINANCIAL SUPPORT FROM STATE

The state government generates fund from its own resources and funds the ITS project. The state can generate funds from following sources:

(a) Allocation of funds from budget
(b) Funds through public institutes like national banks, public insurance companies, etc.
(c) Funds through international financial institutes by providing sovereign guarantee
(d) Special taxes/cess

Allocation of funds from budget

The government makes provision of the fund in the annual/periodical budget of the country/state. Due to limited financial resources available with government especially with developing countries, the budgetary support is only made available to the projects of national interest. The project related to national security, defense are the largest beneficiary of this option.

Funds through public institutes like national banks and public insurance companies

Government can also provide funds through public institutes like national banks and public insurance companies on long-term loan basis. The financial institutes can arrange funds from their own sources or by collecting funds from the open market. The following tools may be used by the financial institutes to collect funds from open market:

(a) ***Equity finance/share capital:*** The institute may float shares in open market which can be bought by the investors. The investor gets return in the form of dividend and increase in price of the shares. The value of shares keep fluctuating so this is considered to be the most risky investment for the investors.

(b) ***Debentures (debt finance):*** The institute issues debentures of fixed value for a fixed term. At the end of the term, the investors receive the principal amount along with interest so fixed return is assured. From investor point of view the debentures are less risky than shares.

(c) ***Long-term tax free bonds:*** The institute issues long-term (5–10–15 years) bonds of fixed value and for fixed term. The investor can buy bonds and after the maturity period, the investor receives the fixed amount (principal plus interest). This method also allows the investor to claim the income tax rebate in that financial year in which the bonds are purchased.

Funds through international financial institutes by providing sovereign guarantee

The government supports the project executing authority in securing funds from international financial intuitions as long-term loan. Often government also provides sovereign guarantee as a security to international financial institutes against any risk of default by the project executing authority. The major international financial institutes which have been helping infrastructure projects in India are:

(a) World Bank

(b) Asian Development Bank

(c) International institutions like Japan International Cooperation Agency (JICA)

The international financing institutes have their own criteria of selecting projects for funding. Usually internal rate of return (IRR) and economic rate return (ERR) are the guiding factors while deciding whether a project qualifies for funding. The IRR takes into account the financial benefits of the project whereas the ERR also considers economic (environmental and social) benefits of the project. Generally IRR is more than the commercial rate of borrowing and the ERR is positive for a project to be eligible for funding. Some projects for which ERR is very high due to socio-economic benefits may be considered for funding even if IRR is relatively low.

Special taxes/cess

Government may levy taxes or cess to generate funds from taxpayers. The Government of India introduced a cess on both petrol and diesel. This amount at time (at 1999 prices) came to a total of approximately ₹2000 crores per annum. Further, Parliament decreed that the fund so collected were to be put aside in a central road fund (CRF) for exclusive utilisation for the development of a modern road network. Today, the cess contributes between ₹5 and 6 thousand crores per annum towards National Highway Development Project (NHDP).

The government also impose taxes on production of vehicles, sale and registration of vehicles to generate funds.

Surcharge on sale tax/vat/service tax/entrainment tax/development charges/vehicle registration and transfer charges; number of times government imposes surcharge on different types of revenue generation sources in order to raise fund.

17.3 PRIVATE INVESTMENT FROM CORPORATES

If the project is financially lucrative, the private companies may be interested in funding the ITS projects. Needless to mention that the IRR of the project must be reasonably more than rate of commercial borrowing to attract the private investors. The project needs to be designed in such a way that a consistent stream of revenue is generated from the project to financially justify it. The private investor may generate funds from tools like shares, debentures, borrowing/loans, etc. from the market or financial institutes. For example, there are many privately operated mobile applications providing information about railways, buses, metro trains, etc. for the benefit of users. The private party generates revenue from advertising on mobile application. Similarly many Websites provide important information about the transport systems in the city freely and generate revenue from advertising, etc. The ITS applications developed by private radio taxi operators for the convenience of the users, to gain competitive advantage on other operators and security reasons are another example of private investment in ITS solutions. OLA cab application is explained in Box 17.1.

Box 17.1 OLA cab application, India

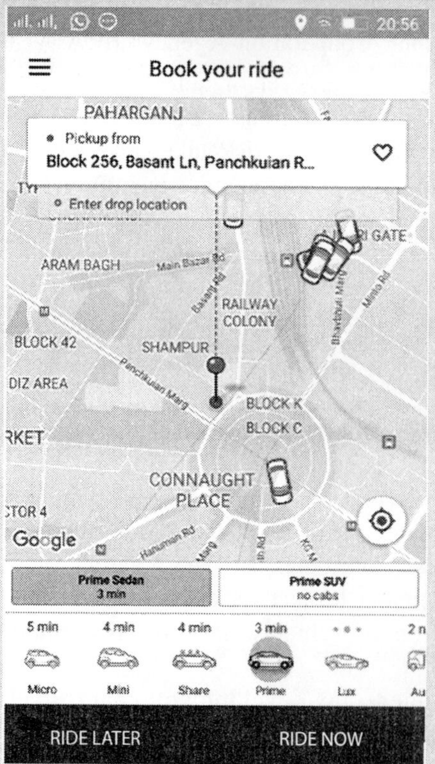

FIGURE 17.1 Mobile app of OLA.

The use of OLA Cab Application (Figure 17.1) for booking a cab is very simple as explained below:

(a) Set your pickup location (e.g.: from airport, railway station, office, home or your current location).

(b) Pick the cab you want and see all available cabs close by.

(c) Tap "ride now" to book your ride.

(d) Get instant confirmation with driver details

(e) See your ride en-route on the map.

(f) Pay only after completion of your ride or use "OLA Money" to ride cash-free!

(g) You want to schedule a pickup for some time later to catch a flight or train. Just use "ride later" and plan ahead. You can also invite friends to use awesome service and earn free "OLA Money".

Source: http://www.9apps.com/android-apps/Ola-cabs-Book-taxi-in-India-1/

17.4 PUBLIC PRIVATE PARTNERSHIP (PPP)

The PPP mode of funding ITS projects envisages investment by private partner as authorised by public authority in which the private party provides a public service or project and assumes substantial financial, technical and operational risk in the project. The private party gets return from the investment in the form of annuity or subsidy from the government, user fee, associated commercial developments, etc. The private investor would only be interested in investing its money if the project has high IRR taking into account high level of risk for private investor involved in PPP projects. The roles and responsibilities of each partner are defined in the agreement usually called *concession agreement*. The strength of PPP is derived from leveraging respective strengths of the partners to create synergy in order to attain common goals. In this model, generally the government provides the licence, facilities and concession to company and company is investing its resources and money in the project. Broadly PPPs can be categorised into two types: a PPP of a purely contractual nature and a PPP of an institutional nature. Typical structure of a PPP project is shown in Figure 17.2.

Different kinds of contractual PPPs exists and innovations emerge continuously as each PPP contract responds to very precise needs. Some of the most frequent types are build, operate and transfer (BOT); that is, the private partner builds and operates the infrastructure, transferring it to the public partner at the end of the contract. Build, own, operate, and transfer (BOOT) is the organisational form when infrastructure ownership is also private during the contract term; DBOT or DBOOT would be the acronyms if arrangements further include the responsibility for the design of the infrastructure project as well.

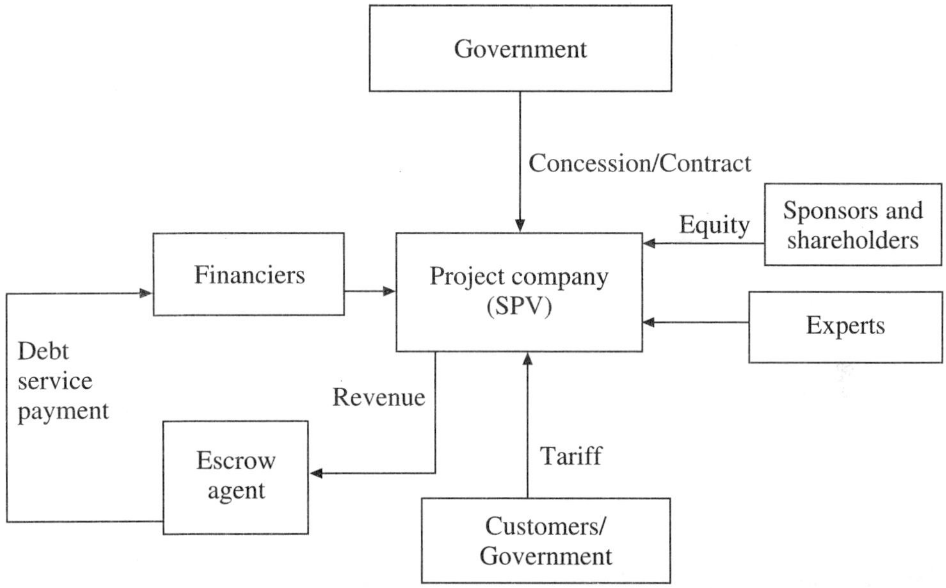

FIGURE 17.2 Typical structure of a PPP project.

The ITS projects involve innovative use of technology to meet the project objectives. An efficient, technically strong and innovative project management is fundamental to the success of ITS projects. PPP mode of funding where the private party is responsible for designing, developing, procuring and delivering the project may be the best suited option for ITS projects which have potential to generate steady stream of revenue. For example, electronic road pricing (ERP) based congestion charging can be implemented through a private party on PPP model as the system is capable of generating assured cash during the operational phase.

17.5 USER FEE TO FUND ITS PROJECTS

The ITS projects can also be funded through the user fee which is levied on the use of the transport services. Some of the possible user fees which can be used for funding are:

Congestion charges

This is other source for fund raising. In this method, the charges are collected from vehicle owners whenever the vehicle is entering in congested area (normally it is CBD). This charge helps in preventing the vehicular traffic in busy area. It also helps in reducing the vehicular pollution in that area. Money generated from this source can be used for road projects.

Special parking charges

This method is used for collecting parking fee from vehicles that are parked in public places (even in front of own house). So the vehicle owner should have own garage to park the vehicle otherwise the parking fee is levied for using public place. This method can generate a huge amount on regular basis.

Licence Fee to use the system

The user may be charged a monthly or annual licence fee to use services such as maps for navigation and other value added services.

Passenger fare

The cost of the ITS application may be built into the passenger fares. For example, Indian Railway Catering and Tourism Corporation (IRCTC) used to collect service charges for online booking of the railway tickets.

17.6 OTHER MEASURES TO PROMOTE ITS SYSTEM

The following measures may also be suitably used to encourage development of ITS systems:

Cross subsidy

The developer of the ITS project may be allowed to generate revenue from other part of the project. For example, a parking operator may be granted advertising rights in its premises in lieu of development of ITS application such as installation of parking management system (PMS), mobile application for users, smart card based payment system, automated parking solutions etc.

Tax rebate on import

To encourage ITS projects, government may offer tax rebates/tax free import of hardware and software required. The companies involved in ITS development may be offered tax holidays say for 5/10 years to encourage them to develop technologies indigenously at an affordable cost.

Mandatory use of ITS equipment like CCTV, GPS for security reasons

Government may make it mandatory for the transporters to use ITS technology such as CCTVs, GPS, etc. for security and safety reasons.

17.7 SUMMARY

Funding is the major challenge in implementation of the ITS technologies especially in the low income group of countries. In long run, the life-cycle cost of the ITS technologies is generally less than the benefits to be accrued, however, the capital investment at procurement stage becomes difficult. Innovative financing options such as public private partnership (PPP), revenue from advertising, affordable user fee, etc. may be deployed to promote ITS without burdening the budget of the government.

Questions

17.1 What are different funding options available for procurement of ITS in a city?

17.2 What are different sources from which state can generate funds for ITS projects?

17.3 Which tools may be used by the financial institutes to collect funds from open market to fund ITS projects?

17.4 Which are the major international financial institutes which have been helping infrastructure projects in India?

17.5 How public private partnership can be leveraged to fund ITS projects? Explain with example(s).

17.6 What are the possible user fees which can be used for funding ITS projects?

Bibliography

A Guidebook on Public-Private Partnership in Infrastructure, UNESCAP Bangkok, January 2011.

Deakin, E. and M. Shirgaokar, "A Compendium of Summaries of Intelligent Transportation Technologies", University of California Transportation Centre, Berkeley, 2003. (Report available over the Website at http://www.uctc.net).

Deakin, E., S. Cheon, J. Ni, A. Leigland, S. Park, and M. Shirgoakar, "Mainstreaming Intelligent Transportation Systems", final report to the California Dept. of Transportation, Institute or Transportation Studies, University of California, Berkeley, 2002. (Report available over the Website at http://www.uctc.net).

Harris, Richard, "Study regarding guidelines for public funding of Intelligent Transport Systems", TREN/G4/FV-2008/475/02, May 2011.

Henk, Russell H., Edward J. Seymour, Robert Harrison, David M. Luskin, and Lisa Loftus-Otway, "An Examination Of Alternative Funding Solutions for ITS Deployment", Technical Report: September 2002–August 2004, 2004.

"Intelligent Transport Systems: EU-funded research for efficient, clean and safe road transport", Luxembourg: Publications Office of the European Union, 2010. (http://www.gppq.fct.pt/h2020/_docs/brochuras/transportes/Intelligent%20transport%20systems.pdf)

"Intelligent Transportation Systems Benefits, Costs, Deployment and Lessons Learned", U.S. Department of Transportation, Research and Innovative Technology Administration, 2008. (http://ntl.bts.gov/lib/30000/30400/30466/14412.pdf)

ITS Toolkit for Traffic Management System, Institute of Urban Transport (IUT), December 2013.

"Study Regarding Guidelines for Public Funding of Intelligent Transport Systems", TREN/G4/FV-2008/475/02, Final Report, Issue 03, May 25, 2011. (http://ec.europa.eu/transport/themes/its/studies/doc/2011_05-its-public-funding-guidelines.pdf)

"The Path Forward: Funding and Financing Our Surface Transportation System", Interim Report of the National Surface Transportation Infrastructure Financing Commission, February 2008.

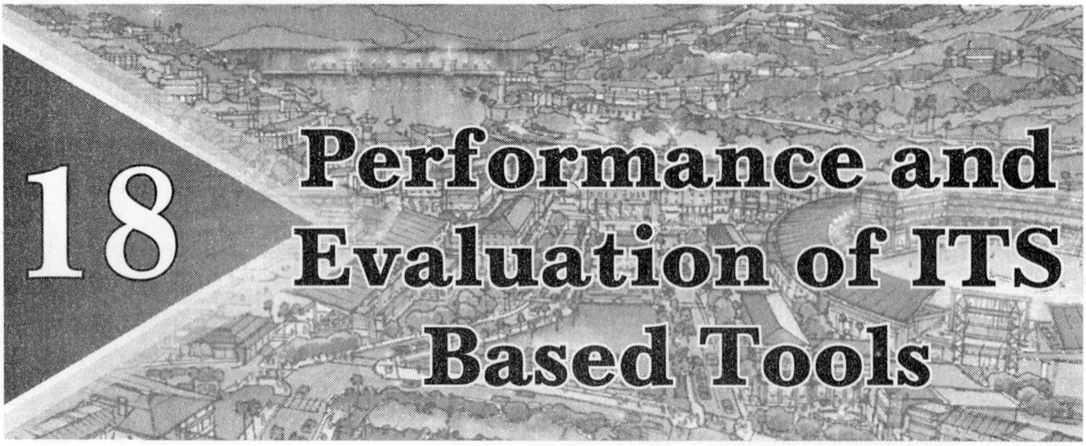

18 Performance and Evaluation of ITS Based Tools

18.1 INTRODUCTION

The ITS technologies are changing at lightning speed. The things that we have not even imagined ten year back are reality today. This aspect makes the whole job very challenging for user in describing the details of each and every thing about the whole system, subsystem, equipment, etc. of its own. Further depending on scope of ITS applications, one requires to define communication system, control room details, IT equipment, cabling, storage system, integration, etc. that are not so easy to freeze.

Thus if ITS based solutions are to be implemented by the city then it is essential to define scope of work, functional specifications, performance parameter and evaluation parameter as clearly as possible. Once it is defined then it is highly recommended to have expression of interest describing scope of work in details, functional specifications and performance parameters. On that basis, the supplier should describe the complete solution showing the fully integrated system, with technical specification and detailed working of that system. Now the user can compare his/her conceived integrated system design and technical details and on that basis the user can finalise the project.

It is important that the user must ask the supplier to show/prove the actual functionalities before releasing the payment to supplier. In that situation, the supplier must show the testing mechanism to prove the supplier's claim about its system.

Thus after functional specification, the next step is to define the performance indicators and evaluation of ITS tools, which are discussed in this chapter along with what user should test from his side is also mentioned. Basically the performance indicators are used or required to compare the product/system/solution provided by different suppliers and on the basis of which one can choose which solution is better for that problem.

The performance and evaluation (P&E) criteria cannot be a standard system applicable to all the ITS tools as P&E criteria depend on the functionality of the tool. The P&E criteria for some of the major ITS tools have been presented here for illustration purpose. The readers are advised to use this chapter as a guide and develop detailed P&E criteria for ITS tools as per their own requirements.

18.2 E-CHALLANING SYSTEM

Performance indicators

The performance indicators are related with the output (or capability of meeting the objective or quantum of output) of system. For e-challaning system, the important performance indicators are average time taken in issuing challans, number of challans issued per officer per day or shift, average time taken in validating the records (driving licence, registration certificate, past crime history, stolen vehicle, pending court case, etc.), number of times the officer caught the repeated offenders, average time taken in receiving e-payment, average time taken in charging the battery of handheld device instrument, average time of working of handheld instrument (battery life) once fully charged. The possibility of customising reports on different platform and transportability of data and information from one source to another and data storage capacity (meta data) in handheld instrument in case of lost connectivity or no connectivity (off-line) are also important parameters for assessing the performance of the system.

Testing—Handheld device

After fixing the performance parameters, the next step is to evaluate the system for these parameters. The testing the process of evaluation of the system for given performance parameters.

The handheld device for e-challan needs to be tested for a number of parameters such as:

- Data management
  - ➢ quantum of data lost when battery shut down due to any reason or connectivity lost or connectivity is very slow or it is not continuous,
  - ➢ login security features and retrieval of password through administrator,
  - ➢ accuracy in maintaining the unique challan number even when some handheld instruments are working off-line and some are working online,
  - ➢ accuracy in automatic recording of location, date, time and other mandatory details at each location of project area
- Speed
  - ➢ average time taken in transferring the details of offenders/challans from field to control room,
- Connectivity
  - ➢ connectivity of instrument with control room at each location of project area,
  - ➢ automatic or manual transfer of data to control room when connectivity restored,
- Storage of data
  - ➢ Capacity of storage of data

- data storage capacity of handheld instrument in case of lost connectivity or no connectivity (off-line),

- Hardware.
    - Battery backup
    - average alert time signal given by instrument for low battery
    - acceptable weather conditions for smooth functioning of the hardware (temperature, humidity etc.)

Further, once the record is locked, then tempering of records should not be possible. A normal eye-sighed person should be able to read and operate the handheld instrument in sunlight or in low lighting conditions. For the printing of challan, printed text on challan paper (in number of days/months/years) should be as per specified requirements. If printer is not attached then maximum range of communication between handheld instrument and printer is to be made available as per specification. If thermal printer is not used then number of characters/ photos (if required) can be printed on paper need to be tested. In addition, the visibility of keypad of handheld device in low light conditions, actual time taken by the instrument for warm up and instrument's validation, resolution of photo taken by handheld instrument as per specification and quantum of data retrieved in case of handheld instrument fall down and broken should also be checked for e-challan system.

Testing—Server

The server of the e-challan system needs to be tested for data transfer/mirroring/back up capability in case of server crash, data security features and data retrieval feature and mechanism, data warehousing capability, response time of various function, reports generated as per specification, data/information transportability on other platform and fault management system and self-diagnostic system.

Evaluation of system

Anybody without using this system can claim that it will give much better results in comparison to manual method.

The best method of evaluating this system is considered in terms of man-hour saved.

The user should calculate the man-hour spent or manpower involved (in completing all stages of challaning) per challan or collecting per ₹100 from compounding fee manual method.

Then the same procedure should be adopted (after all teething problems solved and system stabilised as per specifications) for this new method.

Find out how many times repeated offenders were caught for similar offences or different offences, at same or different locations. Further number of traffic acts used in challaning also gives the versatility of challaning procedure.

The net saving will decide the value of system. If there is no improvement then there is definite some serious problem in configuration or in software or communication or in database or in specifications of equipment. So in that case it needs to check the whole system by the team of multi-disciplinary experts.

18.3 AUTOMATIC SPEED DETECTION SYSTEM (ASDS)

Performance indicators

The performance of automatic speed detection system should be assessed in day/night/bad weather/foggy days (if desired) on the basis of average time taken in measuring the speed with given threshold values, facility/capability of changing the threshold values—manually or remotely fixing different ranges for different corridors, average time taken in recognising the number plate (standard type), revenue collection, reduction in accidents, mean detection time for noticing and processing the alerts for overspeeding vehicles, number/percentage of false alarms raised, actual distance of coverage/capturing range of rule violating vehicle, average time taken in transmitting data by single or by all camera (of area) from field to control room, the storage capacity of data, loss of data in off-line mode, facility of automatic data push as soon as communication restored, number of still frames and size/duration in time of video recording captured for single case, and power back up.

Testing—Field equipment

The equipment needs to be tested to ensure its effective functioning as per the performance indicators. The ASDS should be tested in day/night/bad weather/foggy days (if desired) for accuracy in measuring the speed of vehicle with respect to different distances in defined range zone, accuracy in reading/recognising the standard number plate, accuracy and average time taken in transferring the data (complete meta data—photo and text) from field to control room from different locations of project area, accuracy and time taken when all instruments transmit the data (complete meta data—photo and text) from all location to control room at a time, actual time of power back up for instruments working in field, quantum of data lost at power failure due to any reason or connectivity lost or connectivity is very slow or it is not continuous, connectivity of instruments with control room at each location of project area, actual data storage capacity in field unit in case of lost connectivity or no connectivity (off-line), automatic or manual transfer of data to control room when connectivity restored, accuracy in maintaining the unique ID number for each case even some units are working off-line and some are working online, accuracy in automatic recording of location, date, time and other mandatory details at each location of project area, resolution of photo taken by instrument as per specification, etc.

Testing—Server

The server of the ASDS should be tested for data transfer/mirroring/back up capability in case of server crash, data security features (in house/outside threat, hacking, virus, etc.) and data retrieval feature and mechanism, data warehousing capability, response time of various function, reports generation as per specification, data/information transportability on other platform, fault management system and self-diagnostic system.

Evaluation

The system is to be evaluated by comparing the results of performance indicators measured in pre- and post scenario. This evaluation is very useful in selecting or finding suitability of a particular tool for almost same conditions. Once the ITS tools are selected for particular city and for particular purpose, the architecture is to be defined.

But still

This system works 24×7 in all-weather condition while in practical police force cannot monitor/surveillance the whole area for 24×7 all-weather time. This system is supposed to capture most of the all types of violations (as per the scope of project and capability or configuration of cameras) in 24×7 environments (although some companies claim for 100% accuracy, but in heterogeneous traffic flow as in India, it is not possible). To achieve this in manual system, one requires a large force of enforcement officers, which is not practical also. The 24×7 monitoring manually in a limited area is possible (that is also for special occasions) but as duration and area increases, the chances of pilferages also increase. Further, tempering with data in this system is almost impossible (if computer operator/staff of control room or any hacker (from outside) play mischief, then nobody can help it. Although system can be designed such that any black sheep may not able to do much damage to the system). In manual system, nothing can be guaranteed. With ASDS, it is very easy to identify the regular offenders while in manual system it is very-very difficult. This system provides visual evidence which can be placed in court also while in manual system it is not easy to have independent evidence of violation. Thus the system at the most may also be evaluated on the basis of most important aspect, i.e., number of accidents are reduced, pollution level decreases, etc.

In beginning the number of challans will increase but after a period when drivers develop the sense of responsibility then the number of violation and challans will decline. In that case decrease in the number of fatalities are sufficient to justify the system.

The P&E criteria for mobile surveillance/enforcement system and red light violation detection system can be developed in same way as shown in above systems.

18.4 CCTV SYSTEM

Performance indicator, testing and evaluation for CCTV system can also be drawn in same way as shown in above systems depending upon the purpose for which CCTV system is being used. Some broad parameters for performance and evaluation of the CCTV system have been suggested here. However, the parameters may change according to the intended purpose of the system.

Some more performance indicators

Some more performance indicators applicable to CCTV system are average detection time for detecting the incidence, average detection time of detecting the wrong-way vehicles (or no-entry detection), average detection time in detecting the vehicle crossing the median line, average detection time of detecting the stopped vehicle, parked vehicle, debris, gravel, ash, etc., mean detection time of detecting the vehicle moving for banned turn (i.e., u-tern or right turn, etc.), overall detection rate, percentage of false alarm/alerts, average time taken to transmit the data from single camera and simultaneously by all cameras to control room's server (overall communication network speed), time taken in switching to standby system in case of total crash, and power back up for field equipment.

Some more points for testing

In addition to testing parameters discussed for ASDS, the CCTV system should also be tested for accuracy in measuring the congestion level (in form of level of service), accuracy in measuring the journey time, accuracy in recording from each camera, its coverage distance, covering of all directions as per prescribed angles (by putting some fixed target in field), accuracy in vehicle identification, accuracy in detecting different type of violations, correctness in registering the number plate (standard type) of violating vehicles, etc.

18.5 AREA TRAFFIC CONTROL (ATC)

Performance indicator

It is not easy to judge the performance of this complex system because its efficiency depends on number of feature like level of heterogeneity in traffic (like large number of two-wheeled vehicles—cycles, two-wheeler, motor cycle, and three-wheelers—tanga, cycle rickshaw, etc.). Due to this, it is very difficult to calibrate the system with actual traffic and then to judge the performance. Still, the following performance indicators are generally considered for the evaluation:

- Average waiting time at traffic signals
- Reduction in signal timings
- Average queue length
- Average journey time
- On the average number of green lights (out of total light points) got by priority vehicle (like emergency vehicles, buses)
- Saving in energy (fuel saving)
- Saving in VOT and VOC

Testing

Similarly, for the reasons discussed above, the testing of this system is not so easy job. The accuracy of test results are heavily dependent on traffic flow composition, traffic density, etc. So at the most one can test the system for accuracy in measuring the queue length, traffic count, detecting the vehicles at junctions or in the corridor, providing green light to priority vehicles, and synchronisation of traffic signals. The system should also be checked for number of times false estimation done in a working day, numbers of time the signal are not remotely controlled in a month, in case of connectivity or power failure the number of times the signals fail in coming back to pre-planned cycle time or all red/amber light system and its restoration the signal cycle starts from as planned cycle time, number of signal phases supported by system as claimed, self-testing of controller and system and accuracy in reporting the fault, accuracy of error log records and number of times the green-green conflict reported accurately, etc.

Evaluation of system

The ATC system may be evaluated for average reduction in stop time and delay at signals, average saving in fuel/reduction in pollution, average saving in value of time, average increase in journey time of vehicles, and creditable increase in journey time of priority vehicles.

On these bases, one can, at least, calibrate the system and configured its various components so that the system shows the positive results.

18.6 VARIABLE MESSAGE SIGNAGE (VMS)

Performance indicator

The performance indicators may differ from one country to other. Some of the indicators/ specifications to judge the system as per Wisconsin Department of Transport's guidelines, and Federal Transport Authority (FTA) guidelines are reduction in congestion, number of drivers takes second best option on the basis of information seen, reduction in challans, accidents and incidences, improvement in journey time, pollution level and availability of number of pre-defined stored messages. The system shall be capable for:

- Working in flash on or off mode, capability of writing message at the rate of 60 characters per second
- Incorporating a watchdog timer to detect an out-of-program condition and reset the microprocessor
- Using fail-safe prevention system to prevent the improper information display in case of malfunctioning
- Displaying the blanking feature in case of power failure or break in communication system
- Running the self-diagnostic routines to test the full sign operation
- Running in local and as well as central mode (control room)

The system should be able to be connected with laptop in order to upload/download messages locally. The system housing should be such that any rainy water should not be collected in it. The system should be capable of sensing the background ambient light level and provides a minimum of 7 field—adjustable intensities (dimming). The dimming system should contain 3 photoelectric sensors installed in watertight metal enclosure on the VMS housing. These sensors should be capable of being continually exposed to direct sunlight without impairment of performance. The sensors must be placed so that they view the front, rear and top of the sign. The system should be capable to determine its operating mode (local or central) and select the appropriate method of implementing the dimming. The system should have power backup as per specification and displayed message retained as per required duration. All pivots and bearing shall be non-corrodible and sealed such that dirt, moisture and rainy water may not degrade the strength of system. The complete housing (at least the screen should be water proof) should not stop functioning even in case of rain, winds (up to 160 kmph), storm, etc. The system should be capable of communicating the malfunction to control room or providing facility to monitor it from control room.

Testing

The following parameters of VMS system should be tested to assess the performance with respect to intended purpose:

- Display message
  - ➢ The actual colour, type, size and spacing of fonts and background if any deviation from as per specification than how much?
  - ➢ Is any distortion or gap between character or word or sentence is occurring
  - ➢ Actual time taken in refreshing the message and any deviation from specification
  - ➢ Confirmation of displayed message (activation of lanterns if attached) in control room. Numbers of times fail in confirmation.
  - ➢ Legibility and clarity of message (as per sight distance fixed) in night time or at low visibility time by driver of normal vision
  - ➢ Legibility and clarity of message (as per sight distance fixed), when sun is facing the font or sun is in the back side of font and when sun is at the top of board by driver of normal vision.
  - ➢ Quality/clarity (in multi colour also) of symbols, shapes, figures and video displayed in different sun conditions and in night by driver of normal vision
  - ➢ Display of bilingual and multi colour message and its clarity.
  - ➢ Is the VMS be sited along roadway with a grade less than the specified one (usually 4%.)
- Security
  - ➢ Check the password, and security features so that unauthorised person cannot change the message
  - ➢ Accuracy in reporting and recording the log of messages with time, date and place of VMS system placement.
  - ➢ System capacity
  - ➢ The maximum number of pre-set messages can be stored in this system and the accuracy of these messages in displaying in different situations.
- Power Mnagement
  - ➢ Automatic power backup and the minimum reported duration of working of signage (in case of portable VMS system the duration for vehicle lighting, additional lighting and PA system should also be taken). If there is any deviation, then how much, it would?
  - ➢ Even if power fails then actual time taken in shifting to standby message or blank and its accuracy in displaying the standby message is also to be seen. If power is restored, then how does it come back? Or in which mode or message? Time taken and scope of running the self diagnostic routine and communicating the report to control room. Accuracy in keeping log of failures.
  - ➢ Tolerance limits against voltage fluctuations. Automatic switch to local or safe mode in case of under or over voltage.

- Communication Failure
  - ➢ In case of communication failure, what would be time taken in shifting to local mode or on standby message mode or blank face. Accuracy of display of standby message. At the restoration of communication system than sending a diagnostic report or time taken in running of self diagnostic routine and sending the complete report to control room. Accuracy in keeping log of failures.
- Accuracy in automatic reporting to control room in case of failure in dimming or illumination property, photocell failure, lantern failure, heater or ventilation failure, pixel failure and message failure.
- Is the system's location as per norms (i.e. distance between incidence place and VMS system) with respect to incidence spot, static signage?
- Are the drivers getting sufficient time to manoeuvre their vehicle (can be tested using opinion survey). On the average how many times, it fails to take action.
- The distance between two VMS system is as per norms/contract, if not then why?
- Is the actual vertical and lateral clearance available as per contract?

Evaluation

A number of parameters can be used to evaluate the impact of any VMS system. Some parameters cannot be evaluated directly and for some parameters, one requires to do opinion survey to get the correct picture. The parameters may be the percentage of drivers takes second best action (changing the route, etc.) on the basis of information seen, the percentage of drivers acknowledges the usefulness of information given, improvement in the level of service of concerned corridor, reduction in number of accidents, challans, air and noise pollution, improvement in journey time and average speed.

On the basis of above performance indicators, the system can be evaluated. If results are not positive then the whole process should be checked from designing stage to implementation stage.

18.7 SUMMARY

The P&E of the ITS tools is essential to ensure best output from the ITS tool. The detailed P&E criteria for the ITS tools should be developed at the planning stage itself so that P&E can be made part of the tender and the prospective bidders are fully aware of P&E on which their system will be judged. It is also important to draft P&E very carefully so that desirable but impractical things are not become part of P&E. The P&E should preferably be based on proven systems which have already been used and tested elsewhere. Once system has been installed, it shall undergo P&E test on periodic basis.

Questions

18.1 Why performance and evaluation of ITS tools is important?

18.2 Describe P&E system for following ITS tools:

(a) e-challaning

(b) Automatic speed detection system
(c) CCTV system
(d) Automatic traffic control
(e) Variable message signage

18.3 Develop P&E criteria for automatic number plate recognition system.

Bibliography

Benson, B.G., "Motorist attitude about content of variable message signs", *Transportation Research Record*, Vol. 1550, pp. 48–57, 1996.

Chatterjee, K., N.B. Hounsell, P.E. Firmin, and P.W. Bonsall, "Driver response to Variable Message Sign Information in London", *Transportation Research Part C*, Vol. 10, No. 2, pp. 49–169, 2002.

Evaluation of Advanced Traffic Control Systems, NITC Project Brief, May 2015. (http://ppms.trec.pdx.edu/media/project_files/579_project_brief.pdf)

Intelligent Transport Systems (ITS), Performance Evaluation of Self-Organising TDMA as Medium Access Control Method Applied to ITS, Access Layer Part, Technical Report, ETSI TR 102 862 V1.1.1, 2011–2012. (http://www.etsi.org/deliver/etsi_tr/102800_102899/102862/01.01.01_60/tr_102862v010101p.pdf)

Interfacility Transport Guidelines, State of Wisconsin, Department of Health and Family Services, Division of Public Health, 2006. (https://www.dhs.wisconsin.gov/ems/wi-interfacilitytransportguidelines-2006.pdf)

Nghiem, A.T., F. Bremond, M. Thonnat, and V. Valentin, "ETISEO, performance evaluation for video surveillance systems", Project Orion, INRIA-Sophia Antipolis, France. (https://www-sop.inria.fr/orion/ETISEO/iso_album/AVSS07.pdf)

Payne, Stephen, "Key Performance Indicators for Intelligent Transport Systems", Final Report, 2015. (http://ec.europa.eu/transport/themes/its/studies/doc/its-kpi-final_report_v7_4.pdf)

Raechelle, Newman-Askins, Luis Ferreira, and Jonathan Bunker, "Intelligent Transport Systems Evaluation: From Theory to Practice", *Proceedings 21st ARRB and 11th REAAA Conference*, Cairns, 2003. (http://eprints.qut.edu.au/2392/1/2392.pdf)

Rao, Chinta Sudhakar, M. Parida, and S.S. Jain, "Performance Evaluation of the Intelligent Transport System in Delhi", *Institute of Town Planners, India Journal,* Vol. 8(3), pp. 1–14, July–September 2011.

Ryu, Seungki, Youngtae Jo, Yeohwan Yoon, Gwanho Choi, and Sangman Lee, "Evaluation Methods for Vehicle Speed Detection and Licence Plate Recognition Using Vision-Radar Sensor Systems", *Journal of Emerging Trends in Computing and Information Sciences*, Vol. 7, No. 2, 2016.

Sarkar, P.K. and Vinay Maitri, "ITS Toolkit for Traffic Management System", Institute of Urban Transport, 2012.

19 ITS for Smart Cities

19.1 INTRODUCTION

Although there is no universal definition of smart city, in general, smart city can be defined as a liveable and connected city that provides core infrastructure and offers a decent quality of life, a clean and sustainable environment and application of 'smart' solutions to meet the growing aspiration of its citizens. According to the British Standards Institute (BSI), smart city is "The effective integration of physical, digital and human systems in the built environment to deliver sustainable, prosperous and inclusive future for its citizens". The concept of smart city in itself is unending process. Today's cities are definitely smarter than what they were 100 years ago likewise tomorrow's cities are going to be smarter than today's. Making a city smart is an evolutionary and unending process. Today we are reaping the benefits of IT revolution which have become game changer in making our life smarter at a much faster pace. Smart phones, smart TVs, smart appliances, etc. are not only accessible but have become part of day-to-day life. When everything around is getting smarter even the citizens why not the city itself. This chapter is focused on ITS technologies for 'Smart Cities'. Although, most of the ITS technologies have already been discussed in previous chapters, this chapter presents ITS technologies in a structured fashion in the context of smart cities.

19.2 STRATEGIES AND APPROACHES TO SMART CITY

The various strategies and approaches to develop a city as smart city are shown in Figure 19.1, as spokes on a wheel, with intermediate goals on the inner ring and the resulting smart city at the core. The wheel shows that ICT is a factor contributing to a smart city, but not the defining enabler.

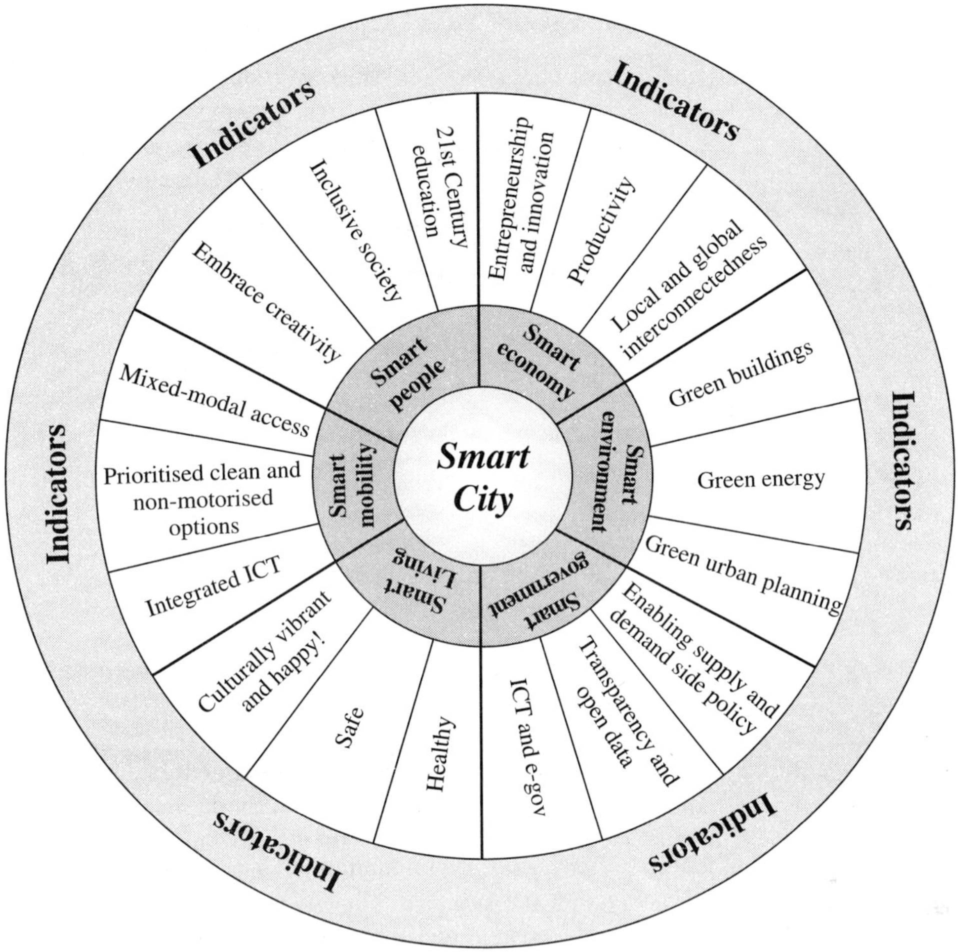

FIGURE 19.1 Smart cities wheel (*Image source:* B. Cohen, reproduced with permission).

19.2.1 Key Components of Smart City

The key components of a smart city are:

(a) Smart people
(b) Smart economy
(c) Smart environment
(d) Smart government
(e) Smart living
(f) Smart mobility

The focus area of this chapter is smart mobility which is fundamental to make a city 'smart'. The intelligent transport system is the foundation of smart mobility in a city.

19.3 SMART CITIES MISSION STATEMENT AND GUIDELINES

Government of India has issued Smart Cities Mission Statement and Guidelines in June 2015. The objective of Smart City Mission is to promote cities that provide core infrastructure and give a decent quality of life to its citizens, a clean and sustainable environment and application of 'smart' solutions. The focus is on sustainable and inclusive development and the idea is to look at compact areas, create a replicable model which will act like a lighthouse to other aspiring cities.

According to Smart City Mission, the core infrastructure elements in a smart city would include:

(a) Adequate water supply
(b) Assured electricity supply
(c) Sanitation, including solid waste management
(d) Efficient urban mobility and public transport
(e) Affordable housing, especially for the poor
(f) Robust IT connectivity and digitalisation
(g) Good governance, especially e-governance and citizen participation
(h) Sustainable environment
(i) Safety and security of citizens, particularly women, children and the elderly
(j) Health and education

In case of transportation, mobility is the key component of the smart cities. There are some important aspects, which need to be considered for mobility in smart cities:

(a) Ease of movement is at the core in the smart city.
(b) Transport system emphasises walking, cycling, public transport (PT) as primary modes of transport while the usages of personalised vehicles are discouraged.
(c) Freight movement network at low cost and high speed.
(d) Improved mobility through three prolonged approaches:
  (i) Improvement of PT—public bus, metro, BRT, LRT, monorail, etc.
  (ii) Improvement in transport infrastructure—ring road, bypass, elevated roads, improvement in existing roadways.
  (iii) Improvements in infrastructure for walking, cycling and waterways.

The National Urban Transport Policy of India focuses on the mobility of people. It encourages people to stop using their personalised vehicles and switch over to PT. It means that the PT should be efficient, effective and economical which can attract commuters. Hence, a successful PT can meet all the aforesaid parameters.

19.4 ITS SOLUTIONS FOR SMART CITIES

Although intelligent transport systems have been around for a long time, the solutions of new generation offer features like traffic prediction, analytics and decision support, traveller information, advisory services, ticketing and fare collection. Innovative tools such as roadside sensors, radio frequency tags and global positioning systems also help in monitoring and managing transport more effectively.

For example, traffic prediction helps multi-modal planning whereby transport services can be integrated to provide seamless and efficient movement of people. Analytics and decision support helps in continuous, system-wide performance measurement that helps to provide optimal solutions to a city's dynamic transport needs. Providing integrated ticketing and fare collection enhance commuter comfort by integrating rail, bus and parking fares.

Sustained research across cities has revealed that an intelligent transport system is much more than just discrete software solutions. It is about adopting customer-centric approaches that create a shift in their attitudes to cost, value and use of transport systems. That is why leading cities worldwide are using these technologies to implement transport strategies and create holistic transport solutions that address three key strategies: governance, transport network optimisation and integrated transport services.

As technologies mature and local administrations become more experienced in optimising their value, one can expect more and more cities to adopt these global practices. Of course, cities can learn from others' experiences and accelerate their own programs for smart transportation.

Traffic management has been a challenge in India and continues to become more difficult with every passing day. It cannot be addressed only by constructing new roads, bridges and underpasses. The solution lies in taking an integrated approach to traffic management and leveraging advanced technologies and intelligent solutions.

19.5 ITS TECHNOLOGIES FOR SMART CITY

The following ITS driven technologies can be deployed to make transportation system smart in a city:

(a) Multi-modal transportation and traveller information systems
(b) Smart ticketing and mobile payments
(c) Intelligent traffic management systems
(d) e-Mobility
(e) Cooperative ITS or cooperative vehicle infrastructure system (CVIS)
(f) Solar power for electric vehicles (EVs) and vehicle-to-grid (V2G)
(g) Freight solutions
(h) Car and bike rental and sharing
(i) On-demand taxis
(j) Congestion zones and road user charging
(k) Automated highway system (AHS)
(l) Driverless public transportation service
(m) Artificial intelligence for smart transportation systems

19.5.1 Multi-modal Transportation and Traveller Information Systems

The transport system in a city is sum total of different modes of transport—metro, bus, monorail, autorikshaw, taxi, private vehicles, etc. A multi-modal integrated transportation system offers several alternatives to users for travelling from one place to other place. For a sustainable, livable, smart city, it is important to promote public transport as an integrated transportation solution to ease congestion and reduce pollution.

As an integral important part of intelligent transport system (ITS), advanced travellers information system (ATIS) provides travellers with pre-trip information about travel options as well as real-time advice on navigating through a dynamic transportation network, where conditions may change rapidly many times in the course of a typical day. Advanced traveller information systems (ATIS) includes following components:

(a) Real-time traffic information provision
(b) Route guidance/navigation systems
(c) Parking information
(d) Roadside weather information systems

19.5.2 Smart Ticketing and Mobile Payments

Smart ticketing

Smart ticketing is where the ticket or product is stored electronically on a microchip rather than printed on a paper ticket. In most existing smart ticketing schemes, this microchip is embedded in a smartcard, but it could also be on a smart phone. Alternatively, a smart product could be linked to a travel account situated in a scheme's back office, so that the chip acts as a token rather holding the ticket product itself.

Smart and integrated ticketing covering most modes of transport and is integral to develop an integrated multi-modal transport system for smart mobility in a city. Smart card based integrated fare product offers flexibility to the users to use it across all major modes of transport like metro, bus, mono rail, taxi, etc. Being very convenient to use, the integrated ticketing encourages use of multiple modes and in turn promotes public transport. Oyster card of London (Figure 19.2), Octopus card of Hong Kong, and ez-Link card of Singapore are some of the successful examples of integrated multi-modal fare products.

FIGURE 19.2 Oyster card, transport for London, UK.

Mobile payment

The mobile payment has further revolutionise the fare collection system for integrated multi-modal transportation in a city. Mobile ticketing (m-ticketing) uses the concept of near field communication (NFC) for electronic ticketing in public transport system. With m-ticketing, a mobile application can provide all information about various modes of transport in a city and can also be used as a convenient fare product.

19.5.3 Intelligent Traffic Management Systems

Intelligent traffic management systems are required not only for smooth management of the traffic but also to reduce traffic congestion. Intelligent traffic management systems are integral part of smart mobility system of a city.

The conventional traffic lights are rigid and do not adapt to the dynamic traffic situation. The intelligent traffic management system takes into consideration the variation in volume of traffic at different point of time and alter the aspect of the signal to ensure optimal green period at the traffic signals. The traffic lights at any one given intersection will be connected to every other traffic light in the city, in order to ensure safe and efficient movement of traffic everywhere.

The green link determination (GLIDE) system of traffic control implemented by Singapore in 1990s and development thereafter is perhaps the best example of intelligent traffic management system. GLIDE system has three main functions. These are the allocation of optimum green-time to each of the approaches at a junction according to the volume of traffic on each of these approaches, the provision of linking for adjacent traffic signals and the ability to monitor the status of the traffic signals in the system centrally. Benefits arising from the implementation of the GLIDE system include higher journey speeds in the city area and lower overall fuel consumption.

19.5.4 E-Mobility

The e-mobility means use of electric vehicles as mode of transportation. As electric vehicles do not use fossil fuels, they do not produce any pollution inside the city. The hybrid vehicles (based on a mix of electric and fossil fuels) have already become quite popular and also being promoted by the government.

Government of India launched the National Electric Mobility Mission Plan (NEMMP) 2020 in 2013. It aims to achieve national fuel security by promoting hybrid and electric vehicles in the country. There is an ambitious target to achieve 6–7 million sales of hybrid and electric vehicles per year from 2020 onwards. Government aims to provide fiscal and monetary incentives to kick start this nascent technology. It is a composite scheme using different policy-levers such as:

- Demand side incentives to facilitate acquisition of hybrid/electric vehicles
- Promoting R&D in technology including battery technology based on lithium ion, power electronics, motors, systems integration, battery management system, testing infrastructure, and ensuring industry participation in the same
- Promoting charging infrastructure
- Supply side incentives for creating manufacturing eco system for production of EVs in the country
- Encouraging retro-fitment of on-road vehicles with hybrid kit

The fully electric vehicles require a network of charging points (Figure 19.3) spread over the city and enhancement in the storage capacity of the batteries to increase the distance travelled in one time charging.

Wireless charging and vehicle-to-grid (V2G) technologies are also under various stages of development. Wireless charging allows the electric vehicle to be charged while in motion. Vehicle-to-Grid (V2G) is a technology that makes clean and efficient electric-powered transportation possible by allowing electric vehicles to power and be powered by the grid.

FIGURE 19.3 Electric car at a charging point (*Source:* https://commons.wikimedia.org/wiki/File:Angouleme_-_Mobilivolt_(1).JPG, Author: Remi Jouan, Free to use).

The following ITS applications are deployed to improve the efficiency and increase the range (distance run per charge) of electric cars:

(a) ***Accurate range prediction:*** Algorithms that calculate the range accurately of an electric vehicle (EV) based on driving style, topology, weather conditions, roads selected, number of passengers (load) and congestion levels, either in real-time or based on predictions.

(b) ***EV route guidance and EV navigation:*** Route guidance and navigation systems that take into account the charging point locations, accurate range predictions based on driving style, state of charge of the battery, topology, weather conditions and congestion levels.

(c) ***EV eco-driving:*** Eco-driving tools (either Web-based or as on-board nomadic device) give advice to the driver on driving an EV efficiently, taking into account regenerative breaking and the use of AC, heater, etc.

Hybrid electric buses

A hybrid electric bus combines a conventional internal combustion engine propulsion system with an electric propulsion system. These types of buses use a diesel-electric or CNG-electric power and are also known as hybrid diesel-electric buses or hybrid CNG-electric buses. These buses claim to cause 30% less carbon emission as compared to conventional diesel buses.

Electric buses

An electric bus is driven by an electric motor and obtains energy from an on-board battery. This bus uses chemical energy stored in rechargeable battery packs. It uses electric motors and motor controllers instead of internal combustion engines (ICEs) for propulsion. It derives all power from battery packs and thus has no internal combustion engine, fuel cell, or fuel tank. The electric buses due to zero emission do not cause any carbon emission in the city where these are operated.

In December 2015, the smart electric bus has been introduced in Delhi. It is an indigenous technology developed by the India headquartered global technology company, KPIT (BSE: 532400, NSE: KPIT). This full electric system for buses can be deployed to create new intelligent electric buses, as well as convert existing conventional fuel buses. It is a versatile system and a variety of existing bus types can be converted to electric buses.

19.5.5 Cooperative ITS or Cooperative Vehicle Infrastructure System (CVIS)

The definition of cooperative systems in road traffic according to the European Commission is

"Road operators, infrastructure, vehicles, their drivers and other road users will cooperate to deliver the most efficient, safe, secure and comfortable journey. The vehicle-vehicle and vehicle-infrastructure cooperative systems will contribute to these objectives beyond the improvements achievable with stand-alone systems."

The stated objectives of CVIS, Europe are:

(a) Create a wireless network between and amongst vehicles and infrastructure
(b) Create an open platform for vehicle-to-vehicle (V2V) and vehicle-to-infrastructure (V2I) cooperative services
(c) Increase road efficiency and safety through vehicle-infrastructure cooperation.

The cooperative ITS permits the vehicles to communicate with each other, with traffic control centres and with the roadside infrastructure. Intersections and information signs also become intelligent and interact with the vehicles to make traffic safer, greener and more efficient. The vehicle driver receives important traffic reports, restriction notices and warnings directly in the vehicle, helping them to avoid accidents or prompting them to take action and this all happens long before the driver can actually see any hazards. The congestion, accidents and pollution are all minimised with the help of cooperative ITS.

Cooperative ITS transmits to drivers necessary real-time information to help realise not only safer transportation and people-friendly environment but also comfortable driving by road infrastructure-to-vehicle (I2V), vehicle-to-vehicle (V2V) and vehicle-to-pedestrian (V2P) communications.

The following applications are part of CVIS, Europe project:

(a) ***Cooperative urban applications*** (CURB subproject): It includes re-routing and on-trip information for individual driver based on accurate area-wide travel time information which is derived from "floating vehicle data collection", while by communicating directly with road users, road operators can improve their traffic management strategies in order to reduce congestion and vehicle emissions. Other examples of this application include priority to emergency vehicles at intersections; speed advisory to drivers for passing through a series of traffic lights without stopping; sharing of bus-lane by other vehicles when not needed for public transport, in order to improve road capacity.

(b) ***Cooperative inter-urban applications*** (CINT subproject): It can improve drivers' awareness of the road environment around their vehicles through alerts and warnings on road incidents, accidents, weather and traffic conditions, speed regulations or wrong-way drivers.

(c) ***Cooperative freight and fleet applications*** (CF&F subproject): It offers substantial benefits for professional drivers and fleet managers. The rest area parking or loading space can be booked in advance to facilitate shorter delivery times, less fuel consumption and emissions while searching for parking, and improved traffic flow by avoiding double-parking. Other applications developed in CVIS may include dangerous goods transport monitoring and controlled access to restricted areas, e.g., bridge, tunnel, environmental zone, etc.

The concept of vehicle-to-infrastructure (V2I) system for CVIS, Europe has been explained in Figure 19.4.

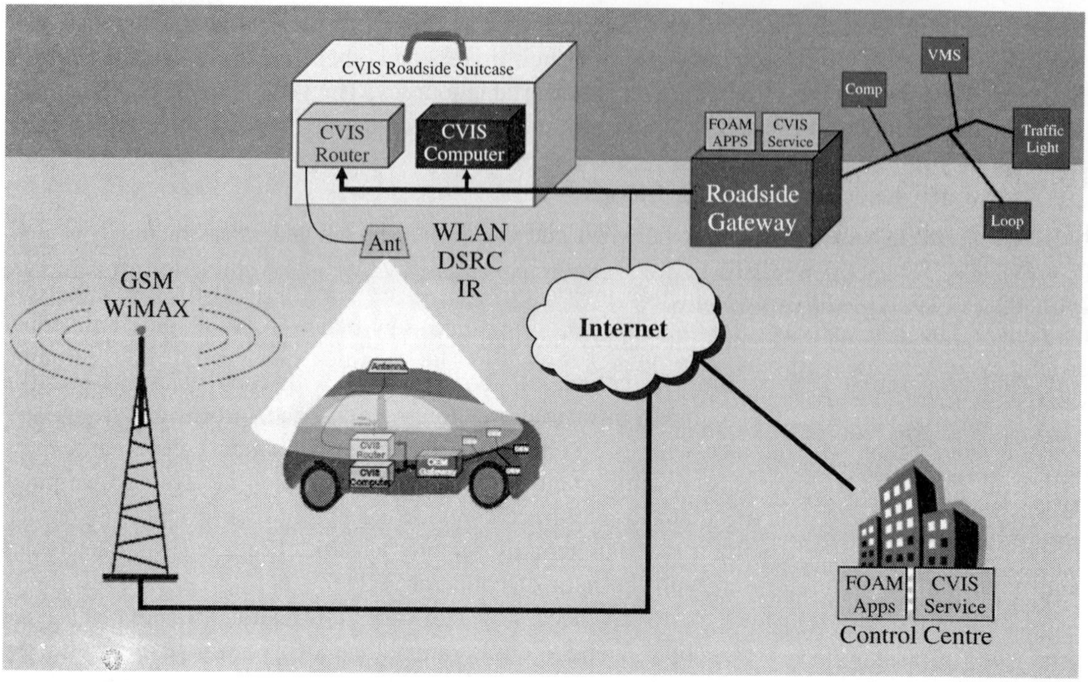

FIGURE 19.4 Vehicle-to-infrastructure (V2I) system for CVIS Project, Europe (*Source:* Prof. Phil Blythe, Professor of Intelligent transport systems, Newcastle University, UK, reproduced with permission).

Similar projects for cooperative ITS are being undertaken by a number of developed countries. ITS Green Safety initiative being implemented by Tokyo envisages a comprehensive cooperative ITS for the city as explained in Figure 19.5.

FIGURE 19.5 Cooperative ITS, Tokyo, Japan (*Source:* http://www.its-jp.org/english/its-green-safety-showcase/, reproduced with permission).

19.5.6 Solar Power for Electric Vehicles (EVs) and Vehicle-to-Grid (V2G)

Renewable sources of energy to power the city is one of the key components of smart city project. The recent advancements in technologies related to solar power and electric vehicles can pave a way for significant reduction in consumption of fossil based fuel for transportation in a city. The productivity of photocells has improved significantly and at the same time efficiency of electric cars has also been increasing. The facilities of charging EVs through solar power are already being created in developed cities on pilot project basis. Further, the renewable energy sources like solar and wind can be used to replace thermal power sources partly or fully. However, the solar power needs to be stored if not being used on real-time basis. The storage of power not only needs capital investment for setting up battery bank but also need precious space. Here, the electric vehicle (EV) can be used as a battery bank to store excess energy from solar power sources such as building with roof top solar panels. The EVs parked in the basement of the buildings can supply power to the building or to the grid called vehicle-to-grid (V2G) system. Such systems require bidirectional capability—the vehicle must be able to send power back to the grid as well as take it from the grid. The requirement and consumption of energy varies building to building. The information about requirement and consumption of energy by different buildings across the city can be collected using information and communication technologies (ICT) and surplus energy from one building can be supplied to meet the shortage elsewhere. This kind of smart network can minimise the reliance on thermal power and use EVs to harness maximum potential of solar power.

Japan is already working on the concept of V-2-G for a smart city. Auto industry is coming up with new models of car which are capable to bidirectional flow of energy (grid-to-vehicle and vehicle-to-grid) to meet the requirement of V2G. The V2G concept is explained in Figure 19.6.

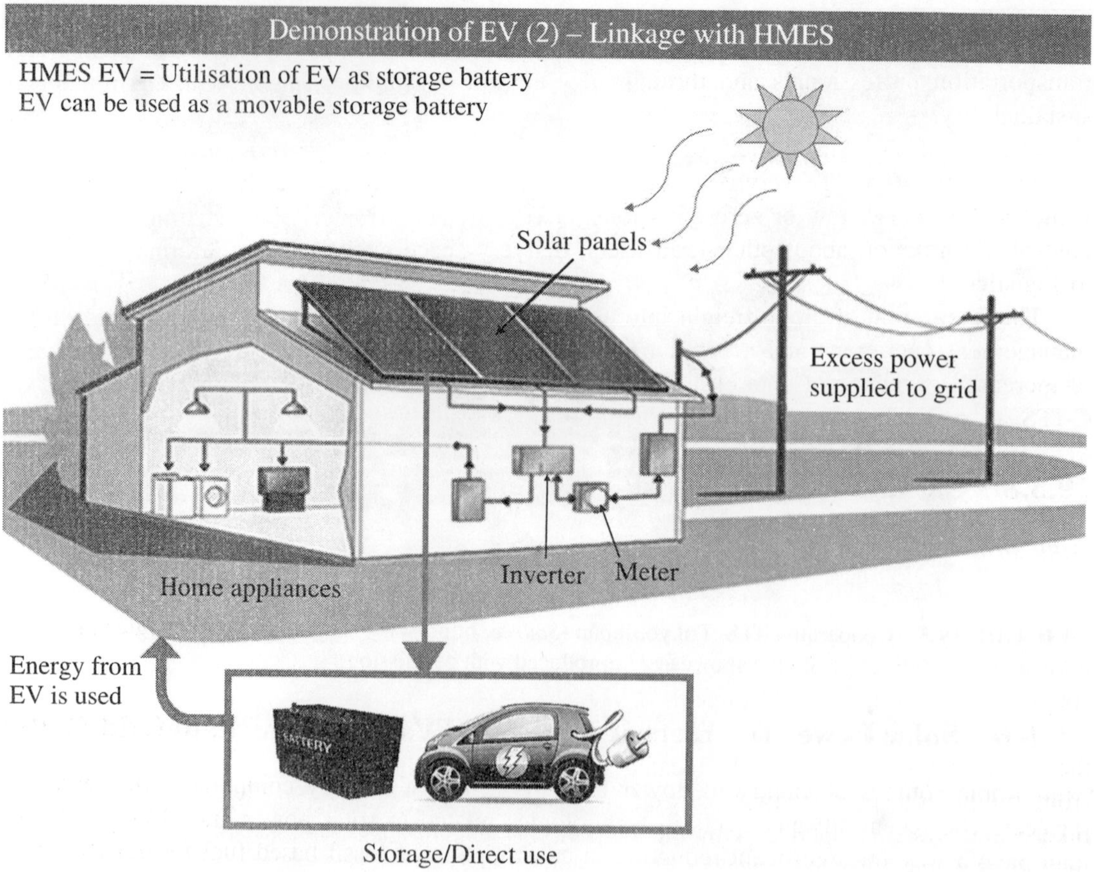

FIGURE 19.6 Concept of V2G, City of Yokohama, Japan.

For further information, please watch https://www.youtube.com/watch?v=z8S_6PLXYog

19.5.7 Freight Solutions

The transportation of freight is inevitable and it indeed expands with the growth of population, economy and transport infrastructure because raw materials for manufacturing industries must be conveyed from the origin to the place of manufacture, and finished goods are required to be transported from factory to market place and delivered to customers for ultimate consumption. All these transportation requirements are met by land-based freight transport system.

With increasing demand, urban freight transport is facing many challenging problems such as high levels of traffic congestion, negative environmental impacts, high energy consumption, inefficient fleet management and operation, insufficient real-time traffic information and poor communication.

The introduction of ITS in freight transportation has offered a balanced solution for maintaining mobility and accessibility, while avoiding and mitigating the undesirable side effects mentioned above. ITS present an opportunity for better management of existing resources and infrastructure, through the provision of information to drivers, logistics companies and transportation professionals and through new control possibilities, along with environmental sustainability.

ITS can offer a wide variety of services which can improve the overall efficiency of urban freight in cities. The ITS services range from access control and privileges granted to specific vehicles (low noise, low or zero emissions, etc.) to dynamic routing, lane sharing, load index control, information about other road users, delivery space availability or information related to logistics.

The integration of urban freight into urban network management relies on advanced traffic management operations and a better use of data on urban freight. This requires the potential of increased exchange of data on urban freight to be explored, and the possible deployment of C-ITS, in particular vehicle-to-infrastructure (V2I).

19.5.8 Car and Bike Rental and Sharing

Bike sharing

A bike rental scheme is a service which makes bicycle available for shared use to individuals on a very short-term basis. Under the bike rental schemes, the users can pick a bike from one bike station and leave it at other bike station as per his convenience. The bike stations are spread across the length and breadth of the city to facilitate easy access to the users.

These bike "pick up" and "return" stations operate on round the clock basis throughout the year. They are strategically located throughout the city to make them easily accessible from public transport stations as well as residential and commercial areas. The advanced bike rental systems operate with the help of IT technologies and provide users with real-time bike availability information through mobile applications. These "smart" bike-sharing systems provide the missing link between existing points of public transportation and desired destinations, offering a new form of mobility that complements the existing public transport systems as feeder service. The smart card based integrated fare product can also be used for bike rental systems.

Taipei, Taiwan (City population 6.8 million): U-Bike, also called Taipei Bike Sharing System, is a public bicycle sharing service offered by Taipei City Department of Transportation in a BOT collaboration with local manufacturer Giant Bicycles. With a number of bicycle parking stations located around the city, many of which are located within walking distance of MRT stations and tourist destinations. U-Bike is perhaps the best way to get around the city, complementing the MRT and public bus systems, giving visitors the option of conveniently renting a bicycle to explore the city or one of Taipei's many bicycle paths. The U-Bike service stations in the bike system are available in Taipei City, New Taipei City, Taoyuan City, Hsinchu City, Taichung City, and Changhua County. In Taipei City, rental is NT$5 for the first 30 min of use and adopts progressive tolls from NT$10 to NT$40 thereafter. The registered users can use metro smart card (easy card) to get the bike from any bike station. There are 230 bike stations in Taipei Core city having 7,000 bikes spread over the city. The Taipei Metropolitan area has 580 stations with 15,800 bikes. The bike sharing scheme is very popular across various sections of the society. The average utilisation of the bikes is 8.6 usage per bike per day.

Encouraged by the success of the system, there is a plan to add around 15,000 more bikes by 2019. The U-Bike system of Taipei is shown in Figure 19.7.

FIGURE 19.7 U-Bike sharing system, Taipei, Taiwan (*Source:* Priya Agarwal, Dy. CSTE, Indian Railways, reproduced with permission).

Paris, France (City population 2.15 million): The Vélib system (Figure 19.8) is highly accessible with bike stations every 300 m and more than 230 miles of cycling lanes. The first half-hour of usage is free of charge (as a result, in the first two months of operation, 92% of trips lasted less than 30 min). After the first half-hour, time is charged by 30 min increments. The aim is to encourage the turnover of bikes, but the rates are also designed to avoid competing with private bike rental companies.

FIGURE 19.8 Vélib bike sharing system, Paris (*Source:* https://commons.wikimedia.org/wiki/File:V%C3%A9libstation.JPG, Author: Fmjwiki, Free to use).

Car sharing

It is a scheme of car rental where people can rent cars for short periods of time, often by the hour. It is a system for sharing cars between a number of users in which each user pays for his share of the operating costs, such as fuel, insurance, parking, and tax, in accordance with their level of use. Car sharing scheme involves allocating vehicle use over shorter periods of time than car rental, with users able to access the service at a cheaper rate each time in the case of brief trips. This scheme also allows users to book car via methods such as the Internet and to pick up and return vehicles at the appointed times through a trouble-free automated and unstaffed process. This scheme is highly appreciated by the customers who make only occasional use of a vehicle, as well as others who would like occasional access to a vehicle of a different type than they use day-to-day.

Autolib: It is an electric car sharing service being operated in Paris to complement the city's bike sharing scheme Velib, since 2007. The Autolib scheme maintains a fleet of all-electric cars for public use on a paid subscription basis, employing a citywide network of parking and charging stations (Figure 19.9).

FIGURE 19.9 Autolib, car sharing, Paris (*Source:* https://commons.wikimedia.org/wiki/File:Autolib%27_rental_station_Paris_01.jpg, Author: Raphael Desrosiers, free to use).

Further, carpooling mobile applications also connect drivers and passengers in real-time, thus enabling dynamic carpooling. Drivers who wish to profit from their journeys can find people situated on the same route via a smartphone app and vice-versa. Passengers can also directly debit his or her fare to app, eliminating the need for any money exchange.

19.5.9 On-demand Taxis

The need of on-demand taxis will remain even with a fully developed integrated smart public transport system, the taxi service in a city for various reasons such as for tourists, medical needs, leisure trips, attending parties, functions, etc. Indeed, the 'smart mobility' in a city is also dependent on well-organised on-demand taxi service which is preferably available online platforms such as Internet and mobile applications. The online applications facilitate the users to check availability of taxi in near vicinity, know applicable fares, book the taxi, pay for the service and submit feedback about the service online. The demand data captured by the online application can be used by service provider to assess demand on different routes at different times of the day and plan the availability of the vehicles accordingly. The system also helps in improving efficiency of taxi operation by real-time monitoring of the vehicle fleet and their utilisation. ITS further helps in improving the availability of the taxis for the users and thereby benefits both operator and the users.

19.5.10 Congestion Zones and Road User Charging

One of the most potential transport management tools to control peak hour traffic congestion in busy areas of the city is to impose congestion charges, i.e., charge the vehicle users for using the road during the peak period. Unlike tolls, where tolls are collected to meet the capital and operational cost of the road, the main objective of congestion charge is to manage the peak hour demand. For implementation of congestion charges, the area where congestion charges would be applicable called *congestion zone* and amount to be charged at different points of time needs to be specified.

The application of ITS for congestion charging facilitates two-way communication between the vehicle and roadside equipment and a charge is deducted from the smart-card when the vehicle passes a charging point or enters a charged area (e.g., Singapore ERP, see Box 19.1).

Box 19.1 Singapore ERP system

What is ERP?

ERP is an electronic road pricing system used in managing road congestion. Based on a pay-as-you-use principle, motorists are charged when they use priced roads during peak hours. ERP rates vary for different roads and time periods depending on local traffic conditions. This encourages motorists to change their mode of transport, travel route or time of travel.

What are its Benefits?

(a) ***Minimises traffic volume:*** It helps in minimising traffic volume in heavily used roads in the CBD and Orchard areas, as well as major expressways.

(b) ***Optimises usage of the road network:*** It helps in optimising the usage of the road network by encouraging motorists to consider alternatives.

(c) ***Provides a fair price for motorists:*** Charges are based on usage—those who use the roads more frequently or during ERP hours pay more; while those who use the roads less frequently or who travel during non-ERP hours pay less or do not need to pay at all.

(d) ***No more monthly/daily licences:*** Motorists no longer need to buy paper licences to drive through high traffic areas in the CBD.

(e) ***No human error:*** ERP's reliable and fully automated system operates 24 hours. Its central computer system ensures gantries are always working properly (Figure 19.10).

Why Do We Need ERP?

Traffic congestion is costly to the individual and society. It results in the loss of productive hours, environmental pollution, wasted fuel and adverse health effects. To keep traffic moving, LTA will continue with a holistic and integrated approach using all the tools available, including building more roads, regulating vehicle growth, implementing traffic engineering solutions and promoting the use of public transport. In addition to the various measures, we also need to manage traffic demand through ERP. With more vehicles on the road, ERP remains effective in addressing current and future traffic conditions and ensuring motorists continue to have a smooth journey.

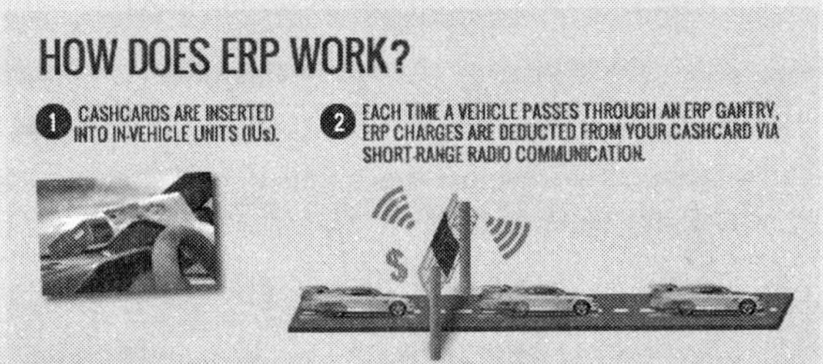

FIGURE 19.10 Working of ERP system, Singapore (reproduced with permission from LTA).

How are ERP Rates Determined?

ERP rates are determined by a **quarterly review of traffic speeds of priced roads and during the June and December school holidays**. Based on an optimal speed range of 20–30 kmph on arterial roads and 45–65 kmph on expressways, ERP rates will be adjusted accordingly. The ERP charges you pay depend on:

(a) ***Passenger car unit (PCU) equivalent:*** Cars, taxis and light goods vehicles are 1 PCU; motorcycles are 0.5 PCU; heavy goods vehicles and small buses are 1.5 PCU. Very heavy goods vehicles and big buses are 2 PCU.

(b) ***Time of entering the restricted zone:*** During peak hours, charges change every half-hour to take into account traffic volumes. This helps to spread the traffic flow over a longer period.

Source: Land and Transport Authority (LTA), Singapore, https://www.lta.gov.sg (reproduced with permission).

The road user charging system collects the charges for the use of the road as per actual distance travelled by the vehicle. The vehicles are charged onetime registration fee and/or some annual road usage fee irrespective of the actual usage of the vehicle. The road users are free to use road as much as they want once they have paid one time/annual fee for road usage. This system does not discourage the road users from excessive use of the road. The road user charging system is a GPS based system which transmits the data of the actual distance travelled by the vehicle to a central server. The vehicle user is charged as per the actual usage of the road.

19.5.11 Automated Highway System (AHS)

The automated highway system is defined as "a lane or set of lanes where specially equipped cars, trucks and buses could travel together under computer control". In an AHS, the vehicle is guided by the road rather than by the driver. Sensors and communication devices link the road and the vehicle to maximise driving performance. Driver error is be reduced and ultimately, with full implementation of AHS, the driver's error is eliminated. It is one of the aspects of intelligent transportation systems (ITS), which apply electronics, computers and control technology developed for aviation, the space programme and defense to the improvement of highways, vehicles and public transportation. The AHS consists of magnetic sensors, computers, digital radio, forward-looking sensors, video cameras, and display technologies. Various combinations of these technologies are being tested in different pilot studies (Table 19.1).

TABLE 19.1 Technologies used in AHS

Technology	*Purpose*
Magnetic sensors	These sensors are imbedded along the highway lanes. Magnetometers in the car's bumpers can sense the magnets and automatically keep the cars in the centre of the lane.
Networked computers	The system does not rely on a central computer to direct the movement of all vehicles. Rather, networks of small computers is installed in vehicles and along the sides of roadways to coordinate the flow of traffic.
Digital radio (DG)	DG equipment in each car allows the on-board computer to communicate with other vehicles in the near vicinity and with the supervisory computers monitoring the roadway.
Forward looking sensors	Using either radar or an infrared laser, these sensors detect dangerous obstructions and other vehicles ahead.
Video cameras (VC)	Video cameras are linked to computers that process images rapidly. The video cameras detects dangerous obstructions and other vehicles ahead. VCs can be used along with or instead of magnets to track lane boundaries.
Visual displays (VD)	VDs are mounted on the dashboard or projected onto the windshield. It informs the driver about the operation of the vehicle.

Benefits of AHS

The benefits of AHS on the performance of the existing transportation system will, over time, be enormous and far-reaching. Over the long term, traffic congestion will be reduced; safety will be enhanced to produce a virtually collision-free environment; driving will be predictable and reliable. More specifically, the following advantages of AHS are expected:

(a) ***Capacity enhancement of highways:*** The number of vehicles per hour per lane can be significantly increased as traffic speeds are standardised and increased and permissible headway distances are decreased.

(b) ***Improve safety:*** Driving safety will be significantly greater than at present. The human error factor will be removed.

(c) ***High-performance driving:*** It can be conducted without regard to weather and environmental conditions.

(d) ***All drivers using AHS can be safe and efficient drivers:*** AHS offers enhanced mobility for people with disabilities, the elderly, and less experienced drivers.

(e) ***Reduce fuel consumption and emissions:*** In the short term, these reductions will be accomplished because start-and-stop driving will be minimised and on-board sensors will be monitored to ensure that the vehicle is operating at top performance.

(f) ***Efficient utilisation of road:*** Land can be used more efficiently. Roads will not need to take up as much room, since AHS facilities should allow for more effective use of the right of way.

(g) ***More efficient commercial operations:*** Commercial trucking can realise better trip reliability to support "just-in-time" delivery.

(h) ***More efficient transit operations:*** Transit operations can be automated, extending the flexibility and convenience of the transit option to increase ridership and service.

19.5.12 Driverless Public Transportation Service

Driverless bus is the next step in the field of autonomous vehicles. The introduction of driverless bus has potential to revolutionise the public transport system in a city. Like driverless (autonomous) cars, the driverless bus will be equipped with multi-sensor technology, computerised driving and will run on batteries. It is believed that the autonomous buses will soon enter into the commercial use and will replace the conventional buses in times to come.

Keolis and Navya has launched the world's first autonomous, driverless public transportation service in Lyon, France in September 2016. The details of the Keolis-Navya autonomous bus are presented in Box 19.2.

Box 19.2 Keolis-Navya autonomous bus

It is an intelligent driverless shuttle that can safely transport 15 passengers at speeds of up to 45 kmph. It has no steering wheel or pedals, run on a battery (100% electric) and has a technologically advanced navigation and sensor system. It is autonomous and flexible as it does not require any driver or specific infrastructure (Figure 19.11). Being Intelligent and reliable, it can adapt to any situation by avoiding the static and dynamic obstacles. It uses electrical energy, its batteries can be recharged by induction and can last from 5 h to 13 h according to the configuration and the traffic conditions. The following equipment make it an intelligent vehicle:

FIGURE 19.11 Keolis-Navya autonomous bus.

(a) ***Multi-sensor technology:*** State-of-the-art sensors capable of communicating between themselves and fusing their data to refine the decision making of the vehicle.

(b) ***LIDAR sensors:*** 3D perception maps the environment, allowing a precise positioning and ensure obstacle detection.

(c) ***GPS RTK:*** Communication between a GPS sensor and a base station to determine at any moment the precise position of the vehicle.

(d) ***Odometer:*** Measures the displacement and the wheels speed to estimate the velocity of the vehicle and confirm its position.

(e) ***Camera stereovision:*** Obstacle detection and estimate their position in regards to the vehicle. Analysis of the environment (road signs, traffic lights) and information extraction.

19.5.13 Artificial Intelligence (AI) for Smart Transportation Systems

Artificial intelligence refers to a system capable of rationally solving complex problems or taking appropriate actions to achieve its goals in whatever real world circumstances it encounters. In other words, AI, a computerized system that exhibits behaviuor, i.e., commonly thought of as requiring intelligence.

AI has already being adopted to design and develop a smart transportation system for a city. Autonomous vehicle are perhaps the best example of AI application in transportation sector. AI-enabled smarter traffic management applications reduce wait times, energy use, and emissions. The smart cities have started leveraging the type of responsive dispatching and routing used by ride-hailing services, and linking it with scheduling and tracking software for public transportation to provide just-in-time access to public transportation that can often be faster, cheaper and, in many cases, more accessible to the public.

Today's new cars have AI-based driver assist features like self-parking and advanced cruise controls that keep a car in its lane and adjust speed based on surrounding vehicles. Fully automated cars monitored by humans can already be seen driving on the roads. The automated surface vehicle technology is likely be safer than human drivers as most of the accidents are attributed to human errors. Automated vehicles also offer the possibility of greater mobility for the elderly and differently abled who may not be able to drive. A well-designed system of automated vehicles able to predict and avoid collisions may also significantly reduce transportation-related emissions and energy consumption.

The future of smart transportation system truly belongs to application of artificial intelligence.

19.6 EXAMPLES OF SMART CITY INITIATIVES IN INDIA

Many Indian cities, viz. Hyderabad, Surat, Coimbatore, Bengaluru, Mangalore, Jamshedpur, Kanpur, Delhi, Mumbai and Chennai, have already begun deploying a few smart technologies to provide transport as well as civic services, which are as follows:

(a) Metro rail system
(b) Traffic management systems
(c) Deployment of advance communication systems
(d) Smart meters
(e) GRPS for solid waste management
(f) GIS to manage property tax
(g) Online water quality monitoring
(g) Online building plan approval schemes, etc.

19.6.1 Lavasa—India's First Fully Planned Hill City

Lavasa is a well-planned hill station in the state of Maharashtra, India. Lavasa has been developed primarily by Hindustan Construction Company (HCC) India. It has been set amidst 7 hills and 60 km of lakefront and spread over 25,000 acres. It is located at a convenient

3 hours drive from Mumbai, an hour drive from Pune. Lavasa has been planned across four town centres. Lavasa city will have a wide range of residences, from sprawling hillside villas, to up to 3 BHK homes and will offer homes which fit budgets across socio-economic classes. It is also expected to facilitate abundant opportunities as global leaders in hospitality, health and wellness, and education. Lavasa Tourism will be setting up their institutions. This hill station in Maharashtra will have a number of diverse work possibilities appealing to wide spectrum of sectors such as IT and biotech industry, KPOs and R&D companies, and the world of art, fashion and animation. Lavasa city is planned for a permanent population of 0.2 million residents and a Lavasa Tourism inflow envisaged at 2 million per annum (Figure 19.12).

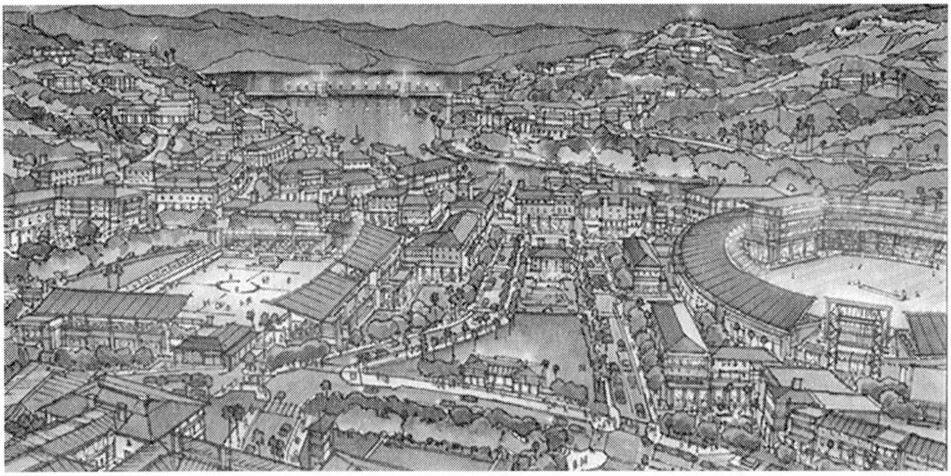

FIGURE 19.12 Development of Lavasa as a smart city (*Source:* https://nimeshpilla.wordpress.com/2016/07/06/re-imagining-a-new-city/).

19.6.2 Gujarat International Finance Tec-City (GIFT), Gujarat

GIFT is planned to be developed as a global financial and IT services hub, a first of its kind in India, designed to be at or above par with globally benchmarked financial centres such as Shinjuku, Tokyo; Lujiazui, Shanghai; La Defense, Paris; London Dockyards, etc. A comparison of GIFT with some of the financial centres of the world is presented in Table 19.2.

TABLE 19.2 Comparison of GIFT with some of the financial centres of the world

	Paris (La Defense)	*Tokyo (Shinjuku)*	*London (Dockyards)*	*Shanghai (Lujiazui)*	*GIFT*
Land area (km^2)	1.6	1.6	1.05	1.7	3.58
Construction scale (in million m^2)	2.5	1.6	1.1	4.5	8.5
Greenbelt (in thousand m^2)	40	120	50	363	1183
Height (m)	200	250	250	490	410

GIFT is a greenfield project to be developed as financial central business district (CBD) located between Ahmedabad and Gandhinagar. GIFT will eventually become a part of the future urban complex of Ahmedabad and Gandhinagar. It is designed as a hub for the global financial services sector and to facilitate these services state-of-the-art connectivity, infrastructure and transportation access have been integrated into the design of the city.

GIFT master plan (Figure 19.13) presents a sophisticated planning approach which combines the intended program into the existing context of both the site and the region. The GIFT development is likely to become a model development in India, furthering the ideas of sustainability and ecology. The project is expected to regenerate the area as high-quality, mixed use district of residential, commercial and open space facilities that optimise land and real estate values.

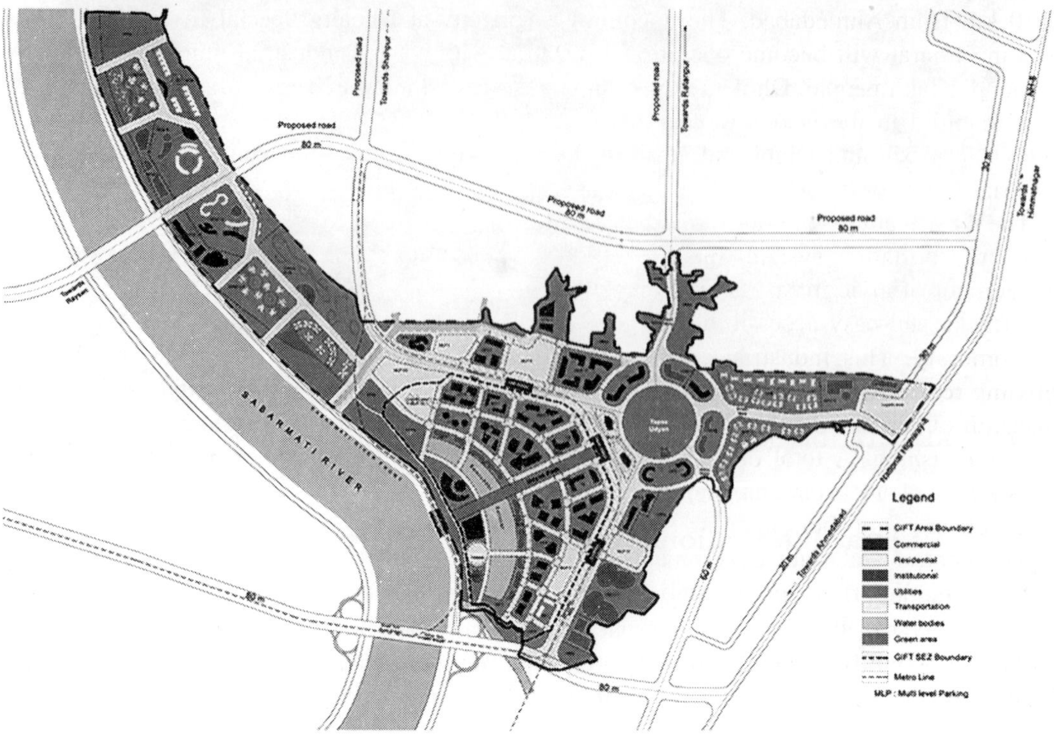

FIGURE 19.13 Master plan of GIFT, Gujarat (*Source:* http://giftgujarat.in/masterplan/zones-delineation.aspx).

The vision for GIFT is achieved through the objectives which are enumerated as follows:

(a) To develop a new format for globally benchmarked integrated city
(b) To propose a road map for fast track development and implementation
(c) To make the city scalable in each and every aspect for a distant future
(d) To derive the city format from fast changing lifestyles and new technologies
(e) To achieve an image of global city, that keeps pace with modern technologies

GIFT is envisaged as an Eco-City which will serve as the Vibrant Hub of Western India and as a habitat showcasing business oriented, environmentally-sensitive growth with equity. The fundamental principles of life and livability lay the foundation for the city.

19.6.3 Delhi-Mumbai Integrated Corridor (DMIC)

The industrial hubs and eco-friendly cities are planned to be developed along the Delhi-Mumbai Industrial Corridor (DMIC). These cities would provide world-class facilities with 24 hours power supply and drinking water, mass rapid urban transportation, with bicycle and walking tracks, complete waste and water recycling, systems for smart grids—digitally managed systems to control energy consumption, and smart metering.

The first of these cities is coming at Dholera investment region in Gujarat which is located at 110 km from Ahmedabad. The upcoming smart city at Dholera Special Investment Region (SIR) in Gujarat will become operational by 2019 with several industrial units and population of around 1 lakh people. Dholera is just one of the five similar-sized greenfield cities that India hopes to build in the next few years. They are Manesar-Bawal in Haryana, Indore-Mhow in Madhya Pradesh, and Dighi and Nasik-Igatpuri in Maharashtra—all along the Delhi-Mumbai Industrial Corridor.

The key features of these smart cities are compact, vertical developments, an efficient public transportation system, the use of digital technology to create smart grids for better management of civic infrastructure, recycling of sewage water for industrial use, green spaces, cycle tracks and easy accessibility to goods, services and activities designed to foster a sense of community. This industrial corridor will include developments of all means of transport including renewing or establishment of airports and seaports in the corridor cities, building and expansion of roadways and rail networks. In future 17 more cities are planned to be developed in similar fashion. A total of 24 such smart cities are being planned for phased development across states of UP, Haryana, Rajasthan, Madhya Pradesh, Gujarat and Maharashtra. The first phase will see 7 of them opening their doors by 2018–2019.

The central and state governments will be responsible for financing trunk infrastructure whereas a public-private partnership model is being explored for the first time to build houses, schools, hospitals and other facilities. Some of the innovative ideas being adopted in these smart cities are—each city will have underground utility corridors for parking, sewage disposal and communication lines to give it a neat look and leave enough space for facilities that are missing in most existing cities, like pavements, parks and cycle tracks. The transportation axis is developed to encourage public transport and to discourage the use of private vehicles. The dedicated bus and light rail corridors are planned to be developed as effective public transport systems in the city. The goal of the planners is to make public transport available within a 10 min walk from home to office.

Eventually, the success of these cities will largely depend on the way they are managed and promoted which will require a strong administrator who works like a city-CEO, the way most mayors are in many big and successful cities in the world. Such cities may eventually become benchmarks for other conventional cities to adapt.

19.6.4 Other Planned Corridor Projects

On the success lines of DMIC, the Chennai-Bangalore Industrial Corridor and Chennai-Hyderabad Industrial Corridors are proposed and are being planned to be developed.

The focus of these corridors will be automobile and ancillaries in Chennai, aerospace in Bangalore and pharmaceuticals in Hyderabad. The Chennai-Bangalore Industrial Corridor is expected to cover the cities of Ranipet and Hosur. Social infrastructure is also being encouraged along this corridor which is an integral part of any industrialisation. Karnataka government wants to extend this corridor to Belgaum and Mangalore with plans to integrate mining, food parks and cements as part of the corridor industries.

19.6.5 Smart City Kochi

Smart City Kochi is an IT Special Economic Zone under construction in Kochi, Kerala in India. Smart City Kochi (SCK) is a joint venture between Smart City Dubai and the Kerala government. It is one of the projects of smart city's vision to build a large network of knowledge-based industry townships across the world. The master plan of the project has already been approved. This project envisions minimum 8.8 million ft^2 of built up space out of which at least 6.21 million ft^2 will be specifically for IT/ITES/allied services. The project is expected to create over 90,000 direct jobs. The project will enjoy a single special economic zone (SEZ) status due to the revision of guidelines for SEZs being initiated by the Union Government. The project will be considered single even though there is a water body separating the land of 246 acres into two parts. The exercise includes final detailed master plan, environmental and sustainability study, urban design landscape guidelines, traffic impact study and plot development guidelines.

19.7 SUMMARY

Smart mobility is one of the pillar to development of a smart city. Ease of movement, well-developed integrated multi-modal transport, walking, cycling, public transport (PT) as primary modes of transport, pollution free freight movement, advanced as well as real-time availability of information are some of the important aspects of smart mobility. There are a number of ITS applications available which can make transportation in the city smart. Advanced travellers information system, Multi-modal integration, and RFID based integrated ticketing help in promoting the use of public transport in a smart city. The schemes of running electric vehicles, bike and car sharing, and on-demand taxi help in providing first and last mile connectivity. Intelligent traffic management system allows to use the smart traffic signaling system for efficient and smooth traffic flow in a city. ITS based congestion charging and road user charging are very useful in easing peak hour congestion and manage the demand. Development of driverless vehicle technology has potential to make transport system intelligent and smarter. However, considering the capital and O&M cost of deployment of these emerging technologies, a progressive adoption of the ITS applications customised to the local needs is the key to success.

Questions

19.1 Define smart city in your own words.

19.2 What is the importance of ITS for smart cities?

19.3 What are the key components of a smart city?

19.4 As per Government of India Smart City Mission, what are the core infrastructure elements in a smart city?

19.5 How multi-modal transportation and traveller information system is important for a smart city?

19.6 Smart and integrated ticketing is fundamental to smart mobility. Explain.

19.7 Explain the role of intelligent traffic management systems in making a city smart.

19.8 How ITS can be used to manage urban freight and to make a city smart?

19.9 Explain car and bike rental and sharing with example(s).

19.10 What do you understand by congestion zones and road user charging?

19.11 Explain the concept of autonomous vehicles and its role in smart mobility in a city.

19.12 Give an example of development of smart city in India.

Bibliography

Alice, "Urban Freight research roadmap", ERTRAC Urban mobility WG, November 2014 (http://www.ertrac.org/uploads/documentsearch/id36/ERTRAC_Alice_Urban_Freight.pdf).

Autolib' from Wikipedia, the free encyclopedia (https://en.wikipedia.org/wiki/Autolib%27).

Cohen, B., "What Exactly is a Smart City?", Fast Company, Co.Design, September 19, 2012 (http://www.fastcoexist.com/1680538/what-exactly-is-a-smart-city).

Intelligent Transportation in Smart Cities, Application Analysis Report, 2014 (https://www.abiresearch.com/market-research/product/1018000-intelligent-transportation-in-smart-cities/)

Luk, Yiu Wah, "Potential applications of intelligent transport systems in urban freight transport", 2004.

Midgley, Peter, "The Role of Smart Bike-sharing Systems in Urban Mobility", published in *Journeys*, May 2009 (http://www.lta.gov.sg/ltaacademy/doc/IS02-p23%20Bike-sharing.pdf).

Sandhu, Sabhijiit Singh, Naman Jain, Aditya Gaurav, and N. Ch. Sriman Narayana Iyengar, "Agent Based Intelligent Traffic Management System for Smart Cities", *International Journal of Smart Home,* Vol. 9, No. 12, pp. 307–316, 2015 (http://www.sersc.org/journals/IJSH/vol9_no12_2015/31.pdf).

Siemens, "Intelligent Infrastructure for the mobility of tomorrow", May 16, 2016 (http://www.siemens.com/press/en/feature/2014/mobility/2014-11-car2x.php).

"The e-Safety Initiative 2004", The 3rd e-Safety Plenary Session, Brussels, March 25, 2004.

Yeomans, Gillian, "Autonomous vehicles—Handing Over Control: Opportunities and Risks for Insurance", 2014. (https://www.lloyds.com/~/media/lloyds/reports/emerging%20risk%20reports/autonomous%20vehicles%20final.pdf)

Zhang, Jianwei, Theo Arentze, and Harry Timmermans, "A Multi-modal Transport Network Model for Advanced Traveller Information System", *Journal of Ubiquitous Systems and Pervasive Networks*, Vol. 4, No. 1, pp. 21–27, 2012 (http://www.iasks.org/sites/default/files/JUSPN20120401021027.pdf).

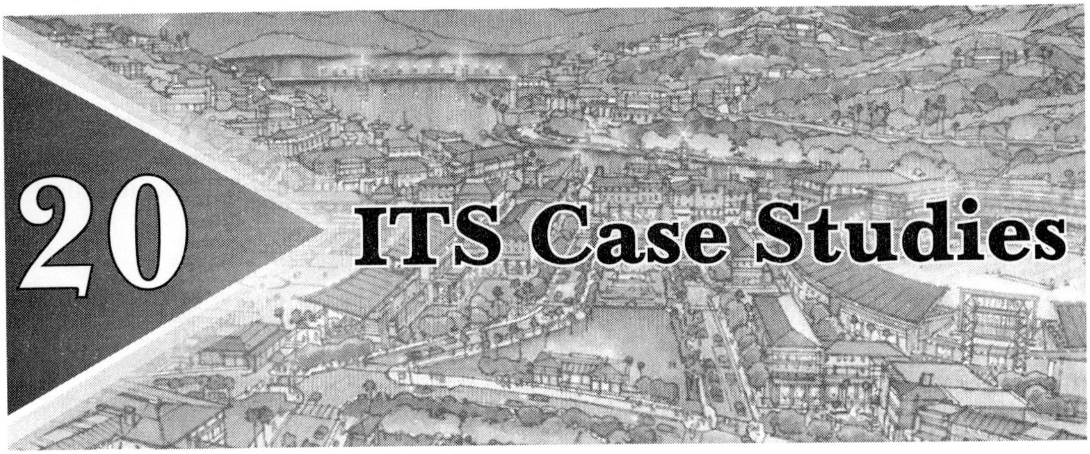

20 ITS Case Studies

20.1 INTRODUCTION

Having studied the theoretical concepts of ITS, it is important to study the real success stories of ITS. In this chapter, various applications of ITS in different cities of the world and in Indian Railways have been presented. A set of Case Studies has been available on 'World Bank Toolkit on Intelligent Transport Systems for Urban Transport'. The case studies for Indian Railways have been picked up from different sources listed in bibliography. The case studies presented in this chapter are as follows:

Roadways

(a) Dublin, Ireland
(b) Florence, Italy
(c) Izmir, Turkey
(d) Zurich, Switzerland
(e) Japan
(f) London, UK
(g) Mysore, India
(h) Indore, India
(i) Delhi, India
(j) Bengaluru, India
(k) Mumbai, India

Railways

(a) Passenger reservation system (PRS)
(b) e-Ticketing

(c) Unreserved ticket system (UTS)
(d) Paperless ticketing
(e) Freight operations information system (FOIS)
(f) Control office application (COA)
(g) Integrated coach management system (ICMS)
(h) National train enquiry system (NTES)
(i) Software aided train scheduling and network governance (SATSaNG)
(j) Integrated train enquiry system (ITES)
(k) Smart phone mobile application

20.2 CASE STUDY: DUBLIN, IRELAND

Dublin Bus is a publicly-owned operator of the bus network in Dublin, Ireland. Around 980 double-decker buses are being operated by Dublin Bus in the city and hinterland of Dublin. It has a 5-year public service obligation (PSO) contract with the National Transport Authority of Ireland. The PSO payment has now being linked to performance measures being monitored by ITS tools. The company has implemented extensive ITS over the course of two decades. Electronic ticket machines were first implemented in 1989 and have provided the backbone of the on-bus intelligent network.

20.2.1 ITS Application by Dublin Bus in Recent Years

Automatic vehicle location and control system (AVLC)

AVLC is implemented since 2009 which involves real-time tracking of vehicles. Buses are available at 20 seconds headway. GPS and odometer are used for getting the location data. The odometer provides the primary means of location, with the GPS supplementing it.

The AVLC has the following features:

(a) Real-time tracking
(b) Efficient route control regime
(c) Centralised control room
(d) Generation of comprehensive reports and statistics
(e) Provides/supports RTPI
(f) Support traffic signal priority

The driver's console shows when the vehicle is ahead of schedule, and hence the driver needs to slow down or wait. It has been decided not to show the driver when the bus is behind schedule, or by how much, since they did not want them to drive over speed in order to catch up and endanger the safety of the passengers.

Automatic fare collection system (AFC)

Smart cards as fare media were implemented by Dublin Bus in 2008. One smart card reader per bus has been installed on the buses. Existing tickets are being migrated to smart cards.

Magnetic strip tickets are still used for common tickets with Irish Rail and Luas (Tram). The current Dublin Bus ticketing system supports the integrated ticketing system.

Real-time passenger information (RTPI)

Dublin City Council (DCC) was entrusted with the responsibility to implement RTPI in Dublin city as it has the authority for any physical works and cabling on the streets. DCC hosts the server that handles the RTPI, and has procured and deployed the at-stop displays. Dublin Bus provides data to DCC to support the RTPI. There was much technical discussion in advance and specifications were agreed together. When DCC went to tender for the RTPI system, it already had the Dublin Bus specifications, so there was not any problem of downstream integration. Now the system is fully bedded in and things are being formalised. Currently, the system can only process real-time information for about 1,000 stops which is planned to be increased to 5,000 stops in due course of time.

Communications

Each bus has been allotted a unique IP address. The AVLC communicates with the DCC/RTPI system by high speed fibre link. Each depot is equipped with its own Wireless LAN and a digital link to the AVLC system.

CCTV

A network of 140 CCTVs provides excellent coverage of most of the key junctions and streets, and most potential disruptions points for the Dublin Bus network. Live CCTV footage is very useful to manage traffic especially in emergency or disruptions. Dublin Bus is now interested to make it available to the dispatchers at the AVLC control centre. Funding had been a problem for the integration, including the provision of a fibre-optic link from DCC to the AVLC control centre in broad stone.

Scheduling system and interface with the AVLC

Dublin Bus is using Microbus scheduling software package for scheduling of vehicle and driver resources, and crew rostering. The schedules are prepared semi-manually which means that an initial set of operational hours, frequencies (by time period) and running times (by time period) are fed in the system manually as an Excel spreadsheet. The duties are imposed on it to generate an initial timetable and duties. This becomes input to Microbus software which is used to optimise it and/or allows some modifications. The final schedule is then exported to other systems such as rostering and AVLC.

20.2.2 Benefits Arising from the ITS Systems

The achieved/perceived benefits of the ITS investments are as follows:

(a) Perceptible improvement in quality of service
(b) Improved reliability of service
(c) Convenience to the passenger with improved services and availability of information through RTPI

(d) Reduction/virtual elimination of buses departing early from terminal

(e) Availability of data on service performance

(f) More precise/appropriate scheduling (sectional journey times, variances), better crew and service management

(g) Use of speed/delay data to identify delay points and supporting case for traffic management improvements or bus priority

20.3 CASE STUDY: FLORENCE, ITALY

The bus services in the City of Florence, Italy is operated by ATAF (Azienda Trasporti dell'Area Fiorentina) which is a municipality-owned bus operator of the City of Florence, serving the metropolitan area and some hinterlands towns. The services are provided under service contract to the province of Florence, for which ATAF joined with Li-nea and some smaller operators in a special purpose entity since 2005. The contract stipulates quantitative and qualitative performance requirements, in particular in relation to volume of service and punctuality. Quite extensive service and event reporting requirements have also been stipulated in the contract agreement.

Automatic vehicle location and control system (AVLC) was first commission by ATAF in 1984 in about 100 of its buses. The AVLC was upgraded in 1996 and further extended to the complete fleet of about 400 buses. AVLC also included GPS, trunked mobile radio, and enhanced functions.

20.3.1 Reasons to Implement ITS

The primary motivations for implementation of ITS are:

(a) To provide the platform for real-time and other information to passengers

(b) To manage the daily operations, on both normal and disrupted state

(c) To assist achievement of the quantity and quality of the service required in the service contract with province of Florence

(d) To generate the trip logs, analysis and reporting required by province of Florence under the service contract

(e) To manage the driver vehicle handovers and shift-changes

(f) To identify vehicle faults and assist rapid response

(g) To provide the platform for e-ticketing

(h) To provide security and incident response on-board vehicles, both for staff and for passengers

(i) To support demand responsive transport and other non-standard mobility services

(j) To generate and manage data for post-event analysis, including running time analysis, scheduling, resource optimisation, and incident investigation

20.3.2 ITS Equipment and Devices

In-vehicle ITS-related equipment consists of:

(a) On-board computer
(b) GPS
(c) Display screens for RTPI
(d) Voice announcer (interior)
(e) External destination board
(f) GPRS/GSM (voice, data) card and antenna
(g) Wi-Fi (wireless LAN) for in-depot data transfer
(h) Driver's console
(i) CCTV
(j) Automatic passenger counters
(k) Forward-facing camera for photographing vehicles violating bus lanes

Station/stop equipment ITS-related equipment consists of:

(a) Displays for RTPI
(b) Self-service ticket vending machines

Depot ITS-related equipment consists of:
Wi-Fi (wireless LAN) for data exchange with vehicles

20.3.3 ITS Applications at ATAF

ITS is used for the following applications at ATAF:

(a) Operations management, including incident management
(b) Real-time passenger information pre-trip, at stations and in vehicles
(c) Vehicle and driver scheduling
(d) Fare collection (for pre-paid tickets only, no on-bus ticket sales)
(e) Data generation for performance monitoring and contract reporting

ITS has been or is being implemented on a limited basis for the following applications, with the possibility to deploy on a wider scale:

(a) In-vehicle video-surveillance
(b) Traffic signal priority
(c) Automatic passenger counting

ITS supports and/or exchanges data with the following:

(a) Driver rostering
(b) Travel and traffic information
(c) Maintenance planning and management

Communication

Due to limited number of radio channels, mobile communication for AVLC has been a challenge for ATAF throughout the history of its ITS deployment.

The private trunked mobile analog radio is the core mode. Only three channels are available for both voice and data communication. In the late-80s/early-90s the national authorities in many of the European countries needed to restructure spectrum allocation to enable new services. So ATAF was allocated different frequencies with narrower bandwidth.

Initially, ATAF had only 100 buses when ATAF commenced AVLC operations in 1984. However, with extension of system to the full fleet of 400 buses in 1996, the constraints were experienced even after adopting a trunked mobile radio approach.

Faced with serious problems with communications based on three radio channels, in 2003, ATAF decided to transfer 100 vehicles to GSM for voice communication, with all data communication staying on the private radio channels. GSM cards were inserted in the existing radios. The transfer between trunked mobile radio and GSM was automatic, and was invisible to the dispatchers and to the driver. Further, in 2009 as GSM/GPRS technologies matured, the commissioning of the new communication system based on the GPRS/GSM network started. This helped the ATAF as new technology brought following benefits:

(a) No infrastructure investment and maintenance costs
(b) High level of service
(c) Rapid-response time for fault repairs, interventions, etc.

Further, close range communication is performed by Wi-Fi (wireless LAN) separately, which is integrated into the on-bus computer. Wi-Fi coverage is available throughout the depots, so vehicles can exchange data wherever they are parked, at maintenance, or at washing/fuelling. Data transfer included:

(a) Upload of reference data (routes, stops, schedule, etc.)
(b) Download of transaction data, statistics, logs, etc.
(c) Real-time passenger information

Passenger information at ATAF provided following functionalities:

(a) Travel planners
(b) Information to passengers on real-time basis
(c) How to use the transport system
(d) Information about incidents and alerts

Real-time passenger information is based on the AVL system. RTPI is provided to the passengers through four main channels:

(a) Internet channels
(b) Mobile personal devices, with a combination of 'push' and 'pull' services
(c) At-stop displays
(d) In-vehicle displays

At stop information consists of two types of device:

(a) Free-standing pole-mounted panel
(b) In-shelter panel

Scrolling-text single-line displays are used in the vehicles for real-time information. The name of the next stop is displayed in the normal mode. The pre-recorded messages or advisories from the dispatcher can also be displayed if required. Further, all buses are fitted with public announcement (PA) system which can be used for stored information (normally announcement of the next stop), for driver messages, or for messages from the dispatcher. A range of SMS, WAP-enabled, and Internet-based traveller information and RTPI services are also available for use in mobiles, tablets, computers, etc. by a dedicated real-time Web portal.

Video surveillance

The buses have been equipped with CCTV cameras for video surveillance. In the standard composition, 3 CCTV cameras are linked to a Video Surveillance Online Repository (VISOR) card that has been added to the on-board computer. In normal mode, the footages are stored in erasable memory, and then overwritten. The driver activate recording footages in hard memory by pressing an alarm button. The CCTV footages are retained until such time as they are recovered or manually erased or transferred over the Wi-Fi link. This option can be used to preserve the evidence when footage may be required for establishment of crime by security agencies or as evidence in the court of law or for any other purpose. The CCCTV footage of any bus can also be accessed on real-time basis. However, a high bandwidth is required to transfer CCTV footage on real-time basis limiting the capability of real-time access. To make it practicable, it would be sent as a sequence of still-shots, sufficient for the dispatcher or other authorised person to be able to discern what is happening on the vehicle. The present day high speed 3G/4G GPRS technology makes it possible to transfer live CCTV footage from the buses without any constraint.

Other imaging technologies

A forward facing camera has been fixed in front of the buses which may capture images of vehicles obstructing bus lanes and other public transport facilities. The licence plate of the vehicle violating the lane can be read from the image and the erring driver can be penalised.

Automatic passenger counters (APC)

APCs have been fitted in around 20 buses. These are used as 'test buses' and are rotated across the fleet to periodically survey all routes. ATAF has not been sufficiently confident in APC technology to instal it as standard across the full fleet.

Scheduling

Micro bus software is used by ATAF for scheduling of the buses. The graphics of the timetable and the actual performance in service can be generated by AVLC. The deviations can be tracked by AVLC to assess the sectional running time between two stops. Any adjustment in run-time can be made as per the observation from AVLC reports. The operation of an individual bus or a whole route for a period of time can also be replayed in order to understand the dynamics of the service and see whether it can be improved through the schedule.

Fare collection

There is no ticket issuing machine available on-board in the buses. The tickets have to be purchased in advance before boarding the bus. The buses are fitted with smart card readers which are provided by AEP, an Italian ticketing validator vendor. The validators are connected to the on-bus computer, which process the transaction, store it and transfer it to the central control.

System integration

The system integration is the backbone of all applications. The on-bus computer provides functionality of full integration and data consistency of all in-vehicle devices. The supplier has developed the system architecture as per ATAF's specifications, and integration has also been done by the supplier as per the specified requirements. The architecture and interfaces reflect European project concepts and CEN TC278 standards, although the actual implementation is the supplier's proprietary approach. The supplier has used its own data model.

20.3.4 Benefits Arising from the ITS Systems

Although cost-benefit analysis (CBA) or any other appraisal of ITS system deployed by ATAF has not been done, ATAF considers that it has received considerable benefits from the ITS implementation, especially the AVLC system. The following benefits of ITS have been perceived by ATAF:

(a) Improvement in operational management capacity
(b) Improvement in operational efficiency
(c) Availability of information for performance analysis and service planning
(d) Passenger convenience in terms of improved services and availability of information
(e) Fulfilment of reporting requirements in relation to the service contract with province of Florence

20.4 CASE STUDY: IZMIR, TURKEY

Elektrik Su Havagazı Otobüs Troleybüs (English: Electricity, Water, Coal Gas, Bus, Trolleybus) or ESHOT is a bus service operating in the Izmir, Turkey. ESHOT is the municipal-owned operator with a fleet of around 1500 buses (including 410 from its associate Izulas). The services are operational on more than 400 routes and carries average 2.8 million passengers per day. The bus services forms part of an integrated transport system in Izmir which includes metro, rail, ferry, and local paratransit. The Izmir metropolitan transport co-ordination centre was established in 2006. It is responsible for integrated planning, co-ordination and determination of routes. ESHOT has been implementing ITS since 1999 starting with introduction on smart card based ticketing system.

20.4.1 Reasons to Implement ITS

ITS has been implemented by ESHOT primarily for following reasons:

(a) To develop a secure and effective revenue collection system which formed the backbone of the subsequent ITS applications
(b) To enable modal and service integration through minimising personal costs of interchange
(c) To develop enhanced operations management capacity to provide reliable services and deal with disruptions to prevent revenue leakages
(d) To provide surveillance for passenger and personnel security
(e) To provide improved passenger information in real-time
(f) To obtain data for planning, resource optimisation and performance monitoring

20.4.2 Intelligent Transport System Applications at ESHOT

Automated Fare Collection

ESHOT introduced electronic ticketing in March 1999, with Akilli Kart as service provider. The Korean technology has been employed for the contactless smart card and validator, and a Motorola communications system. The system was developed as a standalone application to improve revenue integrity and prevent revenue related frauds. However, the system integrator deployed by ESHOT for this program, indigenously world's first integrated automated fare collection and vehicle tracking systems in a single unit. In course of time, the communications were also migrated onto GPRS to get rid of dependency on a dedicated wireless network.

The fare integration encompassing all modes of public transport was launched in February, 2007 as a pilot project. Initially, an application for transferable tickets among modes was launched involving ferries, metro, and 100 buses on 52 routes that provided feeder services. Under the fare integration scheme, a concession of 50% is offered on a second boarding if made within 60 minutes of the first. The new system required time stamping of the transaction, i.e., transaction to be timed and to be written to and read from the smart card. To further promote the use of multiple modes of transport and to encourage people to use public transport, the time period for transfer concession was extended to a 90 min validity in January 2008, and the concession for any subsequent boarding increased to 100% in August 2008. As such a flat-fare structure was introduced for passenger transport across Greater Izmir for all holders of smart cards. Because of flat-fare structure, there is no need of fare validator at exit points. ESHOT buses are only fitted with ticket validators at the entrance door only. Every transaction records the location of the bus at that time using the automatic vehicle location capability of the validator which can be used for further analysis of ridership, occupancy and traffic demand across different routes.

ESHOT has analysed the travel flow data generated from the ticketing transactions to rationalise its route network. As a result of these initiatives, the ridership has been significantly increased, and long and unreliable routes have also been reduced to the benefit of service quality.

Automatic vehicle management

Initially, the system only provided an online vehicle tracking capability. The location of the bus is communicated to the operations control function through GPS after every 10 seconds at a meagre cost €20 per bus per month. The real-time information is used for following purposes:

(a) To monitor adherence with schedule of the service
(b) To control of departure timings, especially at the outer terminals of routes where the level of supervision may not be adequate
(c) To monitor the speed of the bus
(d) To monitor headway between services
(e) To detect any route deviation by cross-checking actual route with schedule route through the co-ordinates of the route to which the bus is currently following
(f) To display information on-board as required in RTPI package
(g) To control bunching of buses by slowing down/stopping the following buses

The AVM system generates a range of reports for management information purposes. These include:

(a) Schedule adherence report
(b) Passenger information systems
(c) Route violation report
(d) Speed violation report
(e) Daily total kilometre report
(f) Speed-band/time-duration graphic

The real-time passenger information systems have been developed at two separate levels

(a) At bus-stops
(b) On the bus itself

There is an enquiry service for the passengers where they can get to know when a bus is expected to arrive at their stop, using an SMS text message via GPRS on their personal cell-phone. The on-bus passenger information system consists of a 19” VGA colour video screen on which the route with its immediate surroundings are displayed. The screen displays the speed and the position of the bus in real-time. The information regarding current and next stop is announced on-board. The system provides on-vehicle route tracking to the passengers to make them ready for alighting at the appropriate stop. The passenger information unit is also capable to provide news feeds and weather forecasts, and hence provides an ‘infotainment’ capability. It also provides point-of-contact details for ESHOT to the passengers if they wish to provide any feedback or lodge complaint about the services including the behaviour of the drivers.

Mobile video surveillance (MVS) system

On-board surveillance in the bus is performed by MVS system. The MVS system comprises a mobile digital video recorder and wide-angle cameras suitably located within the bus to cover the complete area of the bus. The MVS system is also integrated with vehicle tracking and speed data from the AVL capability of the smart card validator. The GPRS communication links the collected data to the operations control centre (OCC) as per the requirement. The recorder is equipped with a smart memory card which can be used to capture visual evidence of an incident and secure it for any investigation purpose. It can also be used as an evidence in the court. It is also equipped with ports to download the data as required for management information purposes, and for direct GPRS wireless transmission as per requirement.

The primary goal of the MVS system is to prevent leakage of revenue through frauds. It is integrated with AFC transactions to facilitate identification of perpetrators of any attempted fraud. The MVS is also helpful in preventing attacks on the bus drivers as visual evidence of any action is captured in the system. In case of any emergency, the driver can alert to OCC by pressing the alarm button to alert the OCC. The AVL supports real-time passenger information on buses and at bus-stops. AVL data is also utilised for fuel issue monitoring and bus maintenance scheduling.

20.4.3 Benefits Arising from the ITS Systems at ESHOT

For the fare collection system, the main benefits reported are:

(a) Better revenue management
  (i) Elimination of cash handling, security and administration costs of traditional paper tickets
  (ii) Improved revenue generation through reduction of fraud and peculation
(b) Operational efficiency
  (i) Greater network efficiency arising from planned interchange, enabled by the through-ticketing capability and elimination of transfer charging
  (ii) Accelerated fleet investment, enabled by improved operating cash flow
(c) Increased ridership as a result of the network integration, though this is of more significance to the rail-based modes

For the other ITS applications, the main benefits reported are:

(a) Improved real-time passenger information, especially on-board the bus
(b) Enhanced security through on-board video surveillance
(c) Improvement in customer satisfaction
(d) Improved operational control, though in a reactive rather than proactive sense
(e) Automation of data collection for management information and preventive maintenance scheduling
(f) Generation of data on service performance and their incorporation into more accurate schedules

20.5 CASE STUDY: ZURICH, SWITZERLAND

Zurich Public Transport (VBZ) is responsible for operation of trams, trolleybuses and buses in Zurich. The network of trams is the primary mode of public transport in the city of Zurich. Since seventies, the City of Zurich has been promoting well-integrated multi-modal public transport in the city by giving it high priority in streets and at junctions. The city offers an excellent operating environment and VBZ is known for operating on-time reliable services. ITS applications have been the backbone to operations management and high-quality, on-time services since its adoption in 1970s. The ITS forms an essential part to the VBZ operation, and to the organisation, business processes, operating procedures, data, and management. VBZ has evolved from using ITS to just know where their vehicles are, to precision operation with on-time running and transfer assurance. The passenger information is extremely well-developed which is able to provide required information to the passengers prior to the trip, at stops and in the vehicles.

20.5.1 Reasons to Implement ITS

The reasons for utilising the ITS include the following:

(a) To offer priority to public transport at traffic signals to minimise delays at junctions and variations in travel time
(b) To provide a reliable service which operates according to the schedule at departure and along the route
(c) To provide real-time and other information to passengers before the trip, at stops and in vehicles
(d) To ensure that planned transfers at interchange points can be achieved
(e) To have the capacity to deal with events, disruptions and emergencies
(f) To achieve integrated and seamless travel across modes
(g) To gather, analyse and utilise extensive data for future planning, route and schedule rationalisation
(h) To optimise the resource utilisation and maximise efficiency

The requirements have evolved over time as VBZ has become familiar with using ITS, and with the data it provides, and as they have built their organisation around it.

20.5.2 ITS Equipment and Devices

In-vehicle ITS-related equipment consists of:

(a) On-board computer
(b) Driver console
(c) Display screens for RTPI
(d) GPS
(e) Radio (voice, data)
(f) Wireless LAN (in-depot data transfer)
(g) Voice announcer (interior)
(h) Voice announcer (external, some vehicles only)

Transponders have been provided to interface with roadside detector (trams only, both for traffic signal pre-emption and for point switching). There is no fare collection equipment on-board the vehicles in the City of Zurich. There are ticket issuing machines and card validators on some buses that operate suburban services in the hinterland. CCTV is not used within the VTZ vehicles.

Station/stop equipment ITS-related equipment consists of:

(a) Displays for RTPI
(b) Self-service ticket vending machines

Depot ITS-related equipment consists of:
Wireless LAN/Ethernet for data exchange with vehicles

20.5.3 Applications for which ITS is Used at VBZ Zurich

The scope, functionality and devices for ITS have evolved over about 40 years since ITS was first utilised at VBZ. ITS is used for the following applications at VBZ Zurich:

(a) Operations management, including incident management
(b) Automatic vehicle location
(c) Real-time passenger information pre-trip, at stations and in vehicles
(d) Fare collection
(e) Traffic signal priority (limited ITS content)
(f) Automatic passenger counting
(g) Timetabling
(h) Vehicle and driver scheduling

ITS supports and/or exchanges data with the following applications:

(a) Driver duty assignment (rostering)
(b) Planning of maintenance of vehicles
(c) Travel and traffic information

AVL and operations management

The ITS for AVL consists of four main elements:

(a) In-vehicle integrated computer
(b) In-vehicle driver console
(c) Control centre workstations
(d) Control centre software

The in-vehicle integrated computer is an IBIS Plus unit. The unit has evolved over time as new technologies are coming up. However, it is essentially the same concept of an on-board unit linked to a driver's console, various in-vehicle devices and a radio. It comprises the following integrated elements:

(a) Computer/core processing unit
(b) Wireless LAN
(c) GPS/Location unit
(d) Reference data store
(e) GSM unit
(f) Transaction store

The unit is the central hub of all the on-board ITS devices, except for ticket issue and ticket validator units on buses where these are installed.

Driver's console

The driver's console comprises a high-definition, multi-function, touch-screen man machine interface (MMI). The MMI is a display unit, providing a user interface to the on-board applications, all processing and data storage functionality are in the on-board computer. When the driver switches on the engine at the time of resuming his duty, the MMI is activated. The on-bus computer is activated from the sleep mode, and the in-vehicle display units are switched on. The driver enters his/her driver identification number and the route/duty data. The on-bus computer retrieves the trip information pertaining to that duty and displays the next scheduled trip. The driver approves that this is the correct trip, or can manually select another trip if necessary. After which, all route and trip information updates automatically.

The normal display mode on the driver's console shows the following:

(a) **Top line:** Route/duty, time, destination, position relative to schedule (+/–, in colour box)

(b) **Right side bar:** Tools including request to speak, alarm, message request and other menu options

(c) **Main body of screen:** Route shown vertically, with current stop at the top, next three stops with their scheduled arrival time, and indicator of whether there is connection data available

(d) **Bottom of screen:** Destination/terminus and scheduled arrival time

The driver's console can display other information on the screen, including:

(a) Real-time information on connections at current and subsequent stops. This allows the driver to check and decide whether to wait for other connections, and also to answer any queries from passengers

(b) Messages from the dispatcher/control centre

(c) Message menu, from which one can be selected for communication to the control centre

(d) Visual of bus occupancy options (empty through to full with passengers left behind) which can either be initiated by the driver or prompted by the dispatcher

Communication

Private analog radio has been used for primary communication for operations management. The radio communication is full duplex, which is rare in the public transport industry. VBZ has an extensive network of base stations spread over not only in the city but also outside. The wide network facilitates all radio communication to be performed on the analog radio system in the city, and along the main service areas outside the city.

Close range communication is performed by wireless LAN. The LAN is also is integrated with on-board computer of the vehicle. The data from the buses can be transferred to the system when they are in the depot for parking, maintenance, fuelling, washing, etc. as wireless LAN coverage is available throughout the depots. Some other critical locations are also equipped with wireless LAN. The exchange of data takes place on an 'opportunistic basis', including for some vehicles if they are passing by the depot. The system keeps track of which data packets have been successfully transferred, and if the transfer of data is not completed, it resumes at the next opportunity.

Data transfer includes:

(a) Upload of reference data—routes, stops, schedule, etc.
(b) Download of transaction data, statistics, logs, etc.
(c) Software upload/update and configuration changes

If necessary, the system can transfer data through the radio system, but this is generally avoided as it uses up the more scarce capacity. The transponders communicate route information for trams to enable automated point switching.

Real-time passenger information

Passenger information consists of multiple strands:

(a) Guide to use the transport system
(b) Travel planners
(c) Real-time information to the passengers
(d) Incident information and alerts

Real-time passenger information is based on the AVL system. The real-time information to the passengers (RTPI) is provided through four main channels:

(a) Internet and mobile channels, delivered to personal devices
(b) In-vehicle displays
(c) At-stop displays
(d) Displays at 3rd party locations

The Internet and mobile channels are increasingly used for providing RTPI, advanced information allows the customer either to choose a better trip alternative or to manage their time better. Real-time information at stops is provided on pole-mounted information displays (Figure 20.1). The display units are LED based with four lines of information, indicating the status of next four vehicles to arrive at the stop.

FIGURE 20.1 Pole-mounted display at bus stop, Zurich, Switzerland (*Source:* VBZ, reproduced with permission).

The information shows:

(a) Line number
(b) Destination
(c) Minutes to arrival
(d) Wheelchair symbol, if the vehicle is accessible

The data is communicated wirelessly to the individual stops from the AVM system. RTPI in the vehicle is displayed on high-definition display screens. The unit is equipped with light sensor which varies the luminance level depending on the ambient light (i.e., turns it down at night, increase in bright sunlight). This functionality not only improves the visibility of the display units but also reduces power consumption and generated heat where possible. The baseline information (line, stops, schedule, transfer information) is pre-fed in the on-bus computer which can be updated periodically as and when required. The real-time information is provided by the AVM system.

There are three main display modes for the on-bus RTPI:

(a) Normal mode which displays information for the next stops
(b) Connection information at interchange points
(c) Messages and alerts from the control centre

In normal mode, the following are displayed (Figure 20.2):

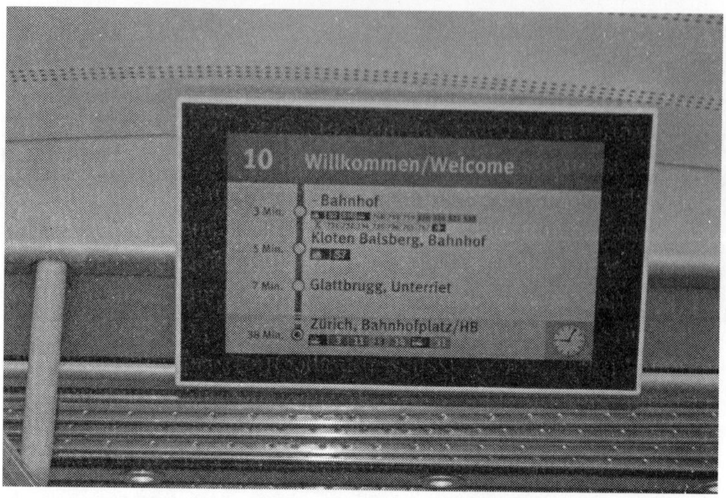

FIGURE 20.2 On-board display (in bus) (*Source:* VBZ, reproduced with permission).

(a) Real-time passenger information
(b) The route number
(c) The current stop
(d) The next three stops, with the number of minutes until arrival
(e) The destination stop, with the number of minutes until arrival
(f) The current time
(g) For each stop, the routes with which connection can be made

Traffic signal priority

To encourage use of public transport, a very high level of priority is accorded to public transport at traffic signals in Zurich. The traffic signal priority is operated from the traffic control centre of the City of Zurich. The traffic control system is responsible for management of the general traffic to minimise congestion, and to accord as much priority as possible to public transport according to the specified rules irrespective of the fact whether the vehicle is on-time, early or late.

Traffic control system

A traffic control centre has been established in Zurich as part of implementing a computer aided traffic control strategy. The traffic control centre possesses complete traffic status information—at network, link and junction level—of various analysis routines, and has CCTV views of key points. Traffic signal priority is triggered by presence of public transport vehicle which is based on the traffic management devices. However, it does not receive any location information from VBZ's AVL system. There are around 400 sets of traffic signals, and about 4,000 traffic detectors which are managed by traffic control system in Zurich. Most of the traffic detectors are induction loops type. The induction loops are able to detect the presence of the vehicle and identify whether it is a public transport vehicle or not. A set of priority rules has been decided and fed in the system for each junction and approach which includes both the priority action to be taken when there is a one-public transport vehicle, and the allocation of priority in case of more than one-public transport vehicles are detected. The underlying principles for traffic signal priority include normally according priority to tram over bus, and peak direction traffic over non-peak direction traffic taking into account the junction configuration. In absence of any applicable rule, the principle of 'first come, first served' is followed. The normal way of giving priority is to curtail the current phase (if required) and to advance the green aspect. This is normally performed by adjusting the phase timekeeping the phase sequence as same. There are fixed rules such as the total cycle time shall not exceed 72 s, the waiting time for pedestrian shall not exceed 30 s. In case of trams, the principle of 'zero wait' is adopted meaning that the traffic signal will always be green when the tram approaches it. So, vehicle can confidently approach the junction without controlling the speed considering that the light will be green before it approaches the signal. The central system is capable to identify faults, and generate an alarm if no signal has been received within a stipulated period of time.

Scheduling

DIVA scheduling package has been deployed by VBZ for generating timetables of the vehicles and driver scheduling. All of the detailed timetable and scheduling development are prepared by the software with except the transitions at the start and end of the peak hours where some smoothing may be done manually. For example, at the end of morning peak, the headway of vehicles is increased from 7.5 min to 10 min. This change in headway is done by manual intervention in the system. The timetable is developed keeping in mind the temporal and spatial variation in demand pattern. The crew schedule is also required to be customised manually sometimes to consider working hour, shift and layover policies of the drivers, however, the timetable is never changed to suit such driver-related constraints.

Fare collection

The fare collection is done before boarding the vehicle. VBZ buses are not equipped with fare collection equipment. The passengers can purchase tickets from self-service vending machines installed at the platform. These offer tickets throughout the Canton of Zurich, and users select how many/which zones they wish, how many trips, any applicable discount, etc. The machines are compatible to both cards and cash. Presently, paper tickets are being issued. Journey and day-related tickets must be validated prior to travel. A mechanical punch removes part of the ticket and makes a time, date and location stamp. There is no need of any further validation, and the ticket can be used on all VBZ modes throughout its validity.

System integration

Even though ITS systems have evolved over four decades in Zurich, the ITS systems are well-integrated. The on-board computer unit provides the primary integration of all on-board ITS-related devices (except fare collection, where they are used). The on-board computer hosts the on-board network and holds the reference data required by the various devices/systems. Within VBZ, the ITS is completely integrated across all the public modes of transport. The trams, trolleybuses and other buses are treated as equal for the purpose of ITS as all of them are public transport. The key interfaces between the AVL and other systems include:

(a) RTPI for smart phone users
(b) Integrated ticketing system
(c) Vehicle scheduling (DIVA)
(d) S-Bahn (SBB) to exchange real-time information and transfer information
(e) SAP, for maintenance scheduling (including unplanned work) 3rd party/external systems

20.5.4 Benefits Arising from the ITS Systems at VBZ

VBZ has been using ITS for operations management for almost 40 years, and has been making available real-time information to passengers for almost 20 years. ITS is an integral part of VBZ's business and operational processes.

Over the time, VBZ has moved far beyond locating its trams and buses, monitoring their running, or looking for data about the day's performance—all these things have long been treated as baseline. With modern ITS tools, VBZ is working on the principle of operating their service according to the schedule, optimising the resources, and guiding the customer through the trip.

The focus now is on:

(a) Maintaining quality of service as general traffic in Zurich approaches the capacity
(b) Integration of the passenger transport services not only in the City of Zurich but also to the neighbouring areas
(c) To ensure closely defined connections to offer seamless and reliable travel to passengers

The expected benefit chain has evolved beyond immediate tangible operational gains. The expected benefit chain offers integrated and seamless travel chains to improve customer satisfaction, which makes passenger transport more attractive, leading to both increased patronage and customer retention, and to the broader strategic objective of shifting passengers from private to public transport. This is required at the policy level as well as at the practical level as traffic in Zurich approaches capacity. The desirable quality of service and reliability become increasingly difficult to maintain when traffic conditions become less free-moving, which causes a drop in the public transport usage and modal share.

20.6 CASE STUDY: ITS IN JAPAN

In Japan, Intelligent Transport Systems (ITS) have played a major role to reduce traffic congestions. ITS is designed to integrate people, roads and vehicles in order to resolve road traffic problems such as congestion, accidents and environmental degradation. The primary goals of Japan ITS are (a) less than 2,500 traffic accident fatalities by (2018) and Japan will have (b) World's safest road traffic society by (2021) leading to zero fatality. There are nine identified areas of ITS implementation in Japan which are.

1. Advances in Navigation Systems
2. Electronic Toll Collection
3. Assistance for Safe Driving
4. Optimization of Traffic Management
5. Increasing Efficiency in Road Management
6. Support for Public Transport
7. Increasing Efficiency in Commercial
8. Support for Pedestrians
9. Support for Emergency Operations

ITS-related infrastructures have been financially supported by the government for thirty years. Vehicle Information and Communication System (VICS), Electric Toll Collection System (ETC) and ITS Spot are major governmental ITS services in Japan.

20.6.1 Widely Deployed ITS Service in Japan—VICS

Vehicle Information and Communication System (VICS) was introduced in Japan in 1996. VICS is a digital data communication system shown in Figure 20.3(a) and (b) that punctually demonstrates the updated necessary road traffic information to drivers through car navigation apparatus. VICS collects traffic information from road operators and police control centers in Japan. The VICS service is available free of charge to the users except the cost of VICS equipment. VICS helps the drivers to find an efficient route to their destination, resulting in the reduction of traffic congestion. The driver is constantly fed with real time data like traffic backup situations or traffic regulations. Japanese drivers are getting the advantage of VICS. It is also a principal solution to lessen problems of traffic congestions. The processing and editing of information regarding traffic congestions, road control and other traffic data is done at the VICS Center. All the data is sent in words and graphics to the navigation devices installed in the vehicles. The VICS is avaialable in around

40 million vehicles in Japan. VICS is a component of the Advanced Mobile Information System (AIMS) in traffic information supply activities carried out by the police, the traffic supervisor.

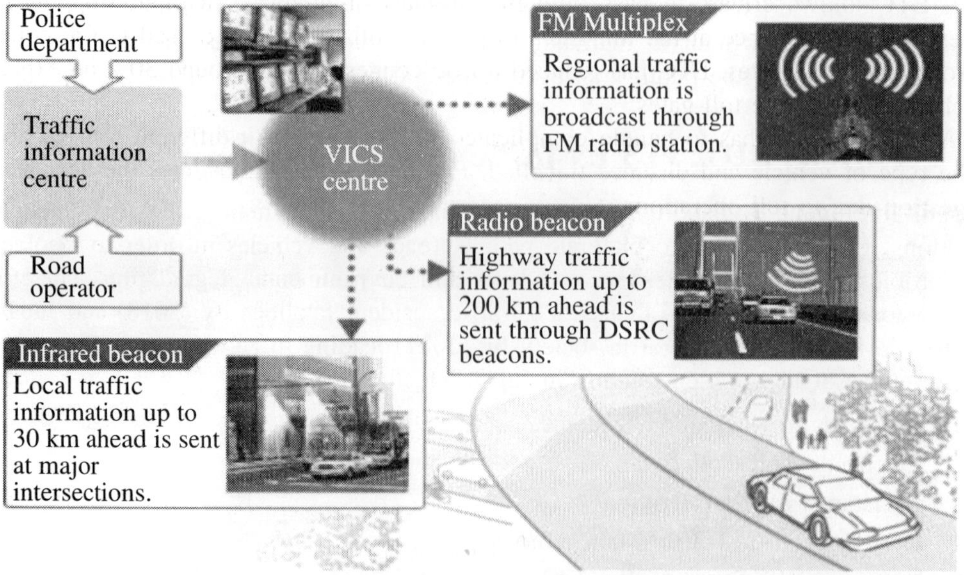

FIGURE 20.3(a) VICS Centre, ITS Japan (*Source:* http://www.vics.or.ip/english/vics/pdf/vics_pamph.pdf).

Display of a VICS compatible car navigation system
(Dotted lines indicate congestion)

FIGURE 20.3(b) Display of VICS showing car navigation system (*Source:* Ministry of Land, Infrastructure, Transport and Tourism, Japan).

20.6.2 ETC—Electronic Toll Collection, Japan

Electronic Toll Collection System (ETC) has been introduced on expressways in Japan since 1997. ETC allows drivers to pass through toll-gates of highways without stopping. Earlier, drivers had to stop once at the toll-gates to pay the toll fee, and it caused traffic congestions in front of the toll-gates. ETC has reduced traffic congestions by around 30% on expressways which occurred at the toll-gates.

An ETC system has to handle complicated toll systems with different toll fee according to the type of vehicle and distance travelled. Figure 20.3(c) demonstrates the decrease traffic congestion due to toll operation.

FIGURE 20.3(c) ETC, Japan (*Source:* https://commons.wikimedia.org/wiki/File:Yokohama_Machida_IC_entrance.jpg, Author: Nikm, Free to use).

ITS Spot

In 2011, ITS Spot service was introduced in Japan which enables the drivers to receive more dedicated traffic information to improve the quality of navigation systems within a city and traffic safety. ITS Spot services consist of three basic services that are made available as an all-in-one system by high speed, high volume road-to-vehicle communications. The first is Wide-Range Road Traffic Information. By transmitting road traffic information in real-time, car navigation systems can search wide-area expressways that cross prefectural borders, urban expressways and other roads, and select the fastest route based on the latest information. The second is safe Driving Support. ITS spots provide wide-area road traffic information on a regular basis including traffic safety issues for specific roads. In emergencies, they provide information that supports safe driving. The third is ETC. ITS spot services utilize the same technologies as ETC, therefore existing ETC services can be enjoyed with a compatible car navigation system. And other services are that information on tourist sites and updated maps are provided over the internet with some car navigation systems. 1600 ITS Spots are installed mainly on expressway in Japan. After released in 2009, totally ten millions of ITS Spot-compatible car navigation systems for automobiles were sold in Japan. For instance, with ITS Spot service, Drivers can receive an alert message in case a congestion occurred at an invisible area beyond a curve as shown in Figure 20.3(d).

FIGURE 20.3(d) ITS Spot system in Japan (*Source:* http://www.its-jp.org/english/files/2013/10/E-GS3_ENG-0802.jpg, reproduced with permission).

20.7 CASE STUDY: LONDON, UK

The need for improved efficiency in bus operations in London has led to the deployment of a wide range of Intelligent Transport System applications, particularly related to bus priority, network management and real time passenger information systems. Some of the ITS tools adopted by London to make transport system more efficient and responsive to customers need are presented here:

A comprehensive bus system management

London, launched a transformation in bus service throughout London in order to promote in increase in the bus ridership that are expected due to the congestion charge, Transport for London. This increased the frequencies of buses (using revenues from the congestion charge), and developed the London Bus Priority Network, an 860 km system of streets managed and enforced to optimize the efficiency of bus service. This initiative has resulted in 70 'Busplus' routes that are featured with advanced passenger information, real-time bus arrival displays, more regular cleaning, low floor buses, modern bus shelters, transit priority traffic signals, and automatic vehicle location and driver instruction systems to decrease busbunching and increase reliability.

By developing 'Quality Incentive Contracts' the city is rejuvinating with increase in bus services, which added financial incentives for operators to improve the efficiency with quality and reliability of the services resulting in additional ridership.

Automated traffic enforcement

Traffic enforcement require a huge amount of commitment of police and judicial resources in case of traditional approaches. Along with this, enforcement actions themselves may give rise to worsening traffic conditions on busy streets. To address these difficulties many cities are resorting to camera-based enforcement of speed limits, red lights, and other traffic laws. Once the violations get detected, these systems opts for license plate recognition to identify and mall citations to vehicle owners. Cases are typically prosecuted less expensively through the civil courts.

This strategy is evolved in London to enforce its congestion charge and to maintain the effectiveness of its bus priority network. Vehicles using bus lanes illegally are being identified by the fixed and bus mounted cameras, generating evidence for enforcement through the civil courts. Along with the introduction of the congestion charge, the adoption of automated bus lane enforcement has dramatically improved the speed and reliability of bus service in central London. This attempts to use smart technology based on RFID, Micro wave and InfraRed system for identifying and detecting the vehicles in the congestion pricing zones in London.

20.8 CASE STUDY: MYSORE, INDIA

Mysore is the 3rd largest city in Karnataka spread geographically over an area of about 128 km^2 having a population of 8,87,446 as per census of 2011. The city is known for tourism, education, medical facilities, etc. Due to setting up of industries such as software, information technology, etc. in Mysore, the number of automobiles has been on the rise in the urban area of Mysore. Mysore city has been grappling with the expected problems of traffic congestion and related issues including delayed and unreliable bus services, limited information about routes, stops, frequency and timetable.

Karnataka State Road Transport Corporation (KSRTC), the State Transport Commission for the State of Karnataka, operates about 7,000 buses on intra and interstate bus services. Mysore City Transport Department (MCTD) is a division of KSRTC, providing urban bus services in the city of Mysore, using a fleet of about 400 buses out of 3 depots and 6 bus stations. Karnataka State Road Transport Corporation (KSRTC) introduced the ITS to deliver high quality services and make the system more passenger friendly through the appropriate use of ICTs.

Launched in 2012, the Mysore ITS has been planned to manage public transport in the city to make it safe, efficient and environment friendly by improving urban public transport in Mysore. The ITS in Mysore was started in 2012. It involved application of information and communication technologies (ICTs) to enable smarter transport networks that help traffic management, ensure real-time control and safety besides curtailing the growth of private vehicles.

The ITS being implemented by KSRTC system in Mysore consists of:

(a) Real-time passenger information at bus stations, bus stops, and in-vehicle

(b) Platform dispatchers at the bus stations

(c) Automatic vehicle location based on GPS and GSM/GPRS

(d) A central control station at the main bus station to manage services along the routes and at outer terminals

(e) An extensive monitoring and evaluation programme to examine the implementation phase, the system and service performance, and the long-term impacts

20.8.1 Reasons to Implement ITS

As a stated policy, the city of Mysore is committed to promote public transport. The city is constantly growing in terms of area and population and as a result the motorisation is also increasing at a faster rate. The city is planning to improve attractiveness of the public transport to encourage people to shift from private to public transport and to encourage migrants to opt for public transport. It is believed that many people would like to use public transport considering that their current means of travel face increasing travel congestion, unpleasant conditions, and increasing out-of-pocket costs as fuel prices rise.

Implementation of effective ITS measures to provide relevant travel information to passengers before and during their journeys is considered as one of the important means to increase attractiveness of the public transport. Further, ITS is helpful in improving the availability and reliability of the services. A stated preference survey conducted for KSRTC revealed that 89% of the people would like to shift to public transport if KSRTC services are reliable. Thanks to proactive approach and dynamic approach, KSRTC has consistently been ranked as one of the leading State Transport Undertaking (STU) in India and received many awards for its quality and innovation. It is natural for KSRTC to lead on new methods to improve service quality and operational efficiency.

20.8.2 ITS Equipment and Devices

In-vehicle ITS-related equipment consists of:

(a) Vehicle mounted unit with inbuilt GPS, GSM and antenna

(b) In-vehicle display screens (1-line)

(c) Bus terminal screens (4-line)

(d) Large display screens for bus stations

(e) Passenger information management system

(f) Bus stop display screens (2-line)

(g) Central control station hardware and software

(h) IT platform, including servers, data storage and peripherals

(i) Application software

20.8.3 Applications for which ITS is Used in MCTD, Mysore

The ITS system installed at MCTD provides the following applications:

(a) Real-time passenger information in-vehicles, at bus stops and bus stations GPS-based automatic vehicle location

(b) Central control station applications (described above)
(c) GSM-based data and voice communication
(d) Real-time passenger information on the Internet and SMS
(e) Digital signage system

A general schematic of the system is shown in Figure 20.4.

Vehicle mounted unit (VMU)

The specified functional requirements for the VMU are:

(a) It generates location databased on GPS
(b) It provides data and voice communication facility using GSM/GPRS
(c) It transmits longitude and latitude position to the central control station system via the GSM/GPRS link, the periodicity (polling cycle) to be user-configurable

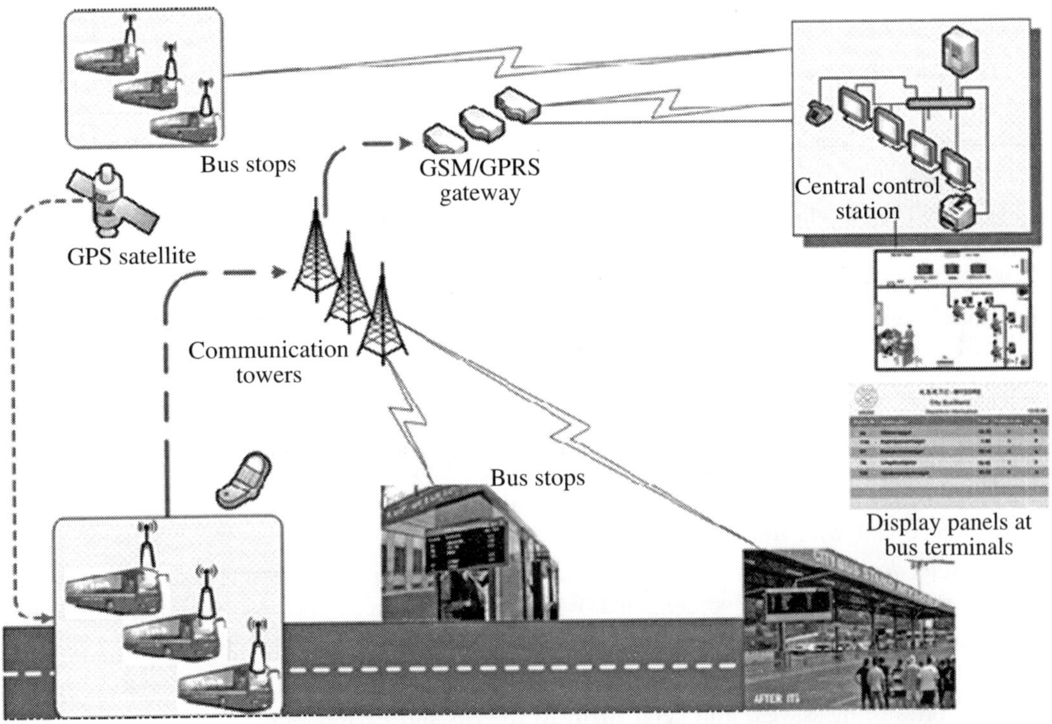

FIGURE 20.4 ITS application in Mysore (*Source:* KSRTC, reproduced with permission).

(d) It generates alert messages for excessive speed, harsh braking or acceleration and other signs of improper driving, the parameters for which should be user-configurable
(e) It hosts two-way communication between the vehicle and the central control station
(f) It is equipped with a microphone/speaker to support voice communication
(g) It has minimum 4 buttons for driver interface, one reserved for SOS/alarm, the others configurable for other message types

(h) It is compatible for integration with other devices such as the display controller over an RS232 interface

The VMU does not function as an on-board computer, i.e., it does not host route data and fare stages for use by other peripherals, nor does it receive transaction data from peripherals for storage and subsequent download.

The AVL system provides location, driving style, incident and alert information to the central control station. A user-configurable polling cycle is set with possibilities to make it distance- or event-based. The AVL supports dynamic trip configuration which allows the CCS to reconfigure the scheduled trips. In particular, it can activate alternate route configurations in case of any temporary diversion of route (due to accident, bandh, etc.) with the revised list of stopping places, and ensures that the in-vehicle next-stop display and announcements are correct.

Communication

The communication is based on GPRS/GSM with following two main streams of data flow:

(a) The Central Control Station (CCS) communicates with the display panels, primarily to provide the real-time information updates

(b) The VMU has data and voice access to the CCS, with the facility to allow voice access to other predetermined numbers

The VMU is having redundancy so that in case of failure of GPRS service, SMS can be activated as an alternate mode.

Voice communication is restricted, and is controlled by the CCS. The driver has following three options to request voice communication:

(a) Select/press a 'request to speak' icon on the driver's console. This is displayed on the dispatcher's screen, the dispatcher may decide whether and when to connect a voice call with the driver

(b) Select/press an alarm icon on the driver's console. This is displayed on the dispatcher's screen with a flashing sign and an audible alarm. The dispatcher shall connect the voice call with the driver

(c) Select/press a hidden emergency button. This is displayed with a special flashing sign on the dispatcher's screen, and with a loud alarm sound. Voice communication is automatically opened up, allowing the dispatcher to hear what is going on in the vehicle

In addition to the request to speak, the driver can select from a menu of status, situation and other pre-fed messages, and send them to the dispatcher.

Real-time passenger information

Real-time information is the main application of the ITS implementation at MCTD. The real-time information system receives input data from the CCS. The real-time information system consists of the following strands:

(a) Bus station display units

(b) Bus-stop display units

(c) In-vehicle display units

(d) Real-time information management
(e) Large display panels
(f) Information management subsystem
(g) Data streaming subsystem
(h) Communications subsystem
(i) Generation of the real-time information and forecast arrival times

Bus location data is obtained by the CCS on real-time basis being generated by the vehicle tracking system. The CCS has a reference database of routes, stops, schedules and scheduled running times, both for operations management and RTPI purposes. For each bus, the CCS calculates the time for the bus to reach all the stops along the route on the basis of current location and any deviation from schedule, and current bus speed. This provides the real-time information set of arrival times of buses for every stop. The CCS transfers the data to the individual displays by GSM for real-time information to the passengers.

Digital signage system

A digital signage system has been installed to manage the content for the digital signs at the various locations. This helps management of the following:

(a) Preparation of content and storage
(b) Allocation of content through the various channels
(c) Facility to organise multiple contents on different areas of the display
(d) Preparation and distribution of the traveller information and alerts
(e) Sales and management of commercial content, including marketing and brand advertising (this is expected to raise significant revenues), and calculation of playtime and preparation of billing

Bus terminal displays

The displays are multi-line (normally 4-line, some 10-line/16-line) with a requirement that their area is readable from at least 30 m, with a good viewing angle, and suitable for use in bright daylight conditions. They support multi-lingual format. The bus terminal displays indicate the routes, destinations and departures times for the routes associated with the related platform or area. These displays receive information from the central server, via wireless or wired communications channels.

Bus stop displays

Bus stop displays have been provided within the bus shelters. They are multi-line (59 No. 2-line, 63 No. 4-line), with dimension of 200 mm × 1000 mm. Similar to bus terminal displays, they are suitable for viewing in bright daylight and support multi-lingual format. The displays present the route, destination and forecast arrival time of the next buses to arrive at the stop. The display units receive their information from the central server via GSM.

Large screen displays

These panels are of approximate dimension 2450 mm × 1525 mm. They will carry both traveller information and commercial advertising.

Internet and mobile device information services

The system is required to support the provision of real-time passenger information and other travel information over the following channels:

(a) Wireless application protocol (WAP)
(b) Internet
(c) Short message service (SMS)
(d) In-vehicle display units
(e) Multi-media messaging services (MMS)

In-vehicles displays

These displays are installed inside the bus, at roof level behind the driver's cab. To allow maximum viewing inside the vehicle, these units are installed facing passenger side. The unit has approximate dimensions 220 × 820 × 150 (all mm), with a single line display. The information received by the unit through serial port. In normal mode, the next stop is displayed. The route and information files are stored in the in-vehicle signage system. GPS location information activates the routes and make announcements about the next stop. Other messages can also be displayed, both from the lists already fed in the system (e.g., travel advice, safety notices) and in real-time from the CCS (e.g., travel alerts, diversions). Information is displayed in both English and Kannada, the local language.

In-vehicle announcement

The in-vehicle announcement system consists of an audio port with amplifier in the display unit which is connected to two speakers located at the front and rear of the bus. The ITS system is capable of three forms of automatic vehicle announcement:

(a) Announcement for next stop, based on the route data (already stored) and cross-checked with the GPS readings
(b) Additional information at designated stops, e.g., historic/tourist information important for first time travellers
(c) Messages which are already fed in the system

The above are pre-recorded files stored in the in-vehicle system. Announcements are made in English and Kannada.

In addition, the driver will be able to make announcement through the VMU microphone system.

Fare collection

The bus conductors have been issued handheld electronic ticket issuing machines (ETMs) to sell tickets. The bus conductors are issued with handheld ETMs at the start of their duty. The ETMs are preloaded with the tariff and stage data for every route. The ETMs are used to issue tickets or recording the pass information (for the pass holders) for each customer. At the end of the shift, the day's transaction data is downloaded from the ETMs to the central system for accounting, MIS and analysis purposes.

Management information and reporting

The management information and reporting requirements can be broadly described in two categories:

(a) ***Operations and MIS reports:*** These are expected to provide a number of reports on the service production, irregularities, performance metrics, etc. The system is further expected to allow quick configuration of new report formats from the available data in consultation with KSRTC.

(b) ***Network management and technical system performance:*** Extensive and detailed requirements have been specified for management and reporting of the IT platform, software systems, network, communications, etc. Service levels management is clearly specified.

System integration

Supplier has carried out integration within the supplied ITS system. At this stage, integration with other existing systems is not specifically provided or defined, e.g., the fare collection system, but it is possible in future.

20.8.4 Benefits Arising from the ITS Systems at MTDC

Implementation of ITS has resulted into following benefits:

(a) Reduction in traffic congestion and pollution levels
(b) Reduction in travel time delays and related uncertainties
(c) Increased commuter satisfaction
(d) Improvement in the reliability of public transport
(e) Real-time information to users
(f) Improvements in safety
(g) Promotion of public transport

This form of extensive and integrated ITS deployment is the first for India. KSRTC leads the rankings of public transport commissions in India, and hence has a high visibility and leadership role within the sector.

20.9 CASE STUDY: INDORE, INDIA

Being the commercial capital of central India, Indore is one of the fastest growing cities in India. The unprecedented growth in population of Indore in recent past is attributed to the rapid industrial and commercial development which has resulted into a large scale increase in traffic on the city roads. This increase in traffic beyond capacity of transport infrastructure has resulted in the manifestation of a number of problems like congestion, delay, accidents, pollution, etc. posing a potential threat to the economic vitality and productive efficiency of the city.

20.9.1 High Capacity Bus Based Rapid Transit System (Indore BRTS)

The Indore bus rapid transit system has been approved at an estimated cost of ₹868.15 crores, by the central sanctioning and monitoring committee through submission Urban Infrastructure and Governance under Jawaharlal Nehru National Urban Renewal Mission (JNNURM). There are 7 corridors identified for developing BRTS in Indore on the basis of travel characteristics, travel impedance matrices, trip assignment transport plan 2025 proposals in CTTS Indore. The 7 corridors are planned to be developed in phases. The feeder routes to BRTS (Figure 20.5(a)) have also been identified as part of the comprehensive mobility plan.

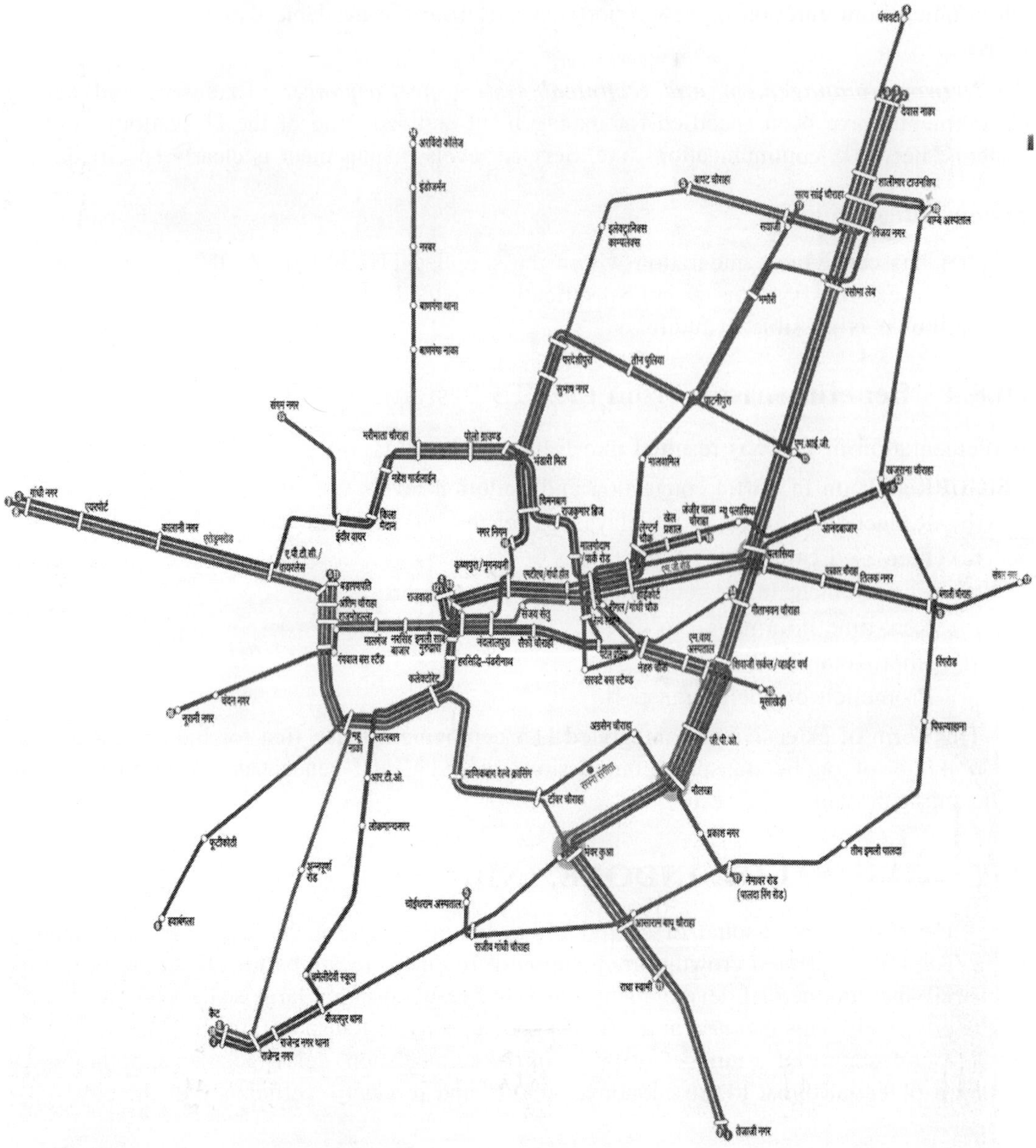

FIGURE 20.5(a) Indore BRT routes, India (*Source:* http://www.citybusindore.com/index.php?page=route-planning).

Indore developed a system approach to implement the BRT system consisting of planning, management, and control through the establishment of Indore City Transport Services Ltd (ICTSL) as presented in Figure 20.5(b).

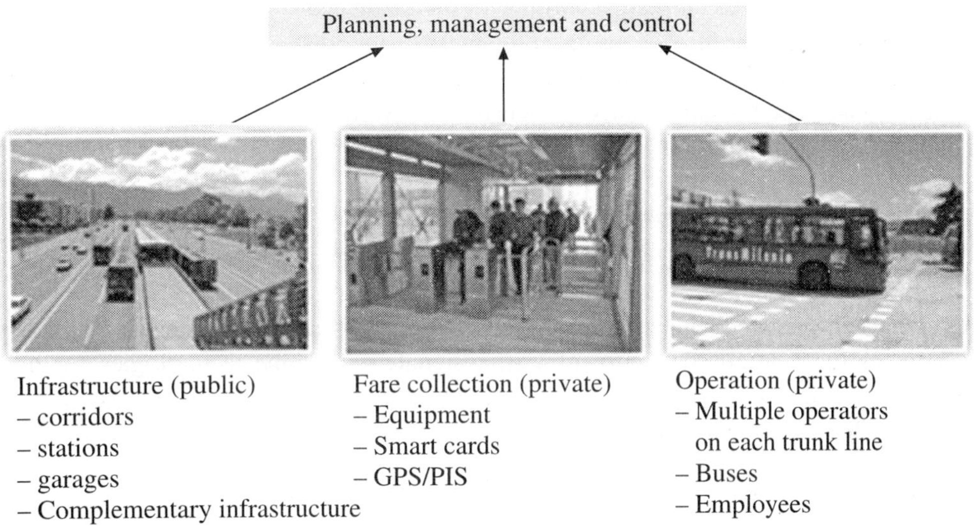

FIGURE 20.5(b) Systems approach to BRT implementation in Indore (*Source:* Detailed Project Report, Indore, reproduced with permission).

20.9.2 Innovative Transport Technologies Used in Indore

ITS application is being planned to improve the safety, efficiency, reliability, dependability and cost-effectiveness of the city's transportation system (Figure 20.6).

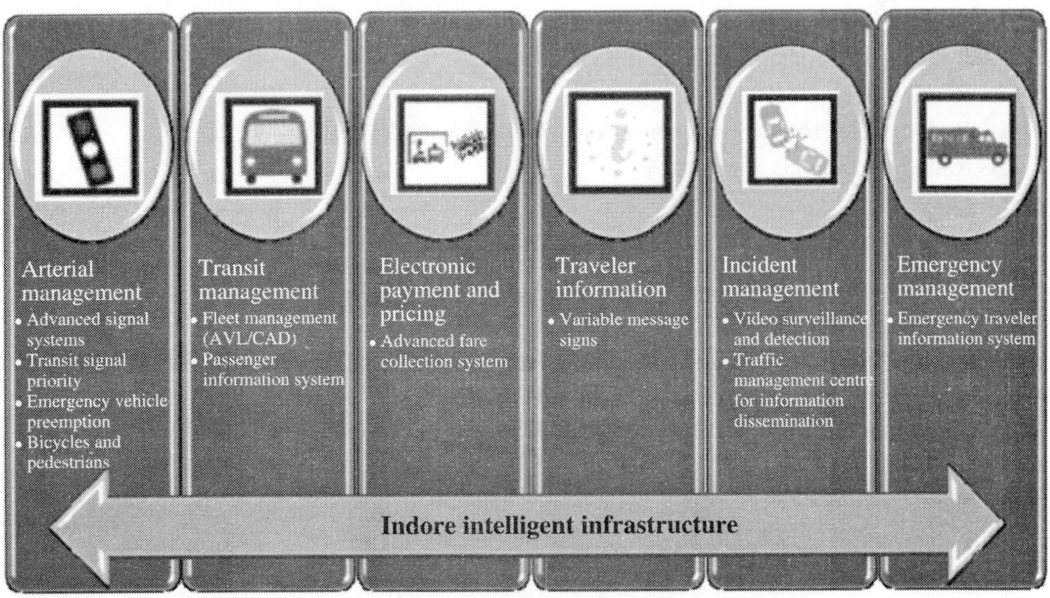

FIGURE 20.6 Intelligent infrastructure applications, Indore (*Source:* Detailed Project Report, Indore, reproduced with permission).

Potential benefits of the proposed ITS system are:

(a) **Increased productivity:** Better operational efficiency of the system.

(b) **Passenger convenience:** Real-time information to users reduces waiting time at the terminals as users can plan according to the estimated time of arrival of the vehicle.

(c) **Increased ridership and mode change:** With improvement in reliability, efficiency and communication mechanism, more and more people would like to use public transport. This would reduce congestion as well as have a beneficial impact on environmental pollution.

20.9.3 Arterial Management

Advanced signal system (ASS)

The ASS has been designed with coordinated signal operations across neighbouring jurisdictions, which includes the centralised control of traffic signals with some technologies for the later development of adaptive signal control. With the help this type of innovative technology, there has been a reduction of cycle lengths, improvement in coordination which has resulted in reduction of:

(a) Motor vehicle delays at junctions
(b) End-to-end travel times in the corridor
(c) Stops/vehicle
(d) Fuel consumption
(e) Emissions, etc.

Quantified benefits

The following benefits are expected from the ITS system:

(a) Reduce stops/vehicle by 15%
(b) Reduce fuel consumption by 2% to 5%
(c) Reduce delays by 15% at junctions
(d) Reduces end to end travel time on road by around 20%
(e) Reduce emissions up to 10%

20.9.4 Transit Signal Priority Systems

TSP is operational strategy to provide priority to the movement of buses through traffic-signal controlled intersections. It has been introduced to improve transit operations, service quality and person mobility and promotes more ridership, reduces traffic congestion, mobile-source emissions and fuel consumption rates. The sensors detect the approaching transit vehicles and alter signal timings to improve the performance of the transit system.

Transit signal priority—strategies

The following strategies are adopted for TSP at a traffic signal:

(a) Green extension
(b) Early green
(c) Phase swapping
(d) Phase skipping
(e) Condition priority

(a) ***Green extension:*** In this strategy, if a request for TSP is received within well in time before the yellow onset, the green aspect of the signal will be extended. This is an important TSP strategy since buses do not have wait a whole cycle before getting the green (Figure 20.7).

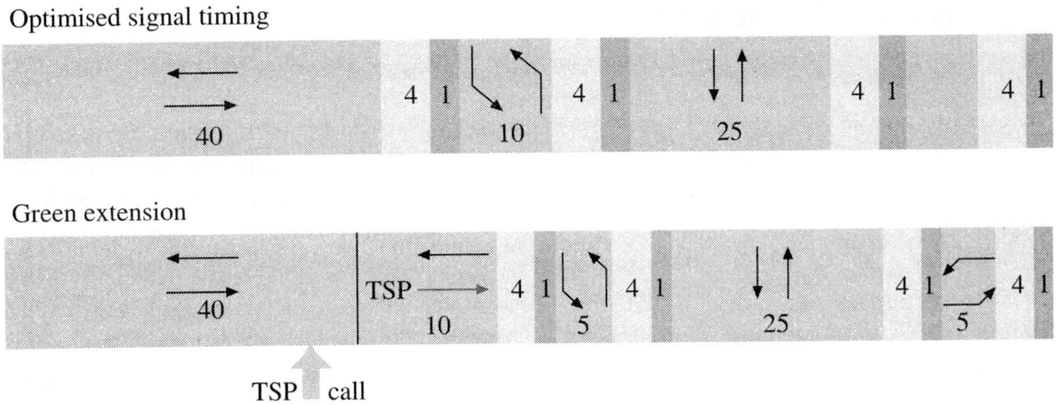

FIGURE 20.7 Transit signal priority (*Source:* DPR, Indore, reproduced with permission).

(b) ***Early green:*** In this strategy, if a request for TSP is received after the TSP phase is timed out, the non-TSP movements are provided only minimum green to provide early green to the bus movement.

(c) ***Phase swapping:*** In this strategy, the transit phase can be introduced immediately after the current phase is time out. However, the road users expect a normal phase sequencing but this strategy get acclimatised with the location, which may result in accidents if users ignore the signal.

(d) ***Phase skipping:*** The phase skipping strategy involves skipping minor phases to provide green to the transit vehicle. However, due to phase skipping all the minor street vehicles have to wait for another cycle to get a green. So this is not the desirable strategy keeping in mind the interest of minor street vehicles.

(e) ***Conditional priority:*** In this strategy, TSP is granted only when a bus is behind schedule with certain time. This is usually a desirable strategy for TSP.

Transit signal priority—operations

TSP operations is generally comprised two primary elements:

(a) Priority request generator (PRG)
(b) Priority request server (PRS)

The bus, through the priority request generator (PRG), submits a request for priority to the PRS. These two elements play a logical process and can be implemented physically in more than one way (Figure 20.8).

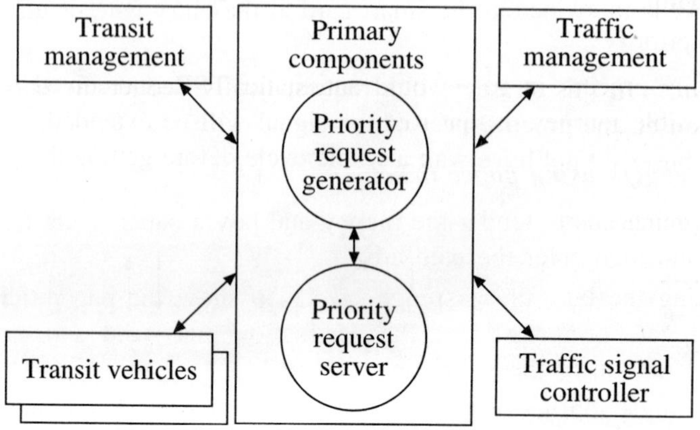

FIGURE 20.8 Transit signal priority—Operations (*Source:* DPR, Indore, reproduced with permission).

The two primary interfaces for TSP functionality are between PRG and PRS and between PRS and the traffic signal controller coordinator. Another element TSP operations in the concept of granting priority directly relates to the traffic signal software, while maintaining coordination with adjacent intersections.

Since the buses in Indore are equipped with AVL system, it is logical to use the technology for granting transit signal priority.

Emergency vehicle pre-emption

This system provides immediate green to emergency vehicles (ambulances, fire trucks, etc.), which can greatly improve the safety of traffic while the emergency vehicle passing through junctions.

20.9.5 Automatic Fare Collection

Indore BRTS bus stand has two gates which acting as the entry and exit (for paid areas). The passenger can enter the station through the unpaid area and has to show his smart card and enter the paid area or take a paper ticket from the TOT and enter the paid area. At the destination station, the passenger would show the smart card again and exit the paid area.

Main components for AFC system

The following are main functional components of the system and interactions during operations.

(a) Ticket operating terminal (TOT)
(b) Passenger information system (PIS)
(c) GPS/GPRS
(d) Smart cards

Passenger work flow process

Work flow of passenger using smart card

(a) The user needs to buy the smart card and deposit required amount at the TOT
(b) The user needs to swipe/tap the smart card at the entry gate or turnstile and enter the paid area
(c) The scanning process in gate would automatically deduct the entire amount for the longest possible journey in that route

Work flow for passengers using paper ticket

(a) The passenger needs to tender the money and buy a paper ticket from the TOT at the bus stand and then enter the paid area
(b) While exiting the bus, the passenger needs to show the paper ticket at space close to the exit gate or turnstile of the destination bus stand and exit paid area. The checking/validation of the paper ticket is manual
(c) Instead of single journey paper ticket, the TOT can used to print return tickets and one-day passes also.

Overlapping of BRTS and non-BRTS systems

In case a passenger carrying a smart card boards on non-BRTS bus from a BRTS bus stand and exists in a non-BRTS bus stand, then the following workflow would exist.

(a) The passenger would display the smart card at the entry turnstile of the BRTS bus stand to enter the paid area
(b) The passenger would board the non-BRTS bus without displaying the smart card
(c) The passenger would exit the bus at the non-BRTS bus stand by displaying the smart card at the exit validator of the bus.

20.10 CASE STUDY: DELHI TRANSPORT CORPORATION DELHI, INDIA

DTC is a public sector undertaking (PSU) under Delhi Government responsible for operation and management of public buses in Delhi. DTC is one of the largest CNG-powered bus service operator in the world. DTC having a fleet of 4020 buses carries 3.5 million passengers per day.

DIMTS

Delhi integrated multi-modal transit system (DIMTS) Ltd. is an urban transport and infrastructure development company instituted in April 2006. It is an equal equity joint venture company between the Government of National Capital Territory of Delhi (GNCTD) and the IDFC Foundation (a not-for-profit organisation).

20.10.1 Reasons for Implementing ITS

The reasons for utilising the ITS include the following:

(a) To provide real-time and other information to passengers before the trip and at stops
(b) To ensure that planned transfers at interchange points can be achieved
(c) To achieve integrated and seamless travel across modes
(d) To gather, analyse and utilise extensive data for future planning, route and schedule rationalisation
(e) To optimise the resource utilisation and maximise efficiency

20.10.2 ITS Applications Adopted by DTC

In order to improve the users experience and enhance operational efficiency, the following ITS applications have been implemented by DTC with the support of DIMTS:

Delhi transit bus info application

The Delhi transit bus info application can be used by the users to get vital information like estimated time of arrival of buses, bus routes, track location of a bus, location of bus stops, etc. The application is available both on the Internet and on Android-based devices. On mobile, the name of the application is 'Next Bus' which can be downloaded from Google Play Store. The screen shot of mobile application is shown in Figure 20.9.

Home page	Route info	Estimated time of arrival

FIGURE 20.9 Screen shot of DIMTS mobile application 'Next Bus'.

The screen shot of the Web portal is shown in Figure 20.10.

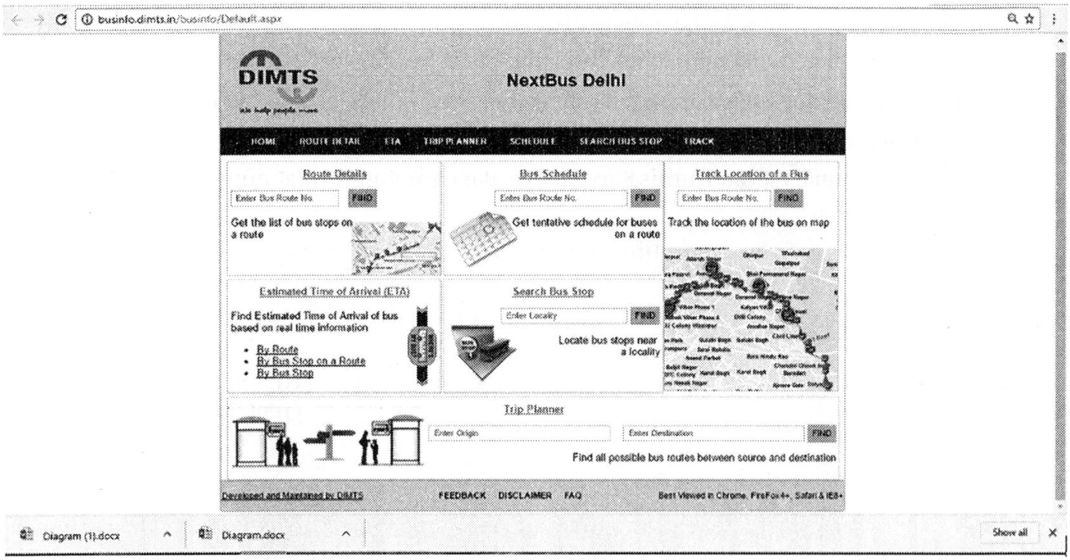

FIGURE 20.10 Screen shot of DIMTS Website (http://businfo.dimts.in).

Automated fare collection system

The automatic fare collection system (AFCS) project consists of designing, developing, implementing, operating and maintaining and facility management of automatic fare collection system and value added services based on smart cards which will be acceptable in multiple travel modes including DTC buses, private buses, metro rail, light rail transit, mono rail and value added services like parking, taxis, autorickshaws, etc.

The main features using AFCS are:

(a) Bring transparency in bus ticketing operations in DTC buses
(b) Cashless travelling
(c) Better control over revenue and precise monitoring
(d) Improve efficiency and decrease operational costs
(e) Conductor less operation using validators
(f) Cost of project can be learnt after opening of financial bid

Soon, the Delhi Metro smart card will be integrated with AFC system of DTC and the same card can be used in metro as well as buses.

Electronic ticketing machines (ETMs)

The ETMs (Figure 20.11), unlike conventional ticket-stub printing machines, are smart card enabled devices with inbuilt GPRS modules. The system is designed to transmit data from ETMs to the central server directly without being connected to the computers at the depots to allow update of master data, configuration data and application in the ETMs over-the-air remotely and in real-time. The ETMs are also capable of validating smart card passes and doing transactions on smart cards. The ETM system has been developed keeping in view, the requirements of a typical automatic fare collection system (AFCS).

These new generation electronic ticketing machines (ETMs) have following features:

(a) Real-time data transfer through GPRS
(b) Smart card enabled
(c) Over the air configuration and update of master data, configuration data and application

FIGURE 20.11 ETMs, DTC, Delhi (*Source:* DTC, reproduced with permission).

The backend system of ETM has following capabilities:

(a) Email operational and revenue to key stakeholders
(b) Performance analysis on route and conductor
(c) Display heath status of the field devices to take proactive action
(d) Day-end revenue reconciliation

After the implementation of ETMs, DTC and Delhi metro are going to launch common mobility card soon through which the commuters can travel in DTC buses and Metro Rail hassle free with the same card.

CCTV in buses

DTC has installed CCTV camera system in buses as a pilot for security of the passengers especially women in view of increasing number of cases of eve teasing. Accordingly, CCTV camera system has been installed in total 200 buses AC and non-AC, i.e., all the 100 buses of Sarojini Nagar Depot and 100 buses of Rajghat Depot. CCTV uses wireless transmission to send the broadcast from video cameras to the monitors or recording devices. Most CCTV used for surveillance which includes:

(a) Security monitoring and safe monitoring
(b) Capturing both gates, face of each commuter, conductor area, driver area and front road view

(c) Camera disconnection alerts
(d) Night vision recoding
(e) Drivers' panic button alert
(f) 24×7 recording including parked buses at terminals or depots
(g) 24×7 service by vendor in both depots for maintenance and providing reports and footage to DTC.
(h) Payment model purely based on uptime of CCTV system
(i) Software for calculating uptime is a part of the project

Operations control centre (OCC)

A state-of-the-art operations control centre has been set up at Kashmiri Gate. The OCC is the key hub for all bus-based transportation activity. The OCC operates and controls the following systems:

(a) Intelligent signaling system (ISS)
(b) Bus rapid transit (BRT) corridor
(c) Automatic vehicle location system (AVL)
(d) Automatic fare collection systems (AFCS)
(e) Private stage carriage in 17 clusters

OCC is equipped with servers that are linked to the ISS, AVL and AFCS components in the field, in real-time through GPRS and optical fibre cable (OFC). The activities are monitored on Barco screens installed in the OCC and LCD panel in addition to the work stations of the respective staff. The command centre is manned 24 h for real-time monitoring of all activities.

20.10.3 Poochh-O Mobile Application

The mobile application called Poochh-O (Figure 20.12) makes it easier for people to get autos and improves their experience of travelling in an auto a richer. The application allows users to locate autos nearest to them and call them on their mobile phones and book a trip. The Android application can be easily downloaded from Google Play store. The app is beneficial for the commuters as well as the auto drivers. While commuters need not hunt for autos, the drivers get assured business without having to move around looking for commuters. Using this app, the users can also find out the estimated fare, travel distance and time of travel, get traffic update of the route they have selected, look for alternative routes as well as track the movement of the autos and even send their own location details to the auto driver, all in real-time. The app also allows the commuters to give feedback as well as share traffic updates, location details and other such details through various platforms such as Facebook, WhatsApp, SMS, Gmail, Google Hangout and more. Poochh-O also provides a link to Tell Tail, another security application developed by DIMTS and already in operation. This enables users to have their auto tracked by their friends and family as well as click on a panic button if there is any emergency.

The key features of the application are:

(a) Locate the autos nearest to you

(b) Call the auto driver on their personal mobile number and book your trip
(c) Calculate estimated fare for your trip
(d) Find out the estimated travel distance
(e) Get information on the estimated travel time
(f) Get traffic update of your route

However the system introduced could not be successful as many a times, the response from the auto rickshaw drivers were not correct as the mobile numbers communicated with the supposed to be drivers did not belong to them.

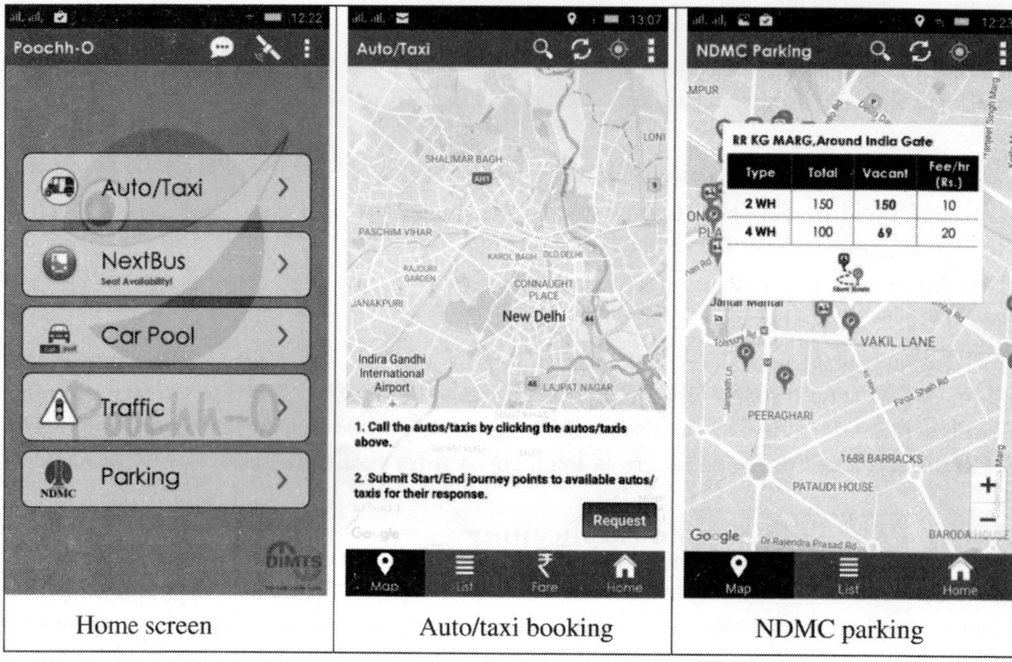

FIGURE 20.12 Poochh-O mobile application.

20.10.4 Future Plans

With successful implementation of above applications, the following ITS applications are planned to be introduced in near future:

(a) VMS
(b) Congestion pricing
(c) On-demand autorickshaw
(d) GreenCAB—Dial-a-Rickshaw Facility

Variable messaging system

(a) A city-wide project will be launched and large LED boards will be installed at various points in the city

(b) Messages will be controlled through a central command centre

(c) Commuters will obtain information to assist their journeys including traffic conditions, traffic advisory, route diversions, etc.

Presently the installation of VMS display board has already been made. The operation is likely to be started shortly.

Congestion pricing

Delhi Government is exploring the option to introduce congestion pricing as a means of managing travel demand on certain roads in Delhi. Delhi Government will identify critically congested arteries where public transport facilities are available for road user charging. In the initial phase, only the CBD area will be covered. A staged roll out will then occur in other parts of the city.

On-demand autorickshaw

Transport department is planning to cover autorickshaws and taxis under the ambit of a vehicle tracking system. A separate call centre will be setup for booking and the dispatch of autorickshaws.

GreenCAB—Dial-a-rickshaw facility

GreenCAB is an idea to provide door-to-door mobility for all classes of people along public transit corridors. In a pilot GreenCAB could use mobile phones to provide a feeder from your doorstep to public transit along the Delhi BRT corridor or other local destinations like schools, shops, etc.

20.11 CASE STUDY: BENGALURU, INDIA

Bengaluru's road network is primarily radial in nature which converges at the centre. The radial network results into unwarranted entry of traffic to the centre. Bengaluru city is grappling with following traffic problems:

(a) Ring roads are inadequate and need more flyovers

(b) Arterial roads are congested

(c) Subarterial roads are congested and have parking pressure

Profile of Bengaluru city

(a) 1000 km^2 of area

(b) 4500 km of road

(c) 40,000 intersections

(d) 330 signalised intersections

(e) 600 manned intersections

The public modes of transport, mainly the buses, are vying for road space and have to compete with other modes of vehicles like autos, two wheelers, cars, etc. With little addition to the infrastructure over the time, it is becoming more and more challenging to manage the congestion levels.

In order to address above issues, the city police have envisaged the Bengaluru Traffic Improvement Project, BTRAC 2010.

20.11.1 BTRAC 2010—Strategy

(a) Set up state-of-the-art traffic management centre (TMC)
(b) Set up nearly 400 signals
(c) Signals to be vehicle actuated, networked, adaptive and controlled/monitored by the traffic management centre (area traffic control)
(d) Install 400 cameras and 20 VMS
(e) Centrally automated traffic enforcement systems
(f) Traffic enforcement cameras
(g) Traffic police mobility, process, communication and modernisation
(h) Capacity building—development of modern traffic training institute

20.11.2 BTRAC 2010—Components

(a) Junction improvement
(b) Intelligent transportation systems
(c) Street furniture
(d) Traffic enforcement cameras
(e) Surveillance/enforcement cameras
(f) Education/awareness and capacity building

The ITS components of BTRAC 2010 have been discussed here.

Variable message signboard (VMS)

Variable message signboards (Figure 20.13) are planned to be installed to inform the motorists about:

(a) Diversions due to natural calamities like flooding, water clogging, fallen trees, electrical poles, etc.
(b) Man-made diversions like strikes, road blocks, agitations, etc.
(c) Safety messages
(d) Road incidents like accidents

Such information will enable motorists to plan their travel route based on current information.

FIGURE 20.13 Variable message signboards (*Source:* http://www.bangaloretrafficpolice.gov.in/, reproduced with permission).

Traffic management centre

The traffic management centre (TMC) has been set up at Ashok Nagar Police Station. TMC is the hub of a transportation management system, where information about the transportation network is collected and integrated with other operational and control data to manage the transportation network and to produce traveller information.

The interim traffic management centre has been developed to house a traffic wireless system, traffic surveillance system, traffic signaling system, traffic help line, traffic planning operations, etc. that are being deployed under BTRAC 2010. The interim TMC shall handle the following components:

(a) Information/data from the 334 traffic signals spread across the city
(b) Live feed of images from the 175 surveillance cameras installed at various junctions in the city
(c) Traffic violation data from the 5 enforcement cameras installed in the city.

The TMC connects various elements of intelligent transportation systems such as vehicle actuated traffic signals, surveillance cameras, enforcement cameras, etc. installed at various locations in the city as a part of BTRAC 2010. TMC enables the decision makers to identify and react to an incident in a timely manner based on real-time data. The TMC also helps in reducing incident response times, lower incident rates (mainly secondary incidents), disseminate traveller information and hence reduce congestion and enhance safety.

Traffic signals

From the traffic signal system, a steady flow of traffic signal timings, traffic flow data, etc. are transmitted and this real-time data can be processed immediately to achieve a synchronised/optimised effect on the chosen corridors.

The TMC receives the real-time data from the field which is being witnessed on a monitor. On the basis of real-time information, optimised plan is generated. The plans so generated are communicated to the local police officers on the field either through the wireless system present in the TMC or by the operations team.

Traffic surveillance cameras

The cameras located at 175 strategic locations across the city would be transmitting live pictures into the TMC. The traffic management team is supposed to look at the live feeds coming in from the field. On the basis of certain parameters like tolerable queue length, congestion or an accident/incident that has occurred on the filed would be able to communicate to the field officers and necessary action to be initiated. Further, the violations happening on the filed can be observed through the system and recorded for further use as evidence in the court and also action initiated through communication to the field officers.

The system collects real-time video from the specified location by using pan tilt zoom (PTZ) camera system transmitting this to the traffic control centre through an IP network. The system therefore comprises outstation equipment installed at the various location in Bengaluru city and complementing equipment installed at the TMC.

Traffic enforcement cameras

The traffic enforcement cameras (Figure 20.14) have been installed at 5 locations across the city which would be recording and beaming the violations into the TMC. The red speed enforcement camera takes photos of motorists who cross stop line or jump traffic signals or exceed speed limit. The enforcement team are required to download the data for cases of violations and then transfer the information to the automated enforcement system. The violators are automatically sent challans through automation enforcement centre (AEC) within 7 days of date of violation. Automated enforcement is considered as a user-friendly and transparent mode of enforcement. The traffic police personnel on duty notes down the numbers of the offending vehicles, the information passed to the automation enforcement centre and computerised challans are generated and sent to the offender. Blackberry and wireless printers have also been provided to traffic subinspectors and inspectors for automated enforcement.

FIGURE 20.14 Traffic enforcement cameras (*Source:* http://www.bangaloretrafficpolice.gov.in/).

20.11.3 Benefits of BTRAC 2010

(a) Substantial compliance of traffic laws and rules will be achieved
(b) Effective trauma care system will be set up
(c) Co-ordinated traffic management will be achieved
(d) Traffic congestion will be reduced by 30% in the central area of Bengaluru City
(e) Level of traffic and road safety awareness will be enhanced
(f) Accidents will be reduced by 30% in the City of Bengaluru
(g) There will be significant reduction in pollution
(h) State-of-the-art traffic policing and regulation will lead to substantial compliance

20.12 CASE STUDY: AREA TRAFFIC CONTROL (ATC), MUMBAI

Area traffic control (ATC) systems are smart real-time dynamic traffic control systems which are designed to effectively react to variations in traffic conditions. ATC is a system responsive to traffic which uses data from vehicle detectors and optimises traffic signal time on real-time basis. The timing plan of traffic controllers is automatically changed. The system employs digital computers for achieving the desired functionality. The ATC Project of the Mumbai Traffic Control Branch is concentrated on synchronising significant intersections. The system has been actualised through the Mumbai Metropolitan Region Development Authority (MMRDA) and Municipal Corporation of Greater Mumbai (MCGM). World Bank has provided necessary budgetary guidance. Modern gadgets such as speed check guns and multi-radar comprising smart cameras, radar sensor, screen, manual control unit, flash generator, flash light, power box and tripod have also utilised as a part of the ATC project.

20.12.1 Working Principle

(a) The vehicle detectors collect the information about the traffic flow and send it to the signal controller.
(b) The signal controller forwards the traffic flow information to the central computer for further processing.
(c) The central computer makes computations on the basis of historical data and present traffic flow and communicate back to the signal controller with processed data.
(d) The signal controller sends timing information to the roadside signal.

The working principle of ATC, Mumbai is shown in Figure 20.15.

20.12.2 Components of ATC, Mumbai

The components of the system procured for Mumbai are as follows:

(a) Fully adaptive ATC system is a software-based system in the control centre. The system which won the bid was intelligent traffic adaptive control area (ITACA), developed and marketed by the Spanish company, Telvent Trafico Y Transporte—now owned by Schneider Electric of France

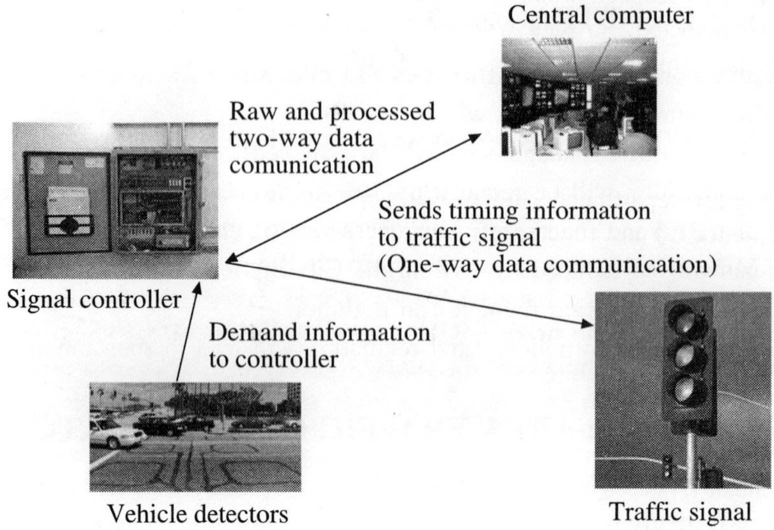

FIGURE 20.15 Working principle of ATC, Mumbai (*Source:* Presentation by Dr. Shankar Vishwanath, Ex Director (ES&P), Municipal Corporation of Greater Mumbai).

(b) New signal posts and signal controllers
(c) Vehicle detectors
(d) Leased data communication network from MTNL
(e) Central control room at traffic police head quarters and information centre at MCGM offices
(f) Civil works including development of a fully ducted network which is expected to ease the maintenance process

20.12.3 Benefits of ATC

The ATC offers following advantages:

(a) **Reduction in journey time for vehicles:** Traffic control system minimises the overall journey time by reducing the number of stop delays, increasing the average travel speed, etc.
(b) **Controlling accidents:** ATC reduces the number of accident by reducing the congestion.
(c) **Saving in fuel:** Reduction in travel time and congestion help in saving the fuel.

However, the ATC system in general is very costly and quite complex in nature. ATC is suitable only for lane following traffic. In area traffic control system, vehicle detector is used to collect the data to find the actual flow and to get signal timing according to the present condition of traffic. These vehicle detectors detect the vehicle on the basis of lane. For example, we are collecting data for two-lane road then the detectors will able to detect the vehicle which will come from their respective lane and the vehicle which is using space other than these two lanes cannot be detected. So data will not be accurate. So we can say that it will give best result only for lane following traffic.

20.13 USE OF ITS IN RAILWAYS: CASE STUDIES

Among government departments, Indian Railways has been in the forefront to use IT systems for the convenience of passengers, efficient operations of trains and better management of its services. The Ministry of Railways set up Centre for Railway Information Systems (CRIS) as a Society in July 1986 with a mandate to leverage IT systems to improve to functioning of Indian Railways. As per Website of CRIS (http://cris.org.in), the following applications have been developed by CRIS to meet its objective:

Passenger reservation system (PRS)

The tickets for reserved accommodation in trains run by Indian Railways are facilitated by the passenger reservation system (PRS). PRS provides reservation services to nearly 1.5 to 2.2 million passengers a day on over 2500 trains running throughout the country. The PRS application country-wide network of computerised enhanced reservation and ticketing (CONCERT) is the largest online reservation application in the world which is developed and maintained by CRIS. The system currently operates from 5 data centres. The server clusters are connected together by a core network that enables universal terminals across country, through which the users can reserve a berth on any train, between any pair of station for any date and class. CONCERT is a complex online distributed transaction application based on client server architecture interconnecting the regional computing system into a national PRS grid.

Dynamic information available on the Website includes enquiry and static information dissemination. Enquiry includes PNR status, train between pair of stations, accommodation availability, train schedule, and station code enquiry. Static information dissemination includes reservation/refund rules, general information, tourist information, special train, and availability of passenger amenities at any station.

E-ticketing

Indian Railways passengers book tickets for reserved accommodation from IRCTC Website or IRCTC mobile application. Centre for Railway Information System (CRIS) has successfully developed the Internet ticketing solution launched by Indian Railway Catering and Tourism Corporation (IRCTC). The effort involved interfacing the IRCTC front end with backend PRS servers, writing procedures for search and queries at the backend, ticket printing on existing clients and accounting software. The system has capacity to book about 15000 tickets per minute. E-ticketing Website for reserved tickets now handles about 60% of total reserved tickets.

Unreserved ticket system (UTS)

Twenty-one million passengers travel daily on Indian Railways in unreserved category. An unreserved ticket allows the passengers to travel in unreserved coach, i.e., without any reserved seats or berths. As such the ticket is general in nature and is not specific to a train service. This facility is predominantly used by commuters and suburban travellers over short distances where reservation is not a necessity. The unreserved tickets can be issued any time of day or night as booking offices remain open 24×7. The unreserved ticketing system (UTS) caters to this segment of the market and seeks to provide a centrally administered computerised ticketing system over the entire Indian Railways. Today more than 95% of the unreserved tickets are sold through this system. The ability of the system to deliver ticketing at remote corners of the country and provide uninterrupted services everywhere is an example of successful implementation of ITS in ticketing system by Indian Railways.

Paperless ticketing

To facilitate convenient booking of unreserved tickets by passengers without having to stand in long queues to purchase railway tickets, Indian Railways has introduced the facility of booking of unreserved tickets, season tickets and platform tickets through mobile phone. Paperless tickets can be booked through smartphones through the app as per geo-fencing limits and the ticket stored in the smartphone in encrypted form which can also be checked by the ticket checking staff through the app. This facility is available at suburban sections of Indian Railways.

In addition, paper tickets can be booked on smartphone app and print taken from the automatic ticket vending machines (ATVM) at all locations where ATVMs are available.

Freight operations information system (FOIS)

The Indian Railways annually carries nearly 1 billion tonnes of freight. This translates to about 9000 freight trains daily. Indian Railways earns two-thirds of the revenues from freight business. The major commodities carried by Indian Railways are coal, iron ore, food grains, iron and steel, cement, petroleum products, fertiliser, containerised traffic and army equipment. There are specialised wagons to handle the transportation needs of the different types of commodities. Unlike passenger carrying trains, freight trains do not run as per the timetable (schedule) which makes freight operations a highly information intensive activity. The information is essentially required for operations manager to make allocation decisions continually, to dynamically optimise utilisation of resources like wagons, locomotives, crew and paths on the network. Real-time information allows effective decision making and thus ensures high levels of mobility within the system. The freight operations information system (FOIS) was the first project which CRIS embarked upon to improve efficiency of freight operation in Indian Railways. FOIS has started as an application to track and monitor the movement of wagons, locomotives and unit trains. Over the time, new features have been added and now it is a complete management module for freight trains handling the billing and revenue collections as well. It has been playing a major role in the improved wagon productivity on Indian Railways and the objective is to use the information to further improve productivity, customer service and thus meet the needs of a rapidly growing economy.

Control office application (COA)

Train operations on the Indian Railways are controlled and monitored by the control rooms in each of the seventy-seven divisional/area control offices. The control room is the brain of the division. The mobility of train movements over the jurisdiction of the division largely depends on the efficiency of the control room operations. The control office works on 24×7 basis. The use of IT based control office application has changed the way the trains are monitored as train movements are recorded in real-time and movement of scheduled and unscheduled trains planned and controlled through the computer aided interface.

Along with freight operations information system (FOIS), COA has led to all information on train operations being computer generated. It is this application that feeds the national train enquiry system (NTES) that provides passengers up-to-date information on train running. COA also provides train operations information to FOIS and integrated coaching management system (ICMS) also.

Integrated coach management system (ICMS)

Over 50 thousand coaches needs to be tracked and monitored by Indian Railways on day-to-day basis. Efficient deployment of coaches help railways in running more services for the benefit of their passengers, while minimising any chances of disruption and disappointment. Further, timely maintenance of rolling stock also need to be ensured by the operator. Depending upon the need, the system can deliver historical records from past years in printed reports, or up-to-the-minute information on an official's cell phone.

The integrated coach management system is able to offer all above functionality and thus helps in improving the efficiency and utilisation of coaching stock (coaches).

National train enquiry system (NTES)

NTES is an integral part of integrated coaching management system developed and maintained by CRIS. In spite best efforts by Indian Railways to run all passenger carrying trains as per their schedules and maintain their punctuality, at times for reasons beyond the control of Indian Railways trains get delayed. Depending upon amount of delay, the trains are rescheduled from their starting stations, cancelled or diverted to another route resulting in change in the actual arrival/departure time from their scheduled time. The national train enquiry system (NTES) provides information to public about expected arrival/departure of trains at each stopping station, train schedule information, information about cancelled trains, diverted trains, and also platform berthing information at major stations. The main purpose of NTES system is to provide timely and reliable information to general public through user-friendly interfaces and all India accessibility has been achieved to a large extent. Now the information is conveniently and reliably available to public all over the country through various delivery channels, i.e., through Web browsing, through mobile phone or landline (voice and SMS) and also in person through face-to-face enquiry and displays at all IR stations. NTES has won National Award for e-Governance 2010–2011 for innovative use of ICT by PSUs for customer's benefit and national train enquiry system (NTES) was adjudged as the finalist in the Computerworld Honors Program instituted by the Computerworld Information Technology Awards Foundation, USA.

Software aided train scheduling and network governance (SATSaNG)

Due to number of inherent constraints like limited capacity, mixed traffic (passenger as well as freight), multiple types of train control systems, development of a timetable for trains on a busy network like the Indian Railways is an extremely complex and challenging task. Conventionally, the timetable controllers in zonal railways work independently and then collaboratively with other zonal level timetable controllers to design the all India timetable. The time tabling activity is governed by two broad objectives: continence of the passengers and operational feasibility. Introduction of new train services and augmenting older ones in already congested railway network is an art and a select group of planners are highly skilled in this task. Indian Railways has decided to provide software tools to aid the planning process. The software aided train scheduling and network governance (SATSaNG) project is entrusted with the task of building such a tool. The entire resource allocation process will now be aided by the tool leading to more efficient allocations and robust timetables.

Integrated train enquiry system (ITES)

Till 2010, Indian Railways was operating numerous telephonic enquiries all over the country manually, i.e., by its own railway staff. Due to regional in nature, these enquiries provided limited information pertaining to stations under their jurisdiction. Moreover, the manual inquiries were not very customer friendly when it comes to status of arrival and departure of trains through stations under their jurisdiction. The project provided an integrated railway inquiry on pan India basis through a unique number (139) which is a category-I code as defined by DoT and is accessible from anywhere on all landline/mobile/WLL phones irrespective of TSP, without dialling the STD code. Both manual and IVRS enquiries of this enquiry system obtain information online from the already existing computerised NTES and PRS of Indian Railways.

Indian Railways has outsourced enquiry services to Bharat BPO on PPP model and a significant part of the latter's earnings comes by sharing revenues with TSPs on incoming calls for 139. With this project, a major business benefit to railways has been in the form of cost saving and gainful deployment of enquiry staff rendered surplus as a result of closure of manual enquiries.

Smart phone mobile application

Delhi Metro has launched smart phone mobile application on 22nd August, 2013. This application provides important information like fare and journey time from one station to other station, nearest station information and additional information regarding metro stations, which assists commuters to plan their journey (Figure 20.16).

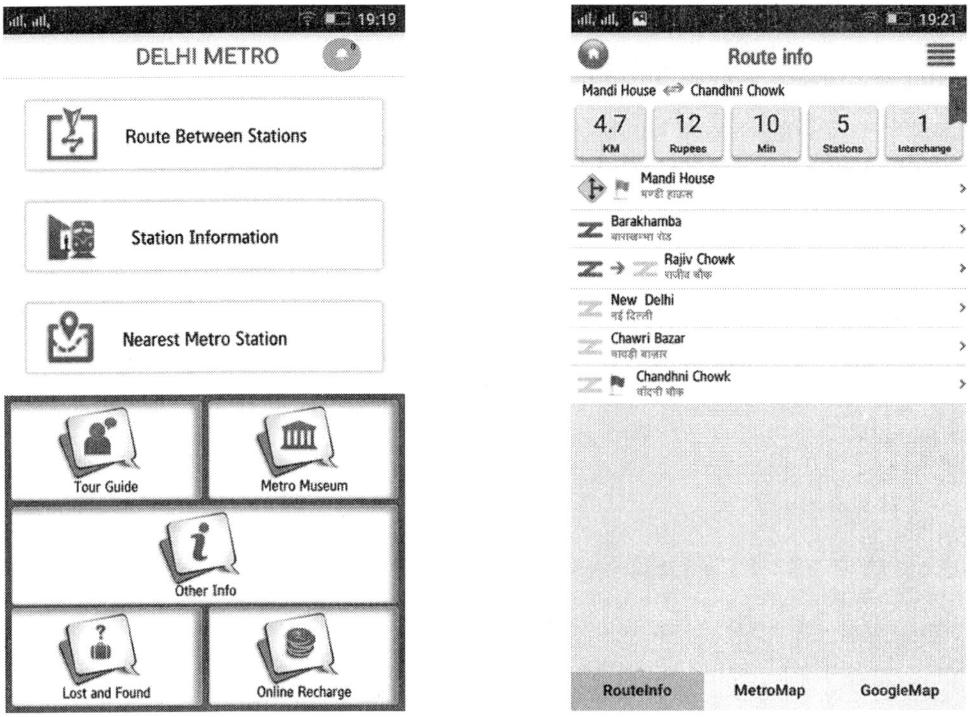

FIGURE 20.16 DMRC mobile application.

Rail SAARTHI – An Integrated Mobile App

With the advent of mobile phones, various apps have been developed enabling passengers to avail themselves of different services of Indian Railways like UTS, PRS, NTES, Air Ticket Booking, Food On Track, Tour Booking, Clean My Coach etc. However, these Apps provided only a single service and the passenger had to search and download each App separately. There was no single App which would allow passenger to plan his journey and book reserved/unreserved tickets and register complaints/ offer suggestions also.

Indian Railways has launched an integrated mobile app 'Rail SAARTHI' in July 2017 for availing all travel-related services such as planning for a journey, ticketing, train related queries, hiring taxis, and porters, e-catering, content on demand, complaints related to the journey, booking retiring rooms, and lounges at stations among other travel related facilities. SAARTHI stands for 'Synergized Advanced Application for Travel, Help and Information. The app which aims to provide a one-stop solution for various services can be used for all travel related services like booking tickets, ordering food from a favourite restaurant, booking tour packaged and finding hotel rooms of passengers' choice outside the stations. The integrated App for Passenger ticketing would help the passengers to book both reserved and unreserved tickets including Season Tickets and as well as purchase Platform tickets by downloading a single Railway App. There is common registration and payment account for all type of ticketing. Expanding the mobile App for unreserved ticketing to all non-suburban stations would further help in reducing queues at the booking offices, eliminating use of paper tickets and promoting digital payment while reducing cash handling at stations. The Integrated Mobile App has replaced the present system where users are required to use multiple Railway apps. The Integrated App offers all user-friendly features and integrating all services provided by Indian Railways. The Integrated App has potential to generate nonfare revenue by utilization of passenger related data available with the Indian Railways as the App will use the affiliate sale partnership model to make available services and products needed by passengers. This mobile app is available on android platform and can be downloaded from Google store. The screen shot of the Rail SAARTHI is shown in Figure 20.17.

FIGURE 20.17 Screen shots of Rail SAARTHI Mobile Application of Indian Railways.

20.14 SUMMARY

There are a number of success stories of application of ITS tools for traffic management, safety, emergency services, real time information to the users and many other emerging areas. This chapter focussed on some of the case studies on ITS applications worldwide. It is important for the planners to assess the efficacy and usability of different ITS applications from success stories so that an efficient ITS system can be designed for the city. As every city has unique features and different challenges, the ITS system needs to be customised to the requirements of the city. However, from the study of the examples in this chapter, it emerges that *Automatic vehicle location and control system (AVLCS), Automatic fare collection system (AFC), Real-time passenger information (RTPI), CCTV, Area Traffic Control (ATC), Variable Message System (VMS) are some of fundamental ITS tools which shall generally be provided in a city depending upon the availability of the funds. The other ITS tools may be progressively added as per the traffic conditions, aspirations of the city dwellers and the availability of funds.*

Today most of the people own a smart phone. A well designed, customer friendly mobile application integrating all travel related services can become very useful for the users. The mobile application shall able to provide real time information of buses, trains, trams etc, fares, offer travel planner and other related services.

We are entering into an era of smart cities where smart travel choices and intelligent transport services would play a major role to make the city smart in real terms. The ITS architecture for the city shall be designed with a vision to develop a smart city where all the services are seamlessly integrated.

Questions

20.1 Give an example of successful implementation of Real-time passenger information (RTPI)?

20.2 What are the main components of ATC Mumbai? Describe each component briefly.

20.3 Briefly describe the ITS infrastructure applications being adopted by Indore City as per DPR?

20.4 What are the different components of BTRAC 2010, Bangalore? Describe each component briefly.

20.5 Write a proposal for ITS tools for a tier two city having a population of 2 million.

20.6 Please short note on following applications used by Railways:

- (a) PRS
- (b) UTS
- (c) FOIS
- (d) COA
- (e) ICMS
- (f) NTES

Bibliography

Bengaluru Traffic Police Website (http://www.bangaloretrafficpolice.gov.in).

Centre of Excellence in Urban Transport, IIT Madras, synthesis report on ITS including Issues and Challenges in India.

Chowdhary, M.A. and A. Sadek, *Fundamentals of Intelligent Transportation Systems Planning*, Artech House Inc., US, 2003.

Detailed Project Report, Indore.

En-route to intelligence mobility, p. 35, November 1999.

http://www.itsindia.org/aitsindia/index.htm

Official Website of Indian Railways:

www.indianrailways.gov.in

www.indianrail.gov.in

http://cris.org.in

Report on Seoul Public Transportation, Seoul Metropolitan Government, 2013.

Sarkar, P.K., Vinay Maitri, G.J. Joshi, *Transportation Planning: Principles, Practices and Policies*, PHI Learning, Delhi, 2014.

Synthesis Report on ITS, IIT Chennai 2010

US DOT, "ITS strategic plan 2015–2019", fact sheet, pp. 2, 15, 48, 54.

US DOT, "Regional ITS architecture", pp. 15, 34, 50.

Williams, Bob, *Intelligent Transportation Systems Standards*, Artech House, London, 2008.

www.nptel.ac.in/cete_49.pdf

www.ssatp.org/sites/ssatp/files/publications/Toolkits/ITS%20Toolkit%20content/case-studies/default.htm

21 ITS Standards and Specifications

21.1 INTRODUCTION

ITS has been recognised as key to address the problems of congestion, accidents, air and noise pollution in urban cities. More recently, concerns about global warming focused our attention on transport as it accounts for about a fifth of all greenhouse gas emissions, mainly carbon dioxide from fuel burnt on the roads by vehicles. The Working Group of Urban Transport for 12th Five Year Plan has recommended for adoption of ITS to overhaul existing urban transport infrastructure in order to meet the increasing demand for mobility. ITS has been recommended to provide high level of service quality and value for money in urban transport sector. Further, the working group has recommended that by 2030, there should be extensive use of technology to manage mobility in the cities. Various forms of road pricing systems such as electronic road pricing, road tolls and congestion pricing which are in use around the world should be used particularly in large cities in the next 20 years. The ITS implementation plan recommended by working group is shown in Figure 21.1

Government of India has instituted the National Urban Transport Policy (NUTP) that lays out the vision for making the cities of India livable and ensures safe, affordable, quick, reliable and sustainable access for the growing number of city residents to jobs, education, recreation and such other needs within the Indian cities.

NUTP 2016 states "ITS is a cost-effective means of increasing the efficiency of the urban transport (UT) network, to manage the crisis of congestion in urban areas, reduce the number of crashes and fatalities, improve safety and security of commuters, promote public transport (PT) usage, and efficient management of freight traffic. Therefore, the Government of India would financially encourage use of ITS for multi-modal integration, traffic management system, freight management system, enforcement of traffic rules, incident or emergency management system, integrated control room, parking information and management system. Also cities should develop a common control centre".

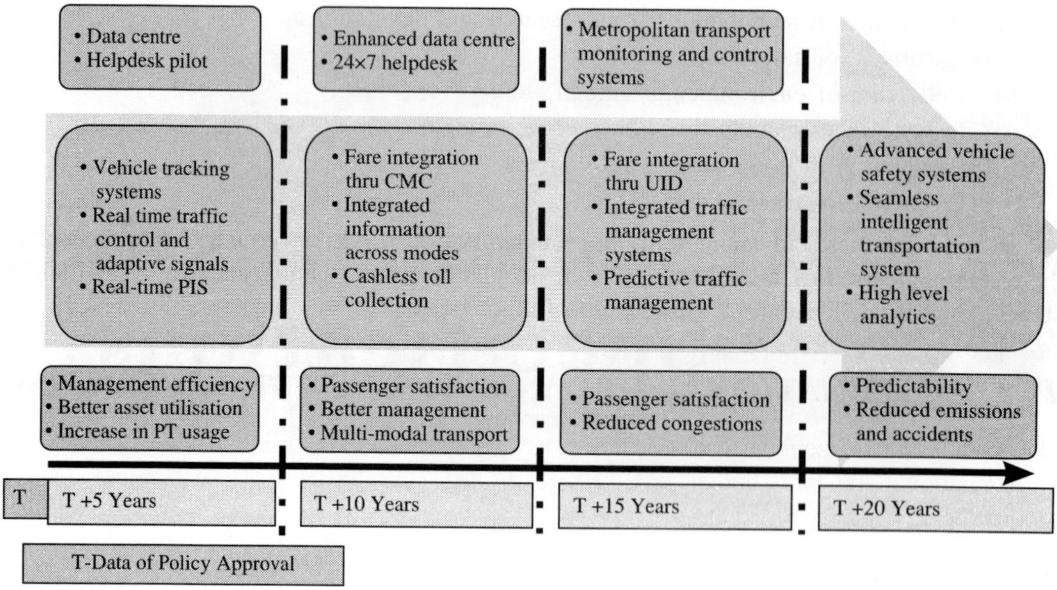

FIGURE 21.1 ITS implementation plan recommended by Working Group of Urban Transport for 12th Five Year Plan.

Although, there has been wise acceptance and recognition of need of ITS to mitigate transportation problems in a city, there is lack of government approved standards and specifications for implementation of ITS in a city. In absence of standards and specifications for different ITS tools, the transport planners are required to study different ITS technologies, select the suitable ones, frame the standards and specifications for the selected ITS tool and invite tenders from the service providers to get the best competitive rates for the ITS tool. Perhaps, the ITS technology are changing too fast to fix the detailed specifications for the ITS tools, equipment, software and hardware system, etc. Further, for an ITS tool, multiple options using different technologies meeting same functional requirements are available in the market. The requirements of ITS tools also vary from city to city. A small city may not require sophisticated adaptive traffic signals whereas an integrated area traffic control (ATC) may be very much required for a big city grappling with the problems of traffic congestion, accidents and pollution. One solution-one technology-one tool cannot serve the needs of all the cities. However, standardisation helps the decision makers to adopt the available technologies without reinventing the wheel every time. Standardisation can help to maximise compatibility, interoperability, safety, repeatability, or quality. The standardisation also optimises the cost as standard products can be mass produced and economy of scale works in favour of standardisation. The key advantages/roles of the standardisation are:

(a) Securing the compatibility of products
(b) Compatibility of interface among different ITS applications
(c) Improvement of production efficiency due to mass scale production of standard products
(d) Assurance of quality
(e) Accurate communication, promotion of mutual understanding

(f) Prevalence of technologies from research and development
(g) Assurance of safety and security
(h) Reduction of environmental burden
(i) Enhancement of industrial competitive strength
(j) Preparation of competitive environment
(k) Promotion of trade and more

In view of these advantages of the standardisation, different countries have developed their standards for ITS also. The approach adopted by some of the countries and India for standardisation of ITS tools are presented in this chapter.

21.2 INTERNATIONAL ORGANISATION OF STANDARDS (ISO)

The ISO is an independent, non-governmental international organisation with a membership of 162 national standards bodies. The headquarter of ISO is in Geneva, Switzerland. Through its members, it brings together experts to share knowledge and develop voluntary, consensus-based, market relevant International Standards that support innovation and provide solutions to global challenges. They give world-class specifications for products, services and systems, to ensure quality, safety and efficiency.

ISO/TC 204 is responsible for the overall system aspects and infrastructure aspects of intelligent transport systems (ITS), as well as the coordination of the overall ISO work programme in this field including the schedule for standards development, taking into account the work of existing international standardisation bodies.

ISO/TC 204 is developing standards broadly for:

(a) System architecture
(b) Interfaces (message sets, etc.)
(c) Frameworks (data dictionaries and message templates)
(d) System performance requirements
(e) Test methods

ISO/TC 204 has already published 246 standards for different ITS tools such as traffic signal controllers, automatic vehicle and equipment identification, lane keeping technologies, RTPI, electronic fee collection, adaptive cruise control systems, in-vehicle navigation systems, vehicle collision warning systems, parking management system, public transport user information, cooperative systems, lane change decision aid systems (LCDAS), emergency evacuation and disaster response and recovery, interoperable fare management system, electronic registration identification (ERI) for vehicles, etc.

These standards can be used for reference purpose and may be adopted as per the local needs.

21.3 EUROPEAN TELECOMMUNICATIONS STANDARDS INSTITUTE (ETSI)

The European Telecommunications Standards Institute (ETSI) is a non-profit organisation that establishes telecommunications standards for Europe. ETSI guidelines are voluntary and almost always comply with standards produced by international bodies. ETSI's Technical Committee for Intelligent Transport Systems (TC ITS) creates and maintains standards and specifications for the use of information and communications technologies in transport systems. Most of the technical committee's ongoing standardisation activities are focused on wireless communications for vehicle-to-vehicle and vehicle-to-roadside communications.

ETSI closely work with ISO to develop standards for different ITS fields.

ETSI TC ITS currently focuses on cooperative systems, electronic fee collection and interoperability of these technologies. Since ITS has a global dimension, great attention is given to the creation of commonly agreed standards for the network architecture, protocols and transmission formats. Having such a set of commonly agreed standards helps lead to a global harmonisation of ITS services and applications. A key issue when working on such standards is ensuring interoperability. Here, ETSI's centre for testing and interoperability (CTI) provides expertise on all aspects of interoperability. The following list shows some potential topics for which ETSI TC ITS is developing standards and technical specifications:

(a) Autonomous driving
(b) Integrated transport supporting smart cities
(c) Roadside platform architecture
(d) Integration of existing infrastructures
(e) Digital maps
(f) In vehicle platform architecture
(g) Urban mobility management
(h) Freight and fleets

ETSI has published two European Standards for intelligent transport systems (ITS): (a) the specification of Cooperative Awareness Basic Service—EN 302 637-2, and (b) the specification of Decentralised Environmental Notification Basic Service— EN 302 637-3. They define the message sets needed for running cooperative ITS safety critical applications.

ETSI standards on ITS can be used for reference purpose to develop and adopt ITS standards suiting local conditions.

21.4 ITS STANDARDISATION COMMITTEE, JAPAN

In Japan, the ITS Standardisation Committee (National Committee) set up under the auspices of the Society of Automotive Engineers of Japan (JSAE) carries out ISO/TC 204 international standardisation activities on behalf of the Japanese Industrial Standards Committee (JISC). Standardisation proposals are set forward concerning items on the ISO/TC 204 activities. The main tasks of the Committee are to:

(a) Act swiftly in response to changes in the standardisation environment
(b) Carry out standardisation activities in accordance with the established strategy
(c) Provide assistance with national standardisation (JIS)
(d) Provide related parties with up-to-date information

The Committee sorted out standardisation trends in and outside Japan and drew up the "Five-year Plan of Strategic International Standardisation 2016", which organises strategies and action plans for each working group (WG). It has also been actively collaborating with the

ITS Info-Communications Forum out of a need to address the WG 10 proposal on Transport Protocol Experts Group (TPEG), traveller information messages, digital media broadcast, and the wide-and medium-area communications, and communications access for land mobiles (CALM) standard proposal by WG 16.

21.5 ITS JOINT PROGRAMME OFFICE, US

The responsibility of developing ITS standards in USA has been entrusted with ITS Joint Programme Office (ITS JPO), and US Department of Transportation (DOT) (https://www.its.dot.gov/about.htm).

For the ITS standards programme, US is working with standards development organisations and public agencies to accelerate the development of open, non-proprietary communications interface standards to support ITS application development and deployment. These standards define how ITS systems and components interconnect and exchange information to deliver ITS services within a multimodal transportation network. The consistent and widespread use of ITS standards permit data and information sharing among public agencies and private organisations, fostering an environment of information sharing and interoperability. Currently nearly 100 standards have been developed under this programme.

Harmonisation of international standards and architecture around the vehicle platform is a process through which various stakeholders, vehicle and equipment manufacturers, standards organisations, and governments work together to agree on common standards. Through the ITS standards and architecture harmonisation programme, the US Department of Transportation (US DOT) participates in international standards harmonisation activities focusing on standards needed to provide connectivity among vehicles, and between vehicles and infrastructure.

The ITS standards programme is a key research programme of the Intelligent Transportation System Joint Programme Office (ITS JPO) within US DOT's Research and Innovative Technology Administration (RITA).

21.6 INTELLIGENT TRANSPORT SOCIETY (ITS), SOUTH KOREA

ITS Korea was set up in 1999 to provide technical advice about ITS policies and to support research and development of ITS technologies by the private sector. In order to effectively establish the ITS, close cooperation between government agencies (the Ministry of Construction and Transportation, the Ministry of Information and Communication, the Ministry of Commerce, Industry and Energy, the Ministry of Science and Technology, and National Police Agency) and the private sector (the industrial sectors, the academia and research institutions related to the ITS) is required. As a result, the government created ITS Korea as a means to strengthen the private sector's capacity for the ITS, to enhance the cooperation between the public and private sectors, and therefore to facilitate the development of the ITS. ITS Korea has been designated by Ministry of Land, Infrastructure and Transport (MOLIT) as national organisation for ITS standards. The main tasks assigned to ITS Korea are:

(a) Standardisation
(b) Standard observance, verification and certification
(c) Performance evaluation
(d) Project consulting
(e) Research and development
(f) Education
(g) Overseas business
(h) Global/domestic alliances

The task of standardisation includes:

(a) Build standard basis environment
(b) Support standard development
(c) Disseminate and maintain standards

ITS Korea supports activities of ISO/TC 204, develops technical regulations and supports the related activities, develops technical regulations conformity tests, administers the standards assembly, publishes ITS standards, operates National ITS Data Registry and ITS Architecture website and carries out competency building programme on ITS.

ITS Korea has published 57 standards on ITS as per the latest information available on its website (August, 2017).

21.7 INDIA

The approach of Government of India is to identify the ITS tools suitable for a city and define the functional requirements of the ITS tool. An ITS tool kit for traffic management system (TMS) has been developed by School of Planning and Architecture (SPA), Delhi and Institute of Urban Transport (IUT), Delhi. The stated aim of this toolkit is to empower city officials to maintain the charter of traffic management system using or applying the "ITS" based tools, to make them understand ITS and benefits thereof, to enable them select ITS tools as per priority for their city. However, the detailed specification and standards for ITS tools have not been developed or defined. Only functional requirements and broad output specifications are defined in the different tender documents of request for proposals (RFPs). The decision to use particular technology, equipment, hardware and software is left to the successful bidder who is responsible to meet the functional requirements only as specified in service level agreement (SLA). An SLA is a contract between a service provider and the public authority that documents what services the provider shall furnish and defines the performance standards that the provider is obligated to meet. A service level agreement (SLA) is signed with the successful bidder. The key performance indicators (KPIs) are defined in SLA to assess the efficacy of the system and payment to the successful bidder is mostly linked to these KPIs.

The functional requirements of the following most commonly used ITS tools as per tenders [request for proposals (RFPs)] issued by different state governments have been presented here for the guidance of the readers:

(a) Intelligent or adaptive traffic control system
(b) Red light violation detection system (RLVDS)

(c) Speed violation detection system (SVDS)
(d) Variable message sign (VMS) boards
(e) Closed-circuit television (CCTV)
(f) Automatic number plate recognition (ANPR)
(g) Automatic vehicle locating system (AVLS)
(h) Passenger information system (PIS)
(i) RFID for tracking of freight containers

The performance and evaluation criteria for some of the ITS tools have already been discussed in Chapter 19. The performance indicators are the parameters on which the performance of the ITS tool needs to be evaluated, whereas the functional requirements are the expected outcome from the ITS tool to meet the objective of installing it.

21.8 ADAPTIVE TRAFFIC CONTROL SYSTEM (ATCS)

Adaptive traffic control is a traffic management strategy in which traffic signal timing changes, or adapts, based on actual traffic demand. This is accomplished using an adaptive traffic control system consisting of both hardware and software. The ATCS consists of the following components:

(a) Traffic signal controller
(b) Vehicle detectors
(c) Communication network

The functional requirements of each component of ATCS are as follows:

21.8.1 Traffic Signal Controller

(a) The traffic signal controller equipment shall be a 32-bit or 64-bit microcontroller with solid state traffic signal lamp switching module with the ability to programme any combination of traffic signal stages, phases and junction groups. The controller shall ideally have a conflict monitoring facility to ensure that any danger is pre-flagged at the programming stage and this is disallowed even during manual override phase.
(b) The traffic signal controller shall be adaptive so that it can be controlled through the central traffic control centre as an individual junction or as part of group of traffic junctions along a corridor or a region. The signal controller design must be flexible for the junction that could be easily configured to be a part of any corridor or group definition and could be changed through central command controller easily.
(c) Site specific configuration data shall be stored in a non-volatile memory device (flash memory) easily programmable at the site through keypad or laptop. A minimum of 512 KB flash memory and 128 KB RAM shall be provided. Volatile memory shall not be used for storing the junction specific plans or signal timings.
(d) All timings generated within a traffic signal controller shall be digitally derived from a crystal clock which shall be accurate to plus or minus 100 ms.

(e) The controller shall provide a real-time clock (RTC) with battery backup that set and update the time, date and day of the week from the GPS. The RTC shall have minimum of 10 years battery backup with maximum time tolerance of 10 s/day.

(f) The controller shall have the facility to update the RTC time from ATCS server, GPS and through manual entry.

(g) The controller shall be capable of communicating with the ATCS server through Ethernet on a managed leased line network or any other appropriate stable communication network.

Police panel

The controller shall provide the following facilities in a separate panel with provision for lock and key arrangements for use by the traffic police.

(a) ***Four hurry call switches:*** The hurry call mode shall provide the means to force the controller to a defined stage, without violating safety clearances. A pre-emption input may be used to demand the 'hurry call' mode to give right of way to emergency vehicles. It should be possible to configure the hurry call switches to any stage as per site requirements.

(b) ***One forced flash switch:*** Activation of this switch should force the signal to flashing amber/flashing red.

(c) ***One auto/manual switch:*** Activation of this switch should enable manual operation of the controller. Deactivation of the manual switch shall continue from the current stage without interruption.

(d) ***One manual advance push button switch:*** In manual operation mode, the stages appear in the sequence specified in the signal plan time table. Activating the push button switch shall terminate the currently running stage and start the next, without violating safety clearances.

(e) ***One junction off switch:*** Activating this switch should put OFF all signal lamps. On deactivation of the switch, the traffic signal controller shall resume its normal operation without violating any safety clearances.

Modes of operation

The traffic signal controller shall have the following modes of operation:

(a) ***Fixed time:*** In fixed time (pre-timed) mode, the traffic signal controller shall execute stage timings according to the site specific time table maintained in the traffic signal controller flash memory. Inputs from vehicle detectors shall be ignored in this mode and no pre-emption shall be made on any stage. Cycle time remains constant in every cycle execution for a given time period.

(b) ***Vehicle actuation with all stages pre-emption:*** In the vehicle actuation with all stages pre-emption mode, the traffic signal controller shall execute stage timings as per demand from vehicle detectors within the constraints of minimum green, maximum green running period for the stage and cycle time stored in the traffic signal controller flash memory. Pre-emption shall be possible for all demand actuated stages. Cycle time may vary in every cycle execution.

(c) ***Semi-actuation:*** In the semi-actuation mode, the traffic signal controller shall execute stage timings in the vehicle actuated stages as per demand from vehicle detectors within the constraints of minimum green, maximum green running period for the stage and cycle time stored in the traffic signal controller flash memory. All other stages shall execute the maximum green time configured for the stage. Pre-emption shall be possible for all demand actuated stages. Cycle time may vary in every cycle execution.

(d) ***Stage skipping:*** The traffic signal controller shall not execute the stage enabled for skipping when there is no vehicle demand registered for the stage till clearance amber time of the previous stage.

(e) ***Transit signal priority (TSP) for BRT buses:*** The traffic signal controller shall provide transit signal priority for buses in dedicated lane to ensure minimum stop delay at the intersection, without violating safety clearances.

(f) ***Vehicle actuation with fixed cycle length:*** In vehicle actuation with fixed cycle length mode, the traffic signal controller shall execute stage timings as per demand from vehicle detectors within the constraints of minimum green, maximum green running period for the stage and cycle time shall be maintained constant during a given timeslot. Pre-emption for all demand actuated stages except for priority stage shall be possible.

(g) ***Full ATCS (FATCS):*** In FATCS mode, the traffic signal controller shall execute stage timings as per demand within the constraints of minimum green, maximum green running period for the stage and cycle time specified by the central computer during every cycle switching. Pre-emption for all demand actuated stages except priority stage shall be possible in this mode. The traffic signal controller shall identify a communication failure with the central computer within a specified time period. In such an event, the signal plan timings shall be executed from the local time table stored in the traffic signal controller FLASH memory. Fallback mode of the traffic signal controller shall be vehicle actuated. On restoration of the communication with central computer, the traffic signal controller shall automatically resort to FATCS mode. The traffic signal controller shall accept commands for remote selection/de-selection of the following from the central computer at traffic management centre (TMC).

(i) Hurry call
(ii) Flashing amber/flashing red
(iii) Junction off

If not reverted to the normal operation within the time period listed below, the traffic signal controllers shall timeout the commands and operate normally.

(i) Hurry call: 5 min
(ii) Flashing amber/flashing red: 30 min
(iii) Junction off: 30 min

The traffic signal controller shall report the following to the central computer through the communication network every cycle or on an event as appropriate:

(i) Green time actually exercised for each approach (stage pre-emption timing) against the green running period set for the approach by the central computer
(ii) Mode of operation

(iii) Lamp failure, if any
(iv) Output short circuit, if any
(v) Detector failure, if any

Traffic signal controller operating parameters

(a) ***Phases:*** The controller shall have facility to configure 32 phases either for vehicular movement, filter green, indicative green, pedestrian movement or a combination thereof.

It shall be possible to operate the filter green (turning right signal) along with a vehicular phase. The filter green signal shall flash for a time period equal to the clearance amber period at timeout when operated with a vehicular phase.

The pedestrian phase signal shall be configured for flashing red or flashing green aspect during pedestrian clearance.

It shall be possible to configure any phase to the given lamp numbers at the site.

(b) ***Stages:*** The controller shall have facility to configure 32 stages.

(c) ***Cycle plans:*** The controller shall have facility to configure 24 cycle plans and the amber flashing/red flashing plan. It shall be possible to define different stage switching sequences in different cycle plans. The controller shall have the capability for a minimum of 32 cycle-switching per day in fixed mode of operation.

(d) ***Day plans:*** The controller shall have facility to configure each day of the week with different day plans. It shall also be possible to set any of the day plans to any day of the week. The controller shall have the capability to configure the plans.

(e) ***Special day plans:*** The controller shall have facility to configure a minimum of 20 days as special days in a calendar year.

(f) ***Starting amber:*** During power up the controller shall initially execute the flashing amber/flashing red plan for a time period of 3–10 s. The default value of this starting amber is 5 s. Facility shall be available to configure the time period of starting amber within the given limits at the site.

(g) ***Inter-green:*** Normally the inter-green period formed by the clearance amber and red extension period shall be common for all stages. However, the controller shall have a facility to programme individual inter-green period from 3–10 s.

(h) ***Minimum green:*** The controller shall allow programming the minimum green period from 5–10 s without violating the safety clearances. It should not be possible to pre-empt the minimum green once the stage start commencing execution.

(i) ***All red:*** Immediately after the starting amber, all the approaches should be given red signal for a few seconds before allowing any right of way, as a safety measure. The controller shall have programmability of 3–10 s for all red signal.

(j) ***Signal lamps monitoring:*** The controller shall have inbuilt circuitry to monitor the lamp status.

(k) ***Green-green conflict monitoring:*** The controller shall have a facility to list all conflicting phases at an intersection. The controller should not allow programming of these conflicting phases in a stage. A hardware failure leading to a conflict condition (due to faulty devices or short circuit in the output) shall force the signal into flashing amber/flashing red.

(l) ***Cableless synchronisation:*** It shall be possible to synchronise the traffic signal controllers installed in a corridor in the following modes of operation, without physically linking them and without communication network. GPS enabled RTC shall be the reference for the cable less synchronisation.

(m) Fixed time mode with fixed offsets.

(n) Vehicle actuated mode with fixed offsets.

Input and output facilities

(a) ***Lamp switching:*** The controller shall have maximum 64 individual outputs for signal lamp switching, configurable from 16 to 32 lamps. The signal lamps shall be operating on appropriate DC/AC voltage of applicable rating

(b) ***Detector interface:*** A minimum of 16 vehicle detector inputs shall be available in the controller. All detector inputs shall be optically isolated and provided with LED indication for detection of vehicle.

(c) ***Communication interface:*** The traffic signal controller shall support Ethernet interface to communicate with the ATCS server.

(d) ***Power saving:*** The traffic signal controller shall have a facility to regulate the intensity of signal lamps during different ambient light conditions thereby saving energy.

(e) ***Real-time clock (RTC):*** The GPS receiver for updating time, date and day of the week information of the traffic signal controller should be an integral part of the traffic signal controller.

The traffic signal controller shall update the date, time and day of the week automatically from GPS during power ON and at scheduled intervals.

Manual entry for date, time and day of week shall be provisioned for setting the traffic signal controller RTC.

It shall be possible to set the RTC from the central server when networked.

(f) ***Keypad (optional):*** The traffic signal controller shall have a custom made keypad or should have provision for plan upload and download using PC/laptop/central server.

(g) ***Operator display (optional):*** The traffic signal controller shall optionally have a LED backlit liquid crystal display (LCD) as the operator interface.

21.8.2 Camera based Vehicle Detector

The detector equipment is a separate logic unit, which may be integrated into the controller, or alternatively mounted in its own housing. The outputs of the detectors indicate the presence of vehicles and are used to influence the operation of the traffic signal controller and shall generate counts, demands and extensions for right-of-way. Means shall be provided so that a detector may be connected to demand and/or extend a phase movement as specified. The functional requirements of vehicle detector are as follows:

(a) The contractor shall clearly specify the placement of the detector (upstream, downstream, stop-line, exit, etc.) for independent straight and right turn signals.
(b) The detector shall be able to count vehicles in non-lane based mixed traffic flow conditions and differentiate between different vehicle types (two-wheeler, three-wheeler, car, HGV). The accuracy of counts shall be bigger than 90% over all light and weather conditions. The contractor shall clearly specify how this is accomplished.
(c) The contractor shall give an estimate of the total number of vehicle presence detection zones and vehicle detectors required and the type of detection system recommended.
(d) A detector that does not change its status at least during a stage execution shall be notified to the central computer (in ATCS mode) at the termination of the associated stage.

21.8.3 Communication Network

(a) Function of the communication network is for remote monitoring of the intersection and its management. Real-time data (like RTC time, stage timing, mode, events, etc.) from the traffic signal controller is required to be sent to the central computer in traffic management centre (TMC).
(b) Central computer running the ATCS application shall calculate and send optimum signal timings to all intersections in the corridor.
(c) The contractor shall clearly specify the bandwidth requirements and the type of network recommended for the ATCS.
(d) The contractor shall specify the networking hardware requirements at the traffic management centre and remote intersections for establishing the communication network.

21.9 FUNCTIONAL REQUIREMENTS OF THE RED LIGHT VIOLATION DETECTION (RLVD) SYSTEM

The RLVD system helps in catching the red light violators and thus improves the safety of road transportation. A typical RLVD system shall meet following functional requirements:

21.9.1 General

(a) The following traffic violations to be automatically detected by the system by using appropriate non-intrusive sensors technology:
    (i) Red light violation
    (ii) Stop line violation
(b) The system should have both provisions to detect red light status by taking the signal feed from the traffic signal controller as well as by video analytics method using

another camera (evidence camera) focused at the red light. The evidence camera should also be used for evidence snap generation.

(c) The system should be capable of capturing multiple infracting vehicles simultaneously in different lanes on each arm at any point of time with relevant infraction data like:
   (i) Type of violation
   (ii) Date, time, site name and location of the infraction
   (iii) Registration number of the vehicle through ANPR camera system for each vehicle identified for infraction

(d) The system should be equipped with a camera system to record a digitised image and video of the violation, covering the violating vehicle with its surrounding and current state of signal (red/green/amber) by which the system should clearly show nature of violation and proof thereof:
   (i) When it violates the stop line
   (ii) When it violates the red signal
   (iii) Besides, a closer view indicating readable registration number plate patch of the violating vehicle for court evidence for each violation

(e) The system must have in-built tool to facilitate the user to compose detail evidence by stitching video clips from any IP camera in the junction (including but not limited to the red light violation detection camera and evidence camera), and any other surveillance cameras in the vicinity of the spot of incidence. The entire evidence should be watermarked and encrypted to stand the court of law.

(f) The system shall be able to detect all vehicles infracting simultaneously in each lane/arm at the junction as per locations provided. It should also be able to detect the vehicles infracting serially one after another in the same lane. The vehicles should be clearly identifiable and demarcated in the image produced by the camera system.

(g) The evidence image produced by the system should be wide enough to give the exact position of the infracting vehicles with respect to the stop line and clearly indicate colour of the traffic light at the instant of infraction even if any other means is being used to report the colour of the light.

(h) The system should interface with the traffic controller to validate the colour of the traffic signal reported at the time of infraction so as to give correct inputs of the signal cycle.

(i) The evidence and ANPR camera should continuously record all footage in its field of view to be stored at the local base station. This should be extractable onto a portable device as and when required. The option of live viewing of evidence cameras from the locations shall be available at the TMC. The network should have the capability to provide the real-time feed of the evidence camera to the TMC at the best resolution possible on the available network.

(j) The system shall be equipped with IR illuminator to ensure clear images including illumination of the number plate and capture the violation image under low light conditions and at night time.

21.9.2 Recording and Display Information Archive Medium

The recording and display of information should be detailed on the snapshot of the infracting vehicle as follows:

(a) Computer generated unique ID of each violation
(b) Date
(c) Time (HH:MM:SS)
(d) Equipment ID
(e) Location ID
(f) Carriageway or direction of violating vehicle
(g) Type of violation (signal/stop line)
(h) Lane number of violating vehicle
(i) Time into red/green/amber
(j) Registration number of violating vehicle

21.9.3 On-site Outstation Processing Unit Communication and Electrical Interface

(a) The system should automatically reset in the event of a programme hang up and restart on a button press. However, the system should start automatically after power failure.
(b) The system should have secure access mechanism for validation of authorised personnel.
(c) Deletion or addition and transfer of data should only be permitted to authorised users.
(d) A log of all user activities should be maintained in the system.
(e) Roles and rights of users should be defined in the system as per the requirement of the client.
(f) All formats of the stored data with respect to the infractions should be non-proprietary.
(g) The communication between the on-site outstation processing unit housed in the junction box and the detection systems mounted on the cantilever shall be through appropriate secured technology.
(h) The system should have the capability to transfer the data to traffic control centre (TCC) through proper encryption in real-time and batch mode for verification of the infraction and processing of challan. Call forwarding architecture shall be followed to avoid any data loss during transfer.
(i) In the event that the connectivity to the TCC is not established due to network/ connectivity failures, then all data pertaining to the infraction shall be stored on-site and shall be transferred once the connectivity is re-established automatically. There shall also be a facility of physical transfer of data on portable device whenever required. There should be a provision to store minimum one week of data at each site on a 24 × 7 basis.

21.9.4 RLVD Application

(a) It should be capable of importing violation data for storage in database server which should also be available to the operator for viewing and retrieving the violation images and data for further processing. The programme should allow for viewing, sorting, transfer and printing of violation data.

(b) It should print the photograph of violations captured by the outstation system which would include a wider view covering the violating vehicle with its surrounding and a closer view indicating readable registration number plate patch of the violating vehicle along with all data.

(c) All outstation units should be configurable using the software at the central location.

(d) Violation retrieval could be sorted by date, time, location and vehicle registration number and the data structure should be compatible with local police database structure. It should also be possible to carry out recursive search and wild card search.

(e) The operator at the back office should be able to get an alarm of all faults occurring at the camera site (e.g., sensor failure, camera failure, failure of linkage with traffic signal, connectivity failure, camera tampering, sensor tampering, etc.).

(f) The automatic number plate recognition software shall be part of the supplied system. Success rate of ANPR shall be taken as 75% or better during the day time and 60% or better during the night time with a standard number plate.

(g) The application software should be integrated with the e-challan software for tracing the ownership details of the violating vehicle and issuing/printing notices. Any updates of the software (OS, application software including any proprietary software) shall be updated free of cost during the contract period by the SI.

(h) Image zoom function for number plate and images should be provided. In case the number plate of the infracting vehicle is readable only through the magnifier then in such cases the printing should be possible along with the magnified image.

(i) Various users should be able to access the system using single sign on and should be role based. Different roles which could be defined (to be finalised at the stage of SRS) could be administrator, supervisor, officer, operator, etc.

(j) Apart from role based access, the system should also be able to define access based on location.

(k) Rights to different modules/sub-modules/functionalities should be role based and proper log report should be maintained by the system for such access.

(l) Components of the architecture must provide redundancy and ensure that there are no single points of failure in the key project components. Considering the high sensitivity of the system, design shall be in such a way as to be resilient to technological sabotage. To take care of remote failure, the systems need to be configured to mask and recover with minimum outage.

(m) The architecture must adopt an end-to-end security model that protects data and the infrastructure from malicious attacks, theft, etc. Provisions for security of field equipment as well as protection of the software system from hackers and other threats shall be a part of the proposed system. Using firewalls and intrusion detection systems such attacks and theft shall be controlled and well supported (and implemented) with the security policy. The virus and worms attacks shall be well defended with gateway level anti-virus system, along with workstation level anti-virus mechanism. There shall also be an endeavour to make use of the SSL/VPN technologies to have secured communication between applications and its end users. Furthermore, all the system logs shall be properly stored and archived for future analysis and forensics whenever desired.

(n) The evidence of infraction should be encrypted and protected so that any tampering can be detected.

(o) Ease of configuration, ongoing health monitoring, and failure detection are vital to the goals of scalability, availability, and security and must be able to match the growth of the environment.

(p) System shall use open standards and protocols to the extent possible and declare the proprietary software wherever used.

(q) The user interface should be user friendly and provide facility to user for viewing, sorting and printing violations. The software should also be capable of generating query based statistical reports on the violation data.

(r) The data provided for authentication of violations should be in an easy to use format as per the requirements of user.

(s) User should be provided with means of listing the invalid violations along with the reason(s) of invalidation without deleting the record(s).

(t) Basic image manipulation tools (zoom etc.) should be provided for the displayed image but the actual recorded image should never change.

(u) Log of user actions be maintained in read only mode. User should be provided with the password and ID to access the system along with user type (admin, user).

(v) Image should have a header/footer depicting the information about the site IP and violation details like date, time, equipment ID, location ID, unique ID of each violation, lane number, registration number of violating vehicle and actual violation of violating vehicle, etc. so that the complete lane wise junction behaviour is recorded including speed of violating vehicle, signal jumping, stop line violation, number plate should be readable automatically by the software/interface. There should be user interface for simultaneous manual authentication/correction and saving as well.

(w) Interface for taking prints of the violations (including image and above details).

21.10 FUNCTIONAL REQUIREMENTS OF THE SPEED VIOLATION DETECTION SYSTEM

The speed violation detection system is used to monitor the cases of over speeding by the road vehicle drivers beyond prescribed speed over the road section. The following functional requirements shall be met by the speed violation detection system:

21.10.1 General

(a) The speed violations should be automatically detected by the system by using appropriate sensors technology.

(b) The system should be capable of capturing multiple infracting vehicles simultaneously in defined lanes at any point of time simultaneously with relevant infraction data like:

(i) Type of violation

(ii) Speed of violating vehicle

(iii) Notified speed limit

(iv) Date, time, site name and location of the infraction

(v) Registration number of the vehicle through ANPR camera system for each vehicle identified for infraction

(c) The system should be equipped with a camera system to record a digitised image or video frames of the violation, covering the violating vehicle with its surrounding.

(d) The system shall provide the number of vehicles infracting simultaneously in each lane. The vehicles shall be clearly identifiable and demarcated in the image produced by the camera system.

(e) The system shall be equipped with IR illuminator to ensure clear images including illumination of the number plate and capture the violation image under low light conditions and at night time.

(f) Speed measurement may be made by using non-intrusive technology such as radar/ sensor/camera/virtual based or any other appropriate certified technology.

21.10.2 On-site Outstation Processing Unit Communication and Electrical Interface

(a) The system should automatically reset in the event of a programme hang up and restart after power failure.

(b) The system should have secure access mechanism for validation of authorised personnel.

(c) Deletion or addition and transfer of data should only be permitted to authorised users.

(d) A log of all user activities should be maintained in the system.

(e) Roles and rights of users should be defined in the system.

(f) The data shall be transferred to the traffic control centre (TCC) in real-time for verification of the infraction and processing of challan.

(g) In the event that the connectivity to the TCC is not established then all data pertaining to the infraction shall be stored on-site and shall be transferred once the connectivity is re-established automatically.

21.10.3 Speed Violation Application

The purpose of the Speed Violation Application (software) is to capture and record the details of speed violation which can be used for imposing penalty, as an evidence in court or for any other law enforcement purpose. The application should be able to capture the violation data for the Operator for viewing and retrieving the violation images, print the photograph of violations including readable registration number plate patch of the violating vehicle. The application should be integrated with APNR system for recording the number plate of the violating vehicle and fetch the address of the user from the vehicle registration database. Similar to the RLVD application, the speed violation application is required to keep a log all the violation data in a structured format. However, unlike RLVD, the speed violation system is not required to detect the aspect of the signal when the vehicle crosses the signal. The speed violation detection system is deployed in the mid-section whereas the red light violation detection system is deployed at signal controlled road intersections. Except, these two differences, the functional requirements of speed violation detection application and RLVD application (as presented in Para 21.9.4) are same.

21.11 FUNCTIONAL REQUIREMENTS OF THE VARIABLE MESSAGE SIGNAGE SYSTEM

Variable message sign (VMS) is an electronic traffic sign used to give travellers information about traffic congestion, accidents, incidents, roadwork zones, or speed limits on a specific highway segment, etc. The VMS system shall meet following functional requirements:

21.11.1 System Requirements

(a) The system should be capable to display warnings, traffic advice, route guidance and emergency messages to motorists from the TCC in real-time.

(b) The system should also be capable to display warnings, traffic advice, route guidance and emergency messages to motorist by using local PC/laptops. The system should display graphical representation of the lanes with directional arrows and colour such as green, yellow, red for depicting density of traffic.

(c) The VMS should display text and graphic messages using light emitting diode (led) arrays.

(d) The system should able to display failure status of any LED at TCC.

(e) The system should support display characters in true type fonts and adjustable based on the operating system requirement.

(f) The VMS workstation at the traffic control centre (TCC) should communicate with the VMS controller through the network. It should send out command data to the variable message sign controller and to confirm normal operation of the signboard. In return, the VMS workstation should receive status data from the VMS controller.

(g) VMS controllers should continuously monitor the operation of the VMS via the provided communication network.

(h) Operating status of the variable message sign should be checked periodically from the TCC.

(i) It shall be capable of setting an individual VMS or group of VMS's to display either one of the pre-set messages or symbols entered into the computer via the control computer keyboard or by another means.

(j) It shall be capable of being programmed to display an individual message to a VMS or a group of VMS's at a pre-set date and time.

(k) A sequence of a minimum of 10 messages/pictures/pre-decided signs or group of signs shall be possible to assign for individual VMS or group of VMSs.

(l) It shall also store information about the time log of message displayed on each VMS. The information stored shall contain the identification number of the VMS, content of the message, date and time at which displayed message/picture starts and ends.

(m) The central control computer shall perform regular tests (pre-set basis) for each individual VMS. Data communication shall be provided with sufficient security check to avoid unauthorised access.

21.11.2 Variable Message Signboard Application

(a) Central control software shall allow controlling multiple VMS from one console.

(b) The system should be capable of programming to display all types of message/advertisement having alphanumeric character in English, Hindi, a regional language and combination of text with pictograms signs.

(c) The system should have feature to manage video/still content for VMS display. The system should have capability to divide VMS screen into multi-parts to display diverse form of information like video, text, still images, advertisements, weather info, city info, etc. The system should also provide airtime management and billing system for paid content management.

(d) The system should be capable of controlling and displaying messages on VMS boards as individual/group.

(e) The system should be capable of controlling and displaying multiple font types with flexible size and picture sizes suitable as per the size of the VMS.

(f) The system should be capable of controlling brightness and contrast through software.

(g) The system should be capable to continuously monitor the operation of the variable message sign board, implemented control commands and communicate information to the traffic monitoring centre via communication network.

(h) Real-time log facility and log file documenting the actual sequence of display to be available at central control system.

(i) Access to system should be available only after the authentication and acceptance of authentication based on hardware dongle with its log.

(j) Location of each VMS shall be plotted on GIS map with their functioning status which can be automatically updated.

(k) The system should have report generation facility for individual/group/all VMSs with date and time which includes summary of messages, dynamic changes, fault/repair report and system accessed logs, link breakage logs, down time reports or any other customised report.

(l) The system should be configurable scheduler on date/day of week basis for transmitting pre-programmed message to any VMS unit.

(m) Various users should access the system using single sign on and should be role based. Different roles which could be defined (to be finalised at the stage of SRS) could be administrator, supervisor, officer, operator, etc.
(n) Apart from role based access, the system should also be able to define access based on location.
(o) Rights to different modules/sub-modules/functionalities should be role based and proper log report should be maintained by the system for such access.
(p) Components of the architecture must provide redundancy and ensure that there are no single points of failure in the key project components. To take care of remote failure, the systems need to be configured to mask and recover with minimum outage.
(q) The architecture must adopt an end-to-end security model that protects data and the infrastructure from malicious attacks, theft, natural disasters, etc. provisions for security of field equipment as well as protection of the software system from hackers and other threats shall be a part of the proposed system. Using firewalls and intrusion detection systems such attacks and theft shall be controlled and well supported (and implemented) with the security policy. The virus and worms attacks shall be well defended with gateway level anti-virus system, along with workstation level anti-virus mechanism. There shall also be an endeavour to make use of the SSL/VPN technologies to have secured communication between applications and its end users. Furthermore, all the system logs shall be properly stored and archived for future analysis and forensics whenever desired.
(r) Ease of configuration, ongoing health monitoring, and failure detection are vital to the goals of scalability, availability, and security and must be able to match the growth of the environment.
(s) System shall use open standards and protocols to the extent possible.
(t) The system should be capable to export reports to excel and PDF formats.

21.11.3 Remote Monitoring

(a) All VMS shall be connected/configured to traffic monitoring centre for remote monitoring through network for two-way communication between VMS and control room to check system failure, power failure and link breakage.
(b) Remote diagnostics to allow identifying reason of failure up to the level of failed individual LED.

21.12 FUNCTIONAL REQUIREMENT OF THE CCTV— TRAFFIC SURVEILLANCE CAMERAS

The CCTV cameras are increasingly being used for traffic surveillance and for traffic management purpose. The functional requirement of the overall traffic surveillance system can be categorised into following components:

(a) Information to be captured by edge devises
(b) Information to be analysed at traffic command centre

(c) Role based access to the entire system
(d) Storage/recording requirements
(e) Other general requirements

21.12.1 Information to be Captured by Edge Devices

Traffic surveillance cameras being one of the core sub-modules of ITS project, it is important that their selection and placement is carefully done to ensure the full coverage of the traffic junction along with all associated junction arms, accuracy of the information captured on the field and they are rugged, durable and compact. These cameras need to work on 24 × 7 basis and transmit quality video feeds to the traffic command centre and would capture the video feeds at 30 frames per second (FPS) during entire duration of day. However, city traffic police may take the regular review of the requirements for video resolution, FPS and may change these numbers to suit certain specific requirements (e.g., there could be a situation when certain cameras are required to be viewed at higher FPS for specific period. It is estimated that not more than 5% of the cameras would be required to be viewed at higher FPS at a given point of time). Video feeds shall be stored at 30 FPS for a minimum of 30 days at the data centre. It is recommended to clearly identify in SLAs that cameras need to transmit quality video feed (appropriately focused, clear, un-blurred, jitter free, properly lit, unobstructed, etc.). Packet loss is to be less than 0.5%.

21.12.2 Information to be Analysed at Traffic Command Centre

The proposed video management system should provide a complete end-to-end solution for security surveillance application. The control centre shall allow an operator to view live/ recorded video from any traffic surveillance camera on the IP network. The combination of control centre and the IP network would create a virtual matrix, which would allow switching of video streams around the system.

Not all the traffic surveillance cameras would be simultaneously viewed at traffic command centre. The viewing shall vary from time to time which shall help to manage traffic at the junctions and coordinate with the field traffic officers. Video management system should have event alert facility over e-map. Alert popups/notification along with live video stream of the camera shall be triggered, if someone activate emergency calling button.

21.12.3 Role Based Access to the Entire System

Various users should have access to the system using single sign on and should be role based. Different roles which could be administrator, supervisor, officer, operator, etc. Apart from role based access, the system should also be able to define access based on location. Other minimum features required in the role based authentication systems are as follows:

(a) The management module should be able to capture basic details (including mobile number and email id of the police personnel and other personnel requiring viewing/ administration rights to the system. There should be interface to change these details after proper authentication.

(b) Rights to different modules/sub-modules/functionalities should be role based and proper log report should be maintained by the system for such access.
(c) Biometric standardised coupled with login name and password should be enabled to ensure that only the concerned personnel are able to login into the system.
(d) There should be provision to specify hierarchy of operators/officers for control of the cameras from various locations.
(e) The number of users shall increase as per phase wise implementation. SI is expected to estimate and provision the same based on the phase wise requirements given in RFP. Windows active directory/LDAP or any such system can be used to design role based access.

21.12.4 Storage/Recording Requirements

The storage solution shall be modular enough to ensure compliance to the changes in storage/recording policy, to be evolved upon initial deployment of the system. The following storage requirements shall be fulfilled by the SI as scope for the project:

(a) There shall be data centre (DC) at traffic control centre (TCC).
(b) 30 days of storage for all traffic surveillance camera feeds.
(c) Data on storage would be over-written automatically by newer data after the stipulated time period. If some data is flagged by police personnel (or by designated personnel) as important data/evidence data due to some reporting of crime or accident in the area or due to court order or due to suspicious activity, it would need to be stored for longer duration, as per requirements. City traffic police would analyse such flagged data every 3 months to take such decisions for preservation of the flagged data beyond 90 days.
(d) Full audit trail of reports to be maintained for 90 days.
(e) Retrieval time for any data stored on secondary storage should be maximum 4 h for critical data and 8 h for other data.
(f) The recording servers/system, once configured, shall run independently of the video management system and continue to operate in the event that the management system is off-line.
(g) The system shall support the use of separate networks, virtual local area networks (VLANs) or switches for connecting the cameras to the recording servers to provide physical network separation from the clients and facilitate the use of static IP addresses for the devices.
(h) The system shall support advanced video coding (AVC) or better, MPEG-4 and MJPEG compression formats for all analog cameras connected to encoders and all IP cameras connected to the system.
(i) The system shall record the native frame rate and resolution supplied by the camera or as configured by the operator from the system administration server.
(j) The system should not limit amount of storage to be allocated for each connected device.
(k) The online archiving capability shall be transparent and allow city traffic police to browse and archive recordings without the need to restore the archive video to a local hard drive for access.

(l) The system shall allow for the frame rate, bit rate and resolution of each camera to be configured independently for recording. The system shall allow the user to configure groups of cameras with the same frame rate, bit rate and resolution for efficient set-up of multiple cameras simultaneously.

(m) The system shall support archiving or the automatic transfer of recordings from a camera's default database to another location on a time programmable basis without the need for user action or initiation of the archiving process. Archiving shall allow the duration of the camera's recordings to exceed the camera's default database capacity. Archives shall be located on either the recording server or on a connected network drive. If the storage area on a network drive becomes unavailable for recording the system should have the ability to trigger actions such as the automatic sending of email alerts and sound alerts to necessary personnel.

(n) The recording server/system shall offer different codec (H.264, MJPEG, MPEG-4, etc.), frame rate and resolution (1080P, 960P, 720P, 4CIF, CIF, QCIF) options for managing the bandwidth utilisation for live viewing on the client systems (through use of multiple systems such as transcoding server).

21.13 AUTOMATIC NUMBER PLATE RECOGNITION (ANPR) SYSTEM

The ANPR system shall enable monitoring of vehicle flow at strategic locations. The system shall support real-time detection of vehicles at the deployed locations, recording each vehicle, reading its number plate, database lookup from central server and triggering of alarms/alerts based on the vehicle status and category as specified by the database. The system usage shall be privilege driven using password authentication. The functional requirements of APNR system are as follows:

21.13.1 Vehicle Detection by Colour

The system shall detect the colour of all vehicles in the camera view during day time and label them as per the pre-defined list of configured system colours. The system shall store the colour information of each vehicle along with the licence plate information for each transaction in the database. The system shall have options to search historical records for post event analysis by the vehicle colour or the vehicle colour with licence plate and date time combinations.

21.13.2 Alert Generation

(a) The system should have option to input certain licence plates according to the hot listed categories like "wanted", "suspicious", "stolen", etc. by authorised personnel.

(b) The system should be able to generate automatic alarms to alert the control room personnel for further action, in the event of detection of any vehicle falling in the hot listed categories.

21.13.3 Vehicle Status Alarm Module

(a) On successful recognition of the number plate, system should be able generate automatic alarm to alert the control room for vehicles which have been marked as "wanted", "suspicious", "stolen", "expired", etc.

(b) In case of identity of vehicle in any category which is define by user, generate instantaneous and automatic alarms.

21.13.4 Vehicle Log Module

(a) The system shall enable easy and quick retrieval of snapshots, video and other data for post incident analysis and investigations.

(b) The system should be able to generate suitable MIS reports that shall provide meaningful data to concerned authorities and facilitate optimum utilisation of resources. These reports shall include:

    (i) Report of vehicle flow at each of the installed locations for last day, last week and last month.

    (ii) Report of vehicles in the detected categories at each of the installed locations for last day, last week and last month.

    (iii) Report of vehicle status change in different vehicle categories.

(c) The system shall have search option to tune the reports based on licence plate number, date and time, site location as per the need of the authorities.

(d) The system shall have option to save custom reports for subsequent use. The system shall have option to export report being viewed to common format for use outside of the ANPRS or exporting into other systems.

(e) The system should provide advanced and smart searching facility of licence plates from the database. There should be an option of searching number plates almost matching with the specific number entered (up to 1 and 2 character distance).

21.13.5 Vehicle Category Editor

(a) The system should have option to input certain licence plates according to category like "wanted", "suspicious", "stolen", "expired", etc. by authorised personnel.

(b) The system should have an option to add new category by authorised personnel.

(c) The system should have option to update vehicle status in specific category by authorised personnel, e.g., on retrieval of stolen vehicle, system entry should be changed from "stolen" to "retrieved".

(d) System should have option to specify maximum time to retain vehicle records in specific categories.

21.13.6 Central Management Module

(a) The central management module shall run on the ANPRS central server in control booth. It should be possible to view records and edit hotlists from the central server.

(b) Base specification of fixed box cameras must be part of the ANPR specifications.
(c) IP66 standard with sunshield vandal proof housing.

21.13.7 Systems Requirement

(a) ***Local server at intersection:*** The system must run on a commercial off the shelf server (COTS). Outdoor IP66 quad core processor based server should be able to cover at least 8 lanes. Temperature rating of the server should be at least 60°.

(b) ***Operating system:*** The system must be based on open platform and should run on Linux or Windows operating system (OS).

(c) ***Workstation:*** Workstation must run on latest available OS.

21.14 AUTOMATED VEHICLE LOCATING SYSTEM (AVLS)

An automatic vehicle locator (AVL) is a device to remotely track the location of its vehicle fleet on real-time basis by using the Internet. Global positioning system (GPS) is mostly used to track the location of the vehicle. The following categories of vehicles are usually required to be tracked:

(a) Urban/municipal buses
(b) Mini buses
(c) Feeder buses
(d) Fire vehicles
(e) Dumpers, trippers, etc.
(f) Other vehicles (e-rickshaws and other private vehicles) based on the use of GPS systems available in smart-phones

The software shall be offered using a web based solution that utilises high resolution digital map, e.g., Google Map or similar to show real-time position of the vehicles. The software shall provide map based tracking and transit route line based tracking of vehicles by the control centre. The software is expected to have enterprise capabilities which enables multiple user type to be enabled to carry out various functions like, alarm management, vehicle schedule tracking, speed management, stoppage management, route replays, vehicle tracking dashboard, etc. as a standard functionality. The software shall enable control centre management staff quick decision making capability, which shall be achieved by providing graphical tools for visualisation. The software shall enable the public authority to drill and analyse information and online data in a multi-dimensional manner. Comprehensive analysis and reporting capabilities are expected to be part of the application delivery which matches the world standard capabilities of AVLS systems.

The software should have capability of multi-screen based tracking system, so as to enable quick analyse and have a better insight operational data of all activities within the system. The solution developed/deployed should be scalable to support at specified number vehicles and bus stops. In addition, unlimited number of geo-tagged locations should be supported by the solution. The solution proposed should also assist in integration with other smart city initiatives, for instance, surveillance, traffic management, etc. and this initiative should support works considering citizen as the centre of activities. The software should also have the provision of tracking (potentially unlimited number) of e-rickshaw vehicles and other public transport vehicles (e.g., auto, CNG rickshaw and other classes of vehicles, doctors, etc.), on whom tracking may be performed using their smartphones. Since the software shall be a web based software, it should be able to run on a video wall/control centre and it should have the provision of feeds available to multiple users and locations based on secure authentication. The video generated should be customised to the requirements of public authority. The bidders must be required to provide for appropriate reporting and matching of this data.

21.15 PASSENGER INFORMATION SYSTEM (PIS)

The PIS shall include all the components that are directed to provide information to passengers/commuters. This PIS should work on the data provided by the AVLS. The following functional requirements shall be met by PIS:

(a) Estimated time of arrival/departure (ETA/D) in real-time application shall be developed by the vendor by consulting all the stakeholders. This real-time information should be available through SMS/Internet/Mobile App/LED boards installed at bus stands including information regarding incidents/cancellation/delays.
(b) Real-time information on the internet and SMS shall be provided on demand. SMS and Internet based real-time information should enable commuters to plan their trips in advance and make informed choices about their travel by bus.
(c) Real information through SMS shall be user defined and friendly including bus stop name, service type, route number, time of arrival at the very minimum. The bidder shall be responsible to integrate with the SMS gateway.
(d) Vendor shall ensure smooth coordination and integration for the above.
(e) The PIS should have the capability of tracking vehicles based on their location, etc.
(f) It is a requirement that the solution developed be integrated with commercially available display boards, which are proposed to be installed at bus stations/bus stands and in vehicle passenger information systems.

21.16 RFID TO TRACK FREIGHT CONTAINERS ON REAL-TIME BASIS

To make India's logistics sector more efficient through the use of information technology, the logistics data bank (LDB) project envisages a real-time container tracking system where every container is attached to a radio frequency identification tag (RFID) tag and then tracked through RFID readers. The system will help importers and exporters in tracking their goods in transit. The project will covers the entire movement (of containers) on end-to-end basis through rail or road till the inland container depot and reached at freight station.

The LDB is being implemented through a special purpose vehicle called Delhi Mumbai Industrial Corridor Development Corporation (DMICDC) Logistics Data Services, that is jointly (50:50) owned by the Delhi Mumbai Industrial Corridor (DMIC) Trust and Japanese IT services major Nippon Electric Company (NEC). The services include providing users the 'average delivery time' as well as notifications through SMS and email.

DMICDC Logistics Data Services (DLDSL) has floated tenders (request for proposal) for procurement of RFID tags and RFID readers in February, 2017. The detailed specifications of RFID tags and readers as per RFP floated by DLDSL are available in Appendix B for reference.

21.17 SUMMARY

Intelligent transport system (ITS) has been developed to address various issues, such as road traffic safety, transport efficiency, and environmental measures, by using communication technologies to exchange information between people, infrastructure and vehicles. A number of ITS tools are available to address these issues. The standardisation of ITS tools refers to activities to minimise, simplify, and put in order things, whenever possible, that is, if left alone, would become diversified, complex, or chaotic. Standardisation provides a number of benefits, such as improvement of technological competence and quality, enhancement of operational efficiency, cost reduction, and many more.

At international level, International Standards Origination (ISO), European Telecommunications Standards Institute (ETSI), Japan ITS Standardisation Committee, ITS Joint Programme Office US Department of Transport, Intelligent Transport Society Korea, etc. are working for standardisation of ITS tools.

In India, Ministry of Urban Development, Government of India has taken lead in developing guidelines and some standards for ITS for different cities. An ITS toolkit for traffic management system for cities has been developed to guide the city officials on ITS tools. Further, tenders (request for proposal) have been floated by different cities which specify the functional requirements to be met by the ITS tools. The approach of Government of India perhaps is to specify functional requirements and broad standards only. The functional requirements for most frequently intelligent signaling system, red light violation detection system, speed detection system, CCTV, automatic number plate recognition (ANPR), automatic vehicle locating system (AVLS) and passenger information system (PIS) as available in different RFPs have been presented in this chapter.

Questions

21.1 What do you understand by standardisation of ITS tools and what are the advantages of standardisation?

21.2 Which are the international bodies working for standardisation of ITS tools?

21.3 What do you understand by functional requirements of an ITS tool?

21.4 Briefly describe functional requirements of following ITS tools:

(a) Red light violation detection system

(b) Speed detection system

(c) CCTV

(d) Automatic number plate recognition (ANPR)

(e) Automatic vehicle locating system (AVLS)

Bibliography

ISO/TC 204 (https://www.iso.org/committee/54706/x/catalogue/p/1/u/0/w/0/d/0).

ITS standardisation activities in Japan 2016 (http://www.jsae.or.jp/01info/its/2016_bro_e.pdf).

ETSI—European Telecommunications Standards Institute (http://www.etsi.org/).

ITS Joint Programme Office, US Department of Transportation (https://www.its.dot.gov).

Intelligent Transport Society, Korea (http://itskorea.kr/eng/eng/01_about.jsp).

Smart City Mission, Government of India (http://smartcities.gov.in).

Smart Net, Ministry of Urban Development, Government of India (https://smartnet.niua.org/).

RFP for selection of implementation agency for integrated traffic control system (ITCS) in Surat city (http://smartcities.gov.in/content/citytender.php?statid=11&disid=141&catid=4).

Request for proposal for design, implementation and operation of smart solutions for public mobility and municipal vehicles in Ujjain (http://smartcities.gov.in/upload/tender/591c1a56daa31RfpdocUjjain.pdf)

ITS standards Sweden (http://its-sweden.se/wp-content/uploads/2013/07/intelligent_transport_system_standards.pdf).

ITS toolkit for traffic management system, India (http://www.sutpindia.com/skin/pdf/Toolkits/ITS%20ToolKit%20SentTo%20IUT_200614.pdf).

Appendix A Typical Metro Train Maintenance System

Typically following tools are being used in the metro train systems to monitor the heath of the equipment and generate advanced information for any maintenance requirement for metro trains:

(a) **Mobile communication gateway** for remotely accessing train's on-board database system (ODBS)

(b) **DCUTerm** for train control management system

(c) **Mona** for HVAC maintenance and testing

(d) **WinMonext** for passenger doors maintenance and testing

(e) **Gemini** to test brake control system of the train

(f) **MAVIS** for analysing event and condition data

(g) **TraceWin** for analysing analog form of train's system data

A.1 MOBILE COMMUNICATION GATEWAY (MCG)

The MCG unit is used as a communication gateway between the train and wayside systems by utilising radio communication technology, GPRS-GSM and Wi-Fi. MCG is a part of overall train control and management system (TCMS). It is having an antenna and a dual band combiner by which it is able to transmit the entire train data at regular intervals or as and when required by maintenance team. This tool is very helpful in getting a feedback on train functionality on real-time basis. This data can be accessed remotely at the server located at train depot without physically going on the train. The data can be retrieved even when train is in operations. It helps in analysing the data well before the train is available for the required action and arranging suitable spares thereby not only avoids idling of manpower but also reduces the dead time of the train.

A.2 DCUTERM

DCUTerm is a terminal used for accessing the various control units of TCMS. It is a software-based tool mainly for the maintenance purpose (Figure A.1). It provides the following functionality:

(a) Views analog and logical signals in real-time in a graphical environment
(b) Analyses the system status
(c) Analyses the operation—recording of signals
(d) Enables test procedures through buttons and scripts

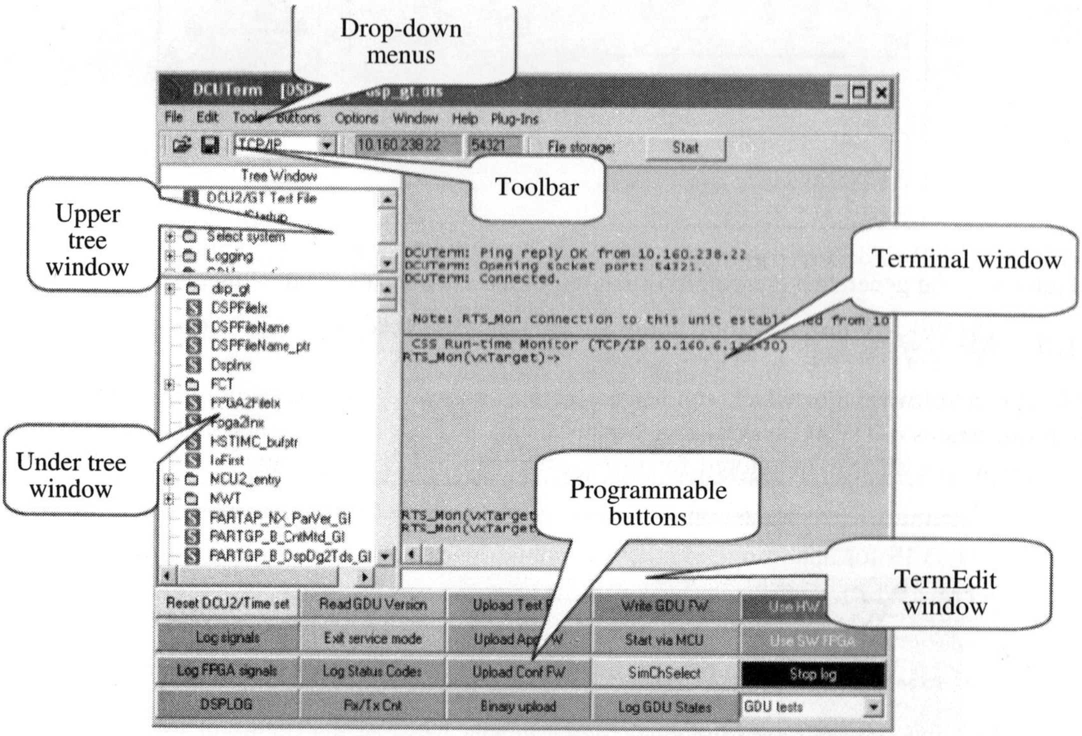

FIGURE A.1 DCUTerm screen shot (reproduced with permission from source).

The various signals in the graph (Figure A.2) provides the vital inputs regarding the functionality of subsystems and the variation of parameters on time scale thereby the true picture of failure scenario can be reconstructed to understand the true cause behind the incidence and likely course of correction can be easily formulated.

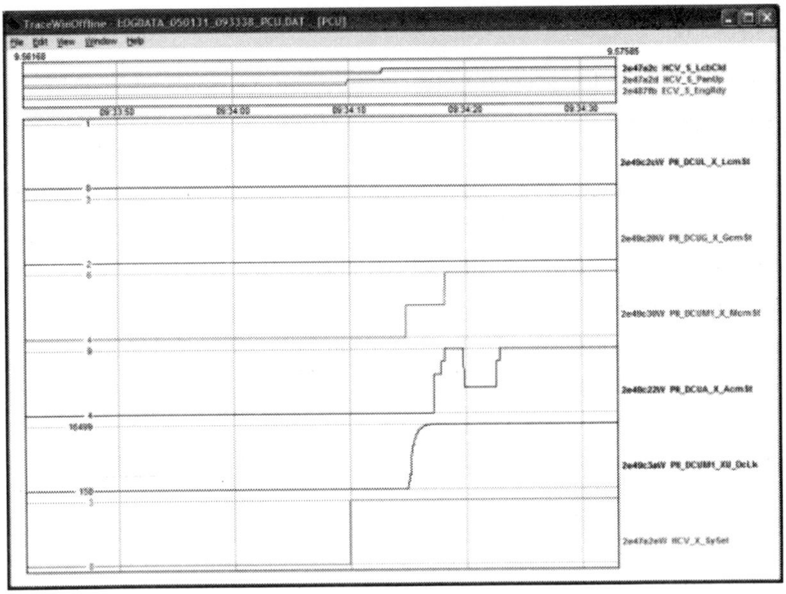

FIGURE A.2 DCUTerm graph for various subsystems.

A.3 MONA

Mona is a software tool which enables the user to monitor, command, control and record the real-time status of HVAC system as per requirement. The status of any subcomponent of HVAC subsystem can also be monitored for its health and functionality (Figure A.3).

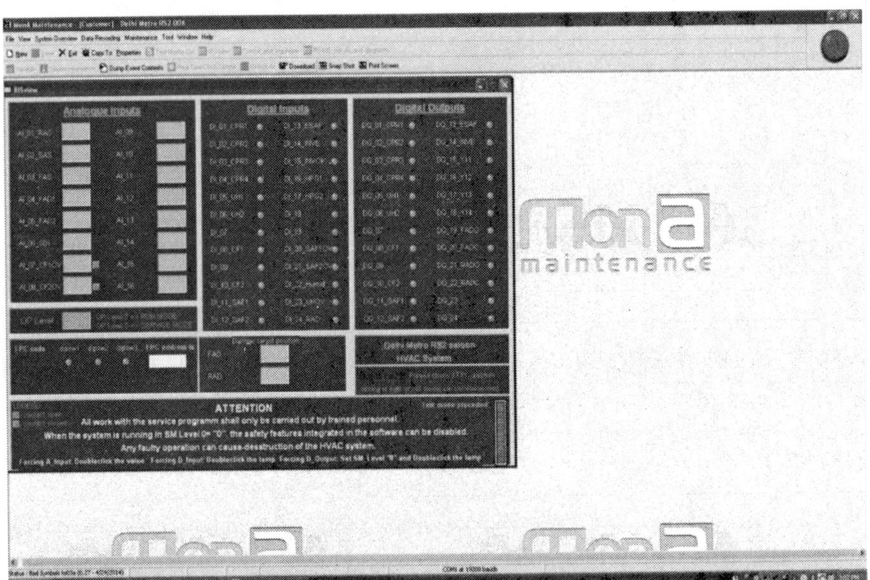

FIGURE A.3 Mona software for HVAC maintenance (reproduced with permission from source).

The real-time monitoring allows advance planning of maintenance activities to be performed during the night.

A.4 WINMONEXT

WinMonext is a software interface available with user of doors system for following actions (Figure A.4):

(a) Monitors (in real-time) all inputs/outputs and failures
(b) Downloads the embedded software
(c) Views dynamic graphic curves depicting various functional states
(d) Modifies the door parameters
(e) Traces events recorder with check-in on hard disk.
(f) Provides the analytic data of doors system in case of failure

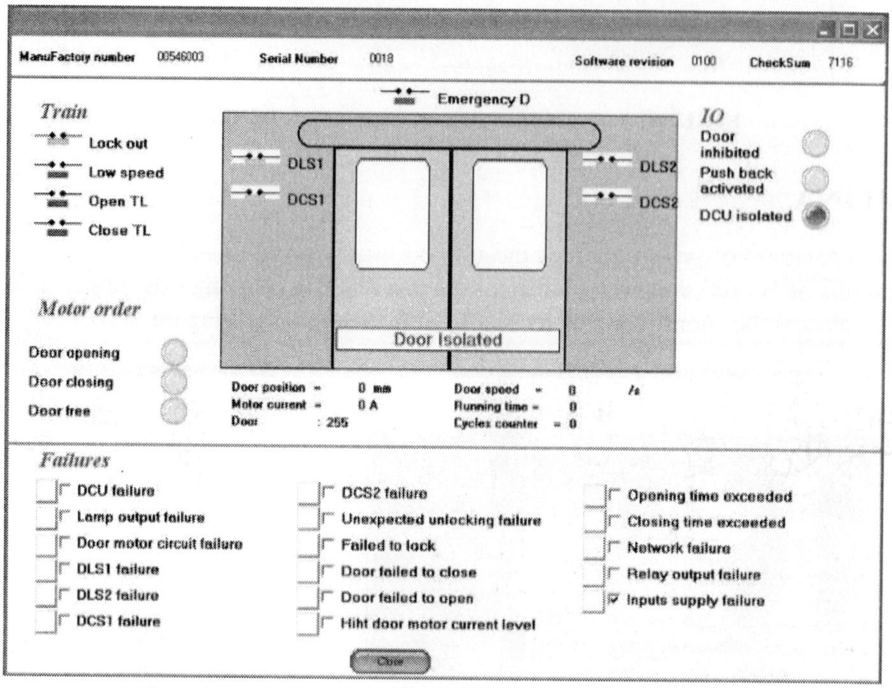

FIGURE A.4 WinMonext software for passenger door maintenance and testing (reproduced with permission from source).

A.5 GEMINI

It provides the real-time status of various subcomponents of brake system at different times. The graphic representation of the signals, variables and controls of BCU are represented by means of objects that change colour dynamically or are animated to indicate a change of status. Values and information are continually updated showing real-time data (Figure A.5).

Parameters monitor by Gemini tool are as follows:

(a) Real-time pulse width modulation (PWM) demand in both traction and braking scenario in the form of percentage

(b) Real-time brake chamber pressure

(c) Real-time load signal (pay load)

(d) Real-time air generation system subcomponent status like air dryer

By above facilities, the various valves and actuators in the brake system can be monitored for any abnormal behaviour, even a minute deviation from normal pattern will be noticed by this tool thereby the Tel Signs can be picked at the incipient level itself and any major failure on mainline could be avoided.

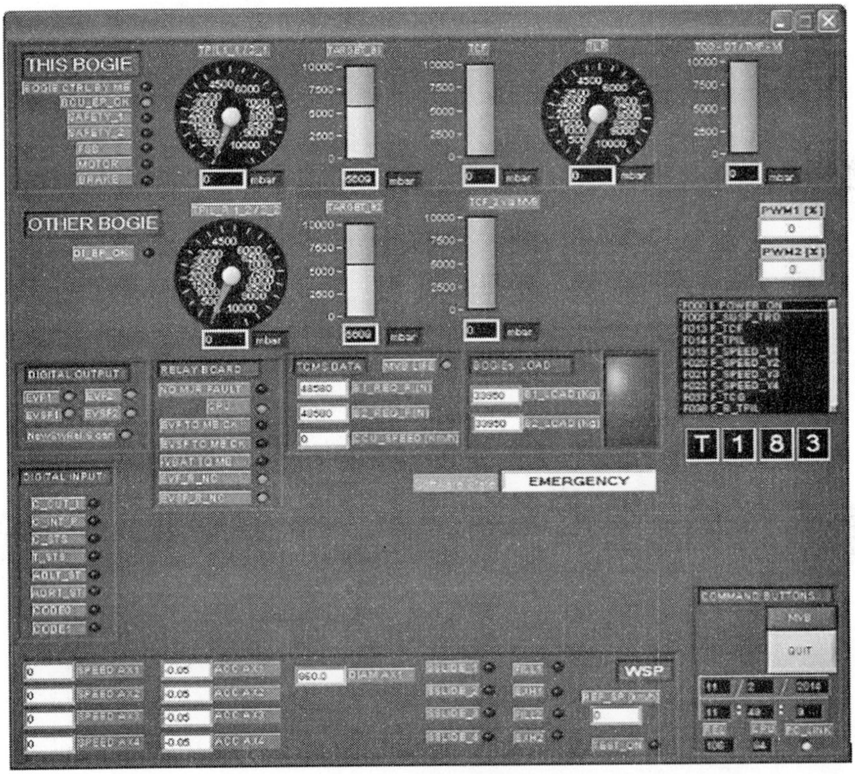

FIGURE A.5 Gemini software to test brake control system of the trains (reproduced with permission from source).

A.6 MAVIS

The off-board maintenance evaluation tool, MAVIS, enables the maintenance staff to view and analyse the information uploaded from the on-board train diagnostic system (TDS). A TDS uploader offers the user an interface for uploading or reading the information stored in the diagnostic system, TDS.

Once this information is retrieved, it can be filtered and sorted in order to analyse trends of events over a period of time. Environment data associated with any failure or event, can be analysed, in tables or graphs (Figure A.6).

The service engineer can see an overview of events search, sort and filter events examine the disturbance, investigate the circumstances, point out possible reasons and give suggestions for repair.

A control is available with user to change/correct the data, e.g., wheel diameter, train energy, kilometre readings, etc.

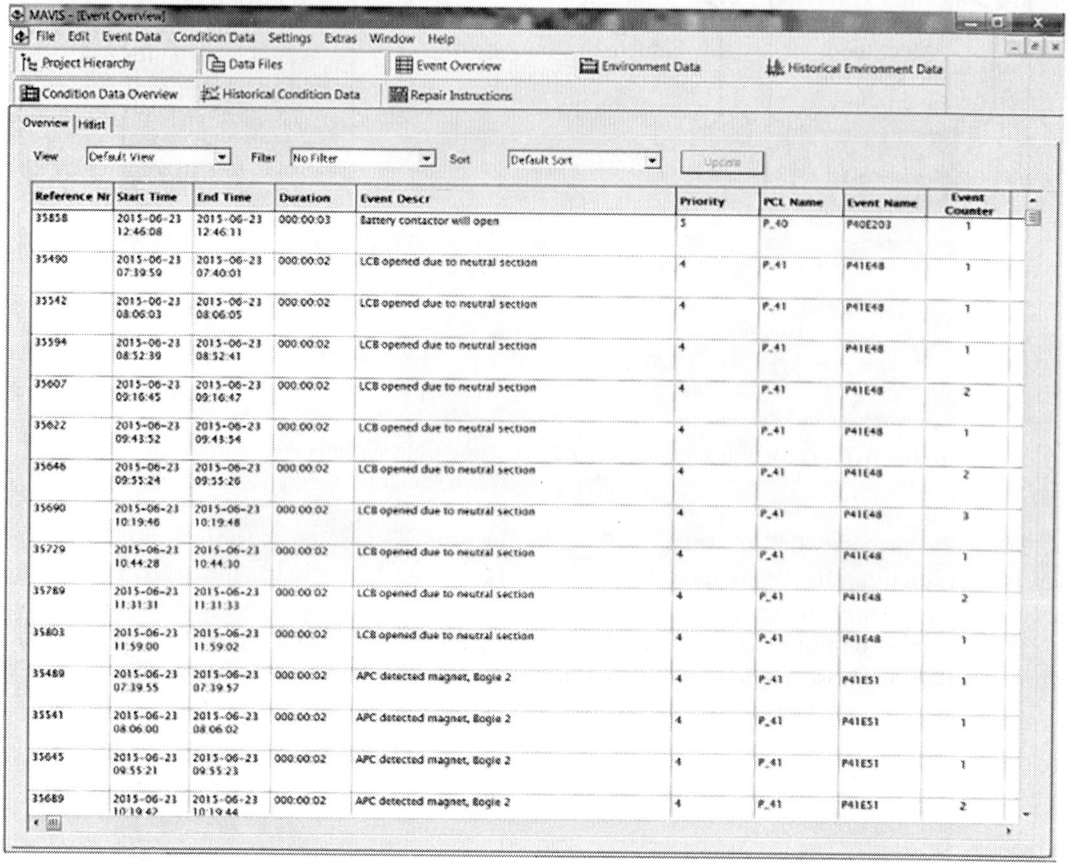

FIGURE A.6 MAVIS software—maintenance evaluation tool (reproduced with permission from source).

A.7 TRACEWIN

This tool is used for viewing logged signals graphically in real-time. For off-line analysis of waveforms, the window is also available as a separate application. Off-line mode helps in analysing various analog signals recorded in on-board database system (ODBS) of the train (Figure A.7). Online mode enables the maintenance engineer to check the status of desired signals whenever required.

When selected, a separate window appears:

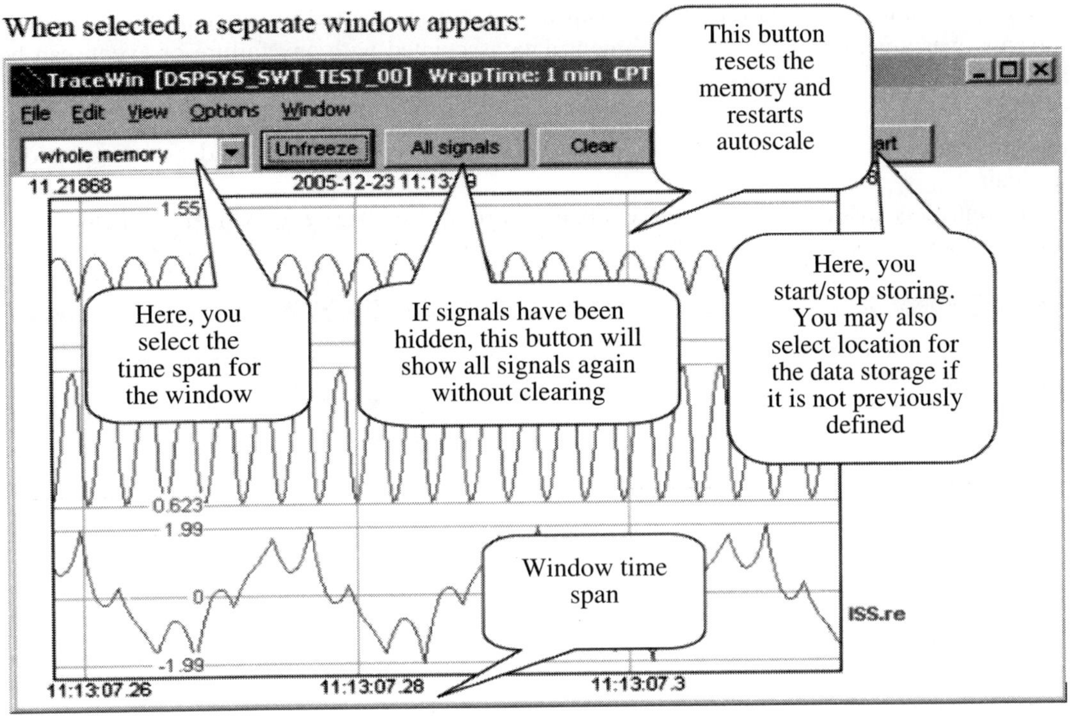

FIGURE A.7 TraceWin software to analyse analog train system data (reproduced with permission from source).

TraceWin provides the information about the pattern of variation of various signals, which may be the tractive effort generated by the train, the actual brake force, the speed profile, the variation of pressures at brake cylinder level, the main reservoir pressures and lot others.

This information not only helps in reconstructing the failure scenario but also provides vital information about those components which are not normally taken up for inspection.

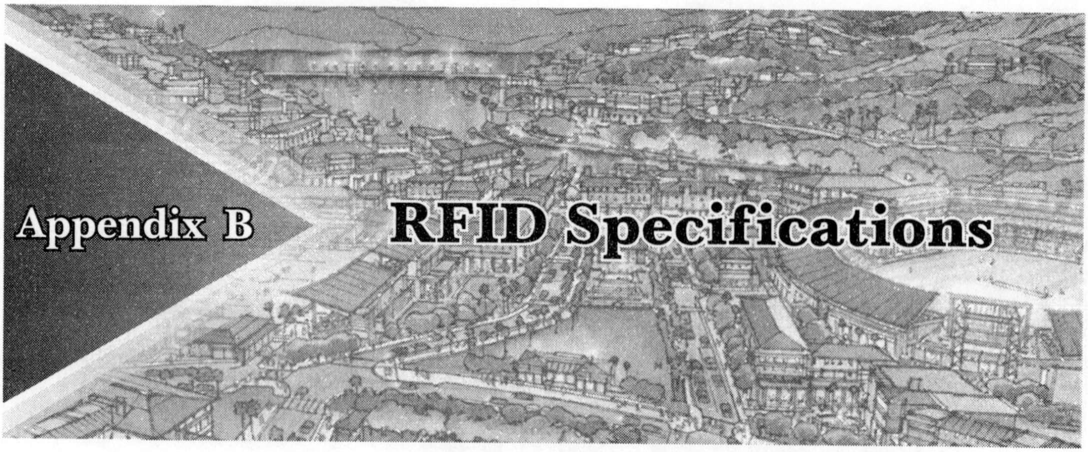

Appendix B RFID Specifications

The details of specifications of RFID used in Mumbai Logistics Park are as follows:

Physical specifications	
Specification for Tag	**Details**
Encasement	Over molded ABS rigid plastic
Size	The tag should be placed easily on the groove of the sea container wall. The size of the tag can be proposed based on this requirement
Weight (g)	Less than 100 g
Environmental specifications	
Storage temperature	–20 °C to +85 °C
Operating temperature	–20 °C to +80 °C
Survivability	–20 °C to +80 °C
IP rating	IP68 or better
Shock and vibration	MIL STD 810 °F
Impact	10 kg from 1 m
Attachment type	Integrated magnets with 7.0 kg or greater pull force considering the attachment on and detachment from the sea container surface. The tag should be re-used any number of times for tagging/de-tagging without impacting or altering the attachment material. The attachment should ensure that the tag should not come out of the container surface during its journey on road or rail
Attachment durability	10,000 cycles of attach/detachment
Attach/detach tool (tag applicator)	For placement and removal of tag from assets several metres above ground level
Certifications	CE approved ROHS approved

RF specifications	
Protocol	EPC Class 1 Gen 2
Frequency range (MHz)	As per the norms in India
Read range using fixed reader (on metal surface)	Contact to 7 m or more
Read range (handheld reader)	Contact to 3 m or more
Material compatibility	Should be optimised for "on metal" use
RFID IC memory capacity	EPC : 96 bits User : 512 bits TID : 64 bits (unique) (Writing capability of all memories should be locked during manufacturing)
EPC memory	The globally unique EPC code will be provided by DLDSL to the supplier for storing the same in the EPC memory during manufacturing of the tags
Customisation	Ease of tagging and de-tagging on the upper side of the container wall using an applicator, while the container is loaded on truck or rail wagon. Approximate height of the attaching/detaching location is between 4 m and 6 m
Project logo	The project name/logo needs to be embossed on the upper surface of the tags

Fixed reader	
Accessories	Reader (indoor unit), power adaptor with Indian standard power plug, and other installation accessories
Ports	Should support 4 or more mono static ports per reader with configurable dynamic channel switching
Antenna mode	Monostatic
Processor	620 MHz or better
Memory	Persistent, flash 128 MB; DRAM 128 MB or better, shall be able to hold at least 100,000 tag data records
Environment	Non-AC, dusty environments such as toll plazas, open highways and state roads, CFS/ICDs
Application	Dense tag environments
Reader housing	Aluminium
	IP53 or better
Dimensions	Small installation footprint
Connectivity	10/100 Base-T Ethernet – RJ45
	DB9 or DB25
Operating temperature	–4 °F to +131 °F
	–20 °C to +55 °C

Firmware and management protocol supported	Remote firmware management capability
API support	.NET and C
Max receive sensitivity ISO standard	80 dbi or more monostatic
	ISO 180006C (EPC Class 1 Gen 2) ISO 180006B
Power output	+10 dBm to +30 dBm
Specific absorption ratings	Complies to FCC 47CFR2:OET Bulletin
	65; EN 50364
Humidity	5% to 95% non-condensing
Safety and health	IEC 60950-1/UL 2043
Service centre	OEM should have in country service centre
Toll free number	OEM should have toll free number easily accessible to partners and customer for online support
Reading accuracy on continuous power "ON"	The device should have designed to be powered on and transmitting throughout its life. The device should not hang or restart automatically due to any reason
Regulatory compliances	Safety: Safety UL 60950-01, UL 2043, IEC 60950-1, EN 60950-1
	Other: ROHS, WEEE

Antenna	
Accessories	Antenna (outdoor), mounting bracket and other installation accessories. (RF cable excluded)
Connectors	2xType N female
Mounting bracket	Integrated mounting holes
Weight	2 kg or less
Casing	UV stable ASA
Frequency and gain	As permitted by regulatory in India
3 db Beamwidth	70° in both phases
Max power	10 W
Axial ratio	1 dB typical
Operating temperatures	–22 °F to +158 °F (–30 °C to +70 °C)
Sealing	IP-67 or better
Environmental	RoHS/WEEE compliant
Vibration	IEC-68-2-6 (10 to 150 Hz, 05 g, 1 h in each of two axes) (Random vibration)
Humidity	MIL-Std 810G, METHOD 507.5, Procedure II–Aggravated

Appendix C

Automatic Number Plate Recognition (ANPR) for Toll Booths and Gate Access Control

An automatic number plate recognition (ANPR) system processes, stores and makes available data collected by the monitored transit lanes. The system can generate an event when a plate is present in a specific list, such as send an alarm when a vehicle is present in a blacklist or can open a barrier when a vehicle is present in a white list. The system can send transit data and images to central system. More in general, the system allows an operator to view the transit and the events generated by the system from one or more control console (called also *workstation*) connected in the network.

ANPR system is structured by the following components:

(a) One or more cameras (at least one for each lane) (Figure C.1)
(b) One or more processors to manage multiple camera
(c) The additional possibility to control a gate barrier or more in general an external device controlled by relay

C.1 CAMERA FOR GATE AND TOLL BOOTH ACCESS CONTROL

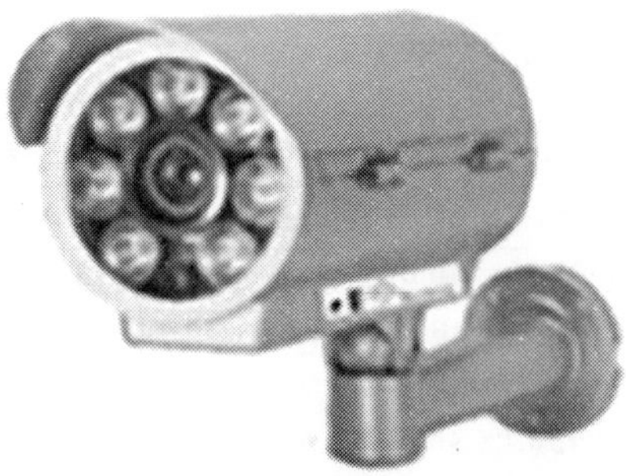

FIGURE C.1 Picture of camera (reproduced with permission from source).

Main camera features:

(a) One lane coverage
(b) Megapixel day/night camera
(c) Recognition of licence at high

The recommended specifications for the cameras to be used for APNR are presented in Table C.1.

TABLE C.1 Recommended specifications for cameras

Recommended specifications*	
Lens	Varifocal
Day and night	Automatically removable infrared-cut filter
Shutter time	1/10000 s to 1/6 s
Video compression	Motion JPEG, H.264 (MPEG4 Part 10)
Resolutions	More than 1 megapixel
Frame rate	Up to 30 fps in all resolutions
Video streaming	Multiple simultaneous individually configurable streams in H.264 and motion JPEG Controllable frame rate and bandwidth
Image settings	Compression, colour, brightness, sharpness, contrast, white balance, exposure control, exposure zones, backlight compensation, fine tuning of behaviour at low light
Supported network protocols	HTTP, FTP, RTSP
Inside camera power	PoE 802.3af, 12 V DC, 24 V AC
Housing power	AC 90 V to 260 V, 32 W
IR wavelength	850 nm
IR power on	Auto sensor control, Adjustable control point
IR distance cover for LPR	Up to 6–20 m
IP rating	IP 66
Housing construction	Aluminium
Weight	4.4 kg
Dimensions	134 mm (diameter) × 318 mm (length) Length bracket: 211.5 mm

*Licence plate recognition software

The software shall be capable of:

(a) Recognition rate of more than 80% for standard number plate in standard font readable by human eye. Indian plates in both day and night condition
(b) For each vehicle transit recognition on multiple frames more frequently in order to reduce the possibility of recognition error
(c) Recognition of licence plates in daylight on colour images

The ANPR process recognises the licence plate of the vehicle and then check if the plate is in a blacklist and store data (ASCII plate, location and timestamp) and image in local database.

Image and data are available to operators through the Web interface; the operator can search for a particular plate in the database or display all the plates in a specific time period. It is also possible to have a live display where the operator can see the live video and plates that are recognised in real-time. Image and transit data may also be transmitted to the central server or any other external software for data processing. (Figures C.2 and C.3).

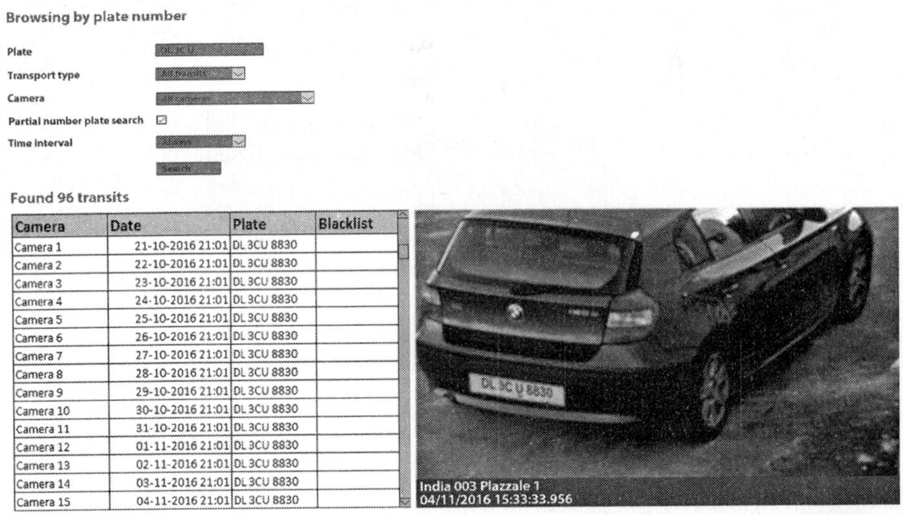

Camera	Date	Plate	Blacklist
Camera 1	21-10-2016 21:01	DL 3CU 8830	
Camera 2	22-10-2016 21:01	DL 3CU 8830	
Camera 3	23-10-2016 21:01	DL 3CU 8830	
Camera 4	24-10-2016 21:01	DL 3CU 8830	
Camera 5	25-10-2016 21:01	DL 3CU 8830	
Camera 6	26-10-2016 21:01	DL 3CU 8830	
Camera 7	27-10-2016 21:01	DL 3CU 8830	
Camera 8	28-10-2016 21:01	DL 3CU 8830	
Camera 9	29-10-2016 21:01	DL 3CU 8830	
Camera 10	30-10-2016 21:01	DL 3CU 8830	
Camera 11	31-10-2016 21:01	DL 3CU 8830	
Camera 12	01-11-2016 21:01	DL 3CU 8830	
Camera 13	02-11-2016 21:01	DL 3CU 8830	
Camera 14	03-11-2016 21:01	DL 3CU 8830	
Camera 15	04-11-2016 21:01	DL 3CU 8830	

FIGURE C.2 Working of licence plate recognition software (reproduced with permission from source).

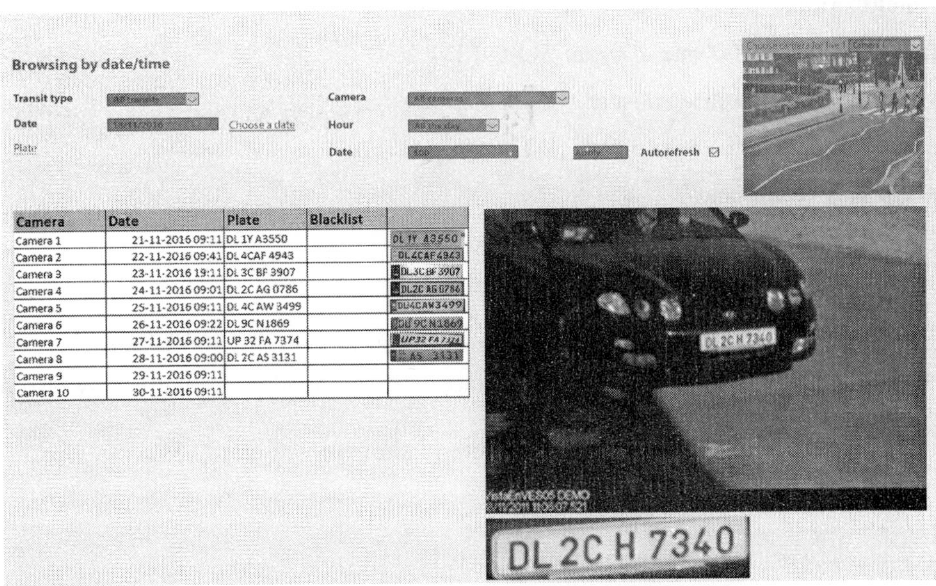

Camera	Date	Plate	Blacklist
Camera 1	21-11-2016 09:11	DL 1Y A3550	
Camera 2	22-11-2016 09:41	DL 4CAF 4943	
Camera 3	23-11-2016 19:11	DL 3C BF 3907	
Camera 4	24-11-2016 09:01	DL 2C AG 0786	
Camera 5	25-11-2016 09:11	DL 4C AW 3499	
Camera 6	26-11-2016 09:22	DL 9C N1869	
Camera 7	27-11-2016 09:11	UP 32 FA 7374	
Camera 8	28-11-2016 09:00	DL 2C AS 3131	
Camera 9	29-11-2016 09:11		
Camera 10	30-11-2016 09:11		

FIGURE C.3 Working of licence plate recognition software (reproduced with permission from source).

C.2 PROCESSOR SERIES FOR GATE AND TOLL BOOTH ACCESS CONTROL

Two types of processors may be used:

Type I processor

One for external environment without any indoor fittings and can be fitted directly in poles/gantries. This processor can be used for maximum of 2 cameras (Figure C.4).

FIGURE C.4 Type 1 processor (typical) (reproduced with permission from source).

Recommended specifications for processor type 1

Chipset: ICH-8M

CPU: Intel® Atom™ D510 Dual Core 1.6

Memory: DDR2 667/800 one 200-pin SO-DIMM 2GB

Construction: Aluminium enclosure with fanless design

Dimensions: 320 mm (L) × 176 mm (W) × 50 mm (H)

IP Rating: IP65 compliant

HD: 1 × 160G 2.5" SATA II HDD

LAN: 3 × 10/100 Ethernet

Power: 8 V ~ 60 V DC thru 3-pin connector (ignition, power and ground)

The type I processor handles up to two cameras. It can be placed comfortably also outdoor as it has IP65 protection.

Diagram of the type I processor architecture is shown in Figure C.5.

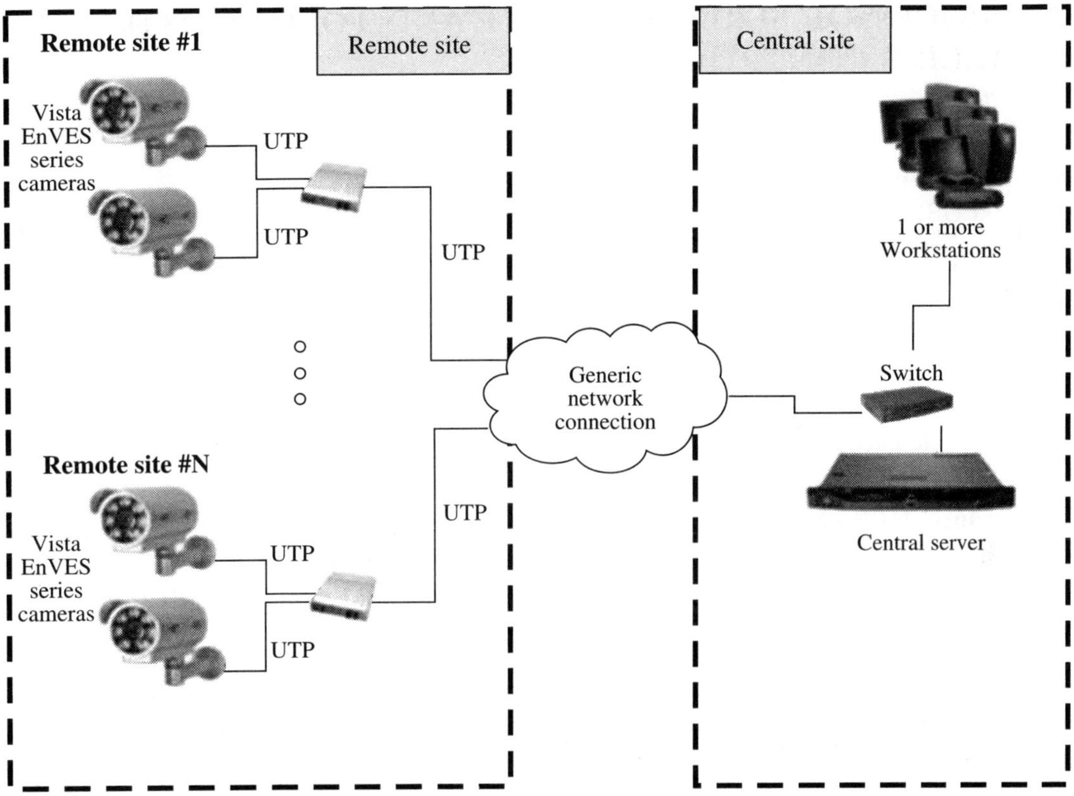

FIGURE C.5 Type 1 Processor architecture (typical) (reproduced with permission from source).

Type II processor

Other processor would be used where indoor cabins are available and can be used with up to 8 lanes at the same time, and the number of cameras depends on the specifications of the camera being used.

This processor is a computer equipment rack-mounted 19” (1-Socket 1U rack server) that offers advanced management capabilities, low wattage power supply and external storage connectivity options in a very small chassis (Figure C.6).

FIGURE C.6 Type II processor (reproduced with permission from source).

The type II processor is able to handle up to 8 devices cameras. It requires a gigabit switch (*not supplied*) to connect to the cameras.

Recommended technical specifications of type II processor

Processor: Intel® Xeon® X3440, 2.53 GHz, 8 MB Cache, Turbo, HT

RAM: 4 GB memory, DDR3, 1333 MHz

HD: 500 GB, SATA, 3.5-in, 7.2 K RPM hard drive

Base: PowerEdge R210 chassis with up to 2 cabled HDs and quad-pack LED diagnostics

Chipset: Intel® 3420

RAID connectivity: C1 MST no raid with on-board SATA controller, Min. 1 Max. 2 SATA only drives

I/O port: n.2 RS/232 port

Power: Single cabled power supply (250 W)

Power cord: Rack power cord 2M (C13/C14 12A)

Front panel: 1U Rack Bezel

Optical device: 16X DVD-ROM drive SATA with SATA cable

Embedded network controller: Broadcom® NetXtreme II 5709 dual port 1GbE NIC with TOE, PCIe-4

Rack chassis: 42.6 H × 431 W × 393.7 D (mm)

C.3 ANPR SYSTEM SCHEME

A typical system flow chart of the ANPR system explained for a simple access control and a toll plaza is shown in Figures C.7.

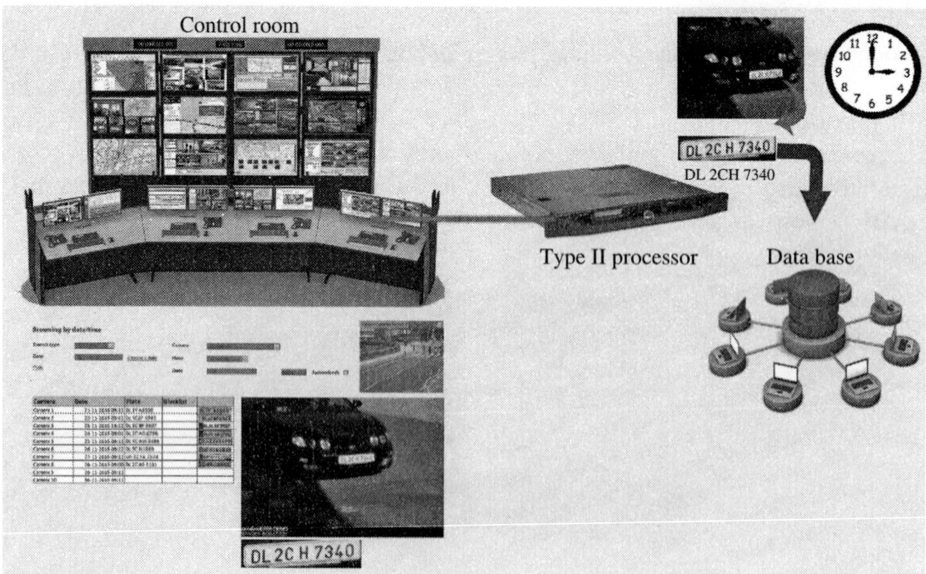

FIGURE C.7 Access control by APNR (reproduced with permission from source).

Optionally the access bar could be controlled via the camera, in this case, depending on the access bar control system, a relay (not provided) should be installed. The camera is connected via IP network to the processor that should be located in a safe place such as a building, a shelter or a control room. The processor is also connected via IP network to one or more workstation located in control rooms from where operator could monitor licence plate that system recognises.

Every processor can be used to manage and store the data of multiple lanes. This feature is suitable for toll plaza installations. As required, additional cameras for each lane can be installed. For example, a series of cameras for the driver image can be added.

For tool booth, installation can be arranged a simple workstation (client) to view transits and to show automatically the last plate read of lane concerned. In any case, all transits will be stored in the internal database system with images (also from any auxiliary cameras), licence plate number and timestamp and could be searched and showed in any time accessing the database via Web interface using a workstation located, for example, in the toll booth.

Appendix D Measuring Traffic Congestion and Speed Using Mobile Network and GPS Probes

The assessment of traffic congestion and speed of vehicles over various sections of the road are important to improve efficiency and safety of the surface transportation. The ITS tools are increasingly being adapted to measure and assess these vital parameters. The ITS infrastructure used for this purpose are traffic sensors (e.g. loop detectors, traffic cameras, etc.), infrastructure for communicating raw data and aggregated traffic information, computation infrastructure for processing raw data from individual traffic sensors to generate aggregated traffic information, and traffic control (e.g., adaptive traffic light control, dynamic message signs, etc.). The researchers have attempted traffic monitoring using variety of sensors. However, the cost deployment and maintenance of traffic sensor is usually very high which makes it difficult to deploy these sensors on large-scale basis. The financial constrains have led to the exploration of alternative solutions by the researchers to assess traffic congestion and speed at minimum cost which has paved way to explore mobile network and GPS system for this purpose.

D.1 MOBILE NETWORK DATA

In India, mobile subscriber has already crossed 1 billion mark at the end of August 2017. In urban areas, the mobile tele-density is more than 145%. Hence, it is safe to assume that every vehicle in a road network is equipped with a cell phone. The location of all the mobile phone users are tracked by the telecom infrastructure for connecting calls efficiently. However, depending upon the density of the mobile communication towers, congestion in communication network and speed of the vehicle, etc., there may be an error of 100–500 m in location of the vehicle assessed from mobile network. This kind of error in location may result in erroneous data of speed and congestion estimated using mobile data only. Several studies have been carried out to assess feasibility of using noisy location data of vehicles collected from mobile network for edge level speed or travel time estimation. The results show that estimates are highly erroneous (mean speed error of 15% or more).

D.2 GPS PROBE DATA

As per the estimates, there are around 300 million smart phone users in India. The density of smart phone users is very high in cities. The penetration of smart phones in cities is witnessing unprecedented growth in India. Most of the private car users in cities possess smart phones. The smart phones are equipped with GPS devices. Further, the number of GPS equipped vehicles such as taxis, autos, buses is also increasing due to instructions issued by government. Due to the increasing penetration of GPS enabled vehicles and smart phone users, the GPS probe data are considered an attractive source for real-time travel speed estimation. The field experiment demonstrated the use of GPS probe vehicles for generating traffic information. The results showed that 2–3% penetration of GPS enabled cell phones in the driver population is necessary to generate the estimation of speed of traffic flow. However, the GPS probe data is not directly useful in Indian traffic scenario due to two major reasons: (a) GPS probe penetration is low in India and (b) GPS is mostly found on cars and public transport buses, whose movement characteristics are very different than that of two wheelers which form approximately 75% of the overall vehicle population on Indian arterial roads.

Chaturvedi and Srivastava (2016) have used mobile network data, GPS data and road infrastructure data to develop a simulation model which can be used to assess the speed and level of congestion of road traffic. The salient features of their work are as follows:

(a) The erroneous location data of vehicles collected from mobile network are processed using a novel map matching algorithm to generate vehicle trajectories and count vehicles. The vehicle count information is used to compute vehicle flow, space occupancy and congestion estimation in real-time. The simulation results show the estimation error of less than 10% in arterial road network scenarios.

(b) To enable edge (road section) level speed estimation, a set of edges for ITS infrastructure deployment are identified using historical data of traffic congestion. The infrastructure requirement depends upon the level of congestion and the road network size. Simulation results show that full coverage of a road network can be assured with infrastructure deployment on less than 30% edges.

(c) As GPS probe speed data are sparse (due to low penetration) and biased (due to deployment on only certain types of vehicles), they are not used in real-time and are collected for every congestion level on every edge. The historical data of GPS probes are used for spatial extrapolation of speed estimation from infrastructure edges to infrastructure less edges. The use of GPS probe data improves accuracy of speed estimation significantly.

Figure D.1 shows the principle of working of the system.

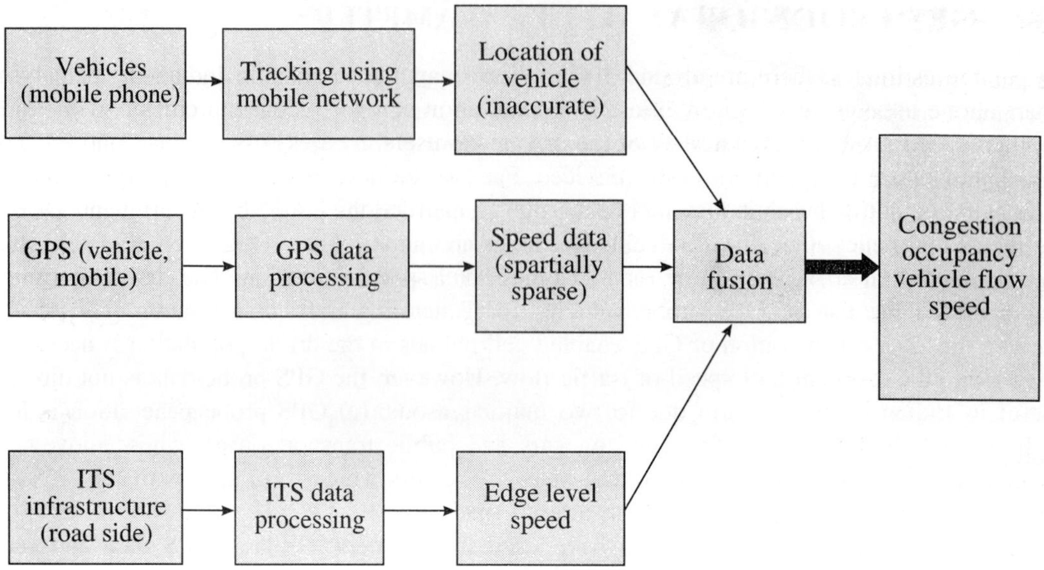

FIGURE D.1 Working principle of measuring traffic congestion and speed using mobile network and GPS probes.

D.3 LOCATION IDENTIFICATION USING MOBILE NETWORK

To assess location error in the cellular network, fingerprinting based localisation can be used that does not require access to the cellular infrastructure elements and is viable using any GPS enabled smart phone. An Android application collects cellular fingerprint [main cell ID and received signal strength indicator (RSSI) values, neighbour cells and RSSI values] and corresponding GPS coordinates every second. The RSSI value indicates the strength of the mobile communication signal. The experiment results show the mean location error of less than 50 m in the regions with very dense cell tower deployment (6–7 neighbour cell towers in addition to the main cell are visible with good signal strength of approximately –80 dbm). In the regions with relatively sparse cell deployment (2–3 neighbour cells), the mean location error of less than 200 m is observed.

D.4 MAP MATCHING FOR ESTIMATING VEHICLE PATHS

After examining location error in the mobile network, a map matching algorithm analyses the location estimates of a vehicle generated using mobile network to compute the vehicle path. A location point along with three subsequent location points, separated by at least 2 × maximum error from one another, is analysed to estimate probable path traversed by a vehicle. This adds some amount of overlapped processing of location estimates and gives overlapping probable path. A unique path estimation is computed by considering only common edges which are part of all the predicted probable paths. This assures the correct edge selection independent of the amount of location error.

D.5 ESTIMATION OF TRAFFIC PARAMETERS

The map matching algorithm processes a series of position estimates for each vehicle to determine a probable set of edges, the vehicle may be traversing at a time step. The trajectory of vehicles are used to count the vehicles detected on every edge in real-time. The vehicle count data is used to determine the vehicle flow and space occupancy. The vehicle count (*N*) on an edge is defined as the count of vehicles detected on the edge during the aggregation period. The vehicle flow (*F*) on an edge is defined as the count of vehicles exiting from the edge during the aggregation period. The exit time of a vehicle on an edge is the time when the vehicle was last detected on the edge. The space occupancy on edge is defined as:

$$\text{Space occupancy } (O) = \frac{\text{Number of vehicles present on edge } (N - F)}{\text{Capacity of the edge } (C)} \tag{D.1}$$

where,

N is the vehicle count on an edge

F is the vehicle flow on an edge = Number of vehicles exiting from the edge during stipulated period

Capacity of an edge (*C*) is the maximum number of vehicles that can be accommodated on the edge and is proportional to the length of the edge and the number of lanes on the edge

The space occupancy for a stipulated time can be used to assess level of congestion using capacity ration (volume/capacity ratio). The *V*/*C* for an edge for a given period is defined as:

$$V/C \text{ ratio} = \frac{\text{Estimated vehicle flow during the period } (F)}{\text{Maximul flow on the edge for the given period } (F_{max})} \tag{D.2}$$

The level of congestion may be assessed using Table D.1.

TABLE D.1 Level of congestion for *V*/*C* ratio

V/C range	*Level of congestion*
<0.125	A
0.125 – 0.276	B
0.276 – 0.479	C
0.479 – 0.715	D
0.715 – 1.00	E
>1.00	F

A indicates no congestion and *F* indicates maximum congestion beyond capacity of the edge. The level congestion on every edge is measured every minute. Using the historical data of congestion, the congestion profile of every edge is estimated. The congestion profile of an edge presents the percentage of overall time the edge remains under given congestion level.

The congestion profile of edges is used for selecting edges for deployment of ITS infrastructure and edge level speed estimation. An edge that remains congested more than 25% time of the day may be considered for deployment of ITS infrastructure.

D.6 USE OF GPS DATA

Only a few numbers of vehicles in India are equipped with GPS devices. The periodic position updates from GPS devices in vehicles are used to compute GPS probe speed on an edge during an aggregation period and is recorded every minute. However, the GPS probe speed data are not directly useful for real-time edge level speed estimation due to low penetration of GPS equipped vehicles.

GPS data is used when the GPS probe data is available for an edge for a given period, it is recorded along with the congestion level data. The historical data of GPS are used to compute speed transition value for extrapolating vehicle speed estimation from the infrastructure edge to the infrastructure without edge.

D.7 ESTIMATION OF SPEED FOR ROAD EDGES WITHOUT ITS INFRASTRUCTURE

The speed for the road edges equipped with ITS infrastructure can be estimated using the ITS tools. The infrastructure edges assumed to collect correct speed information, in addition to the flow, space occupancy and congestion information estimated using the mobile network, every minute. Every infrastructure edge learns occupancy-speed relationship using historical data.

A relationship can be derived between speed (V) and space occupancy (O), the regression analysis of historical data is shown as follows:

$$\text{Speed } V = a + (b \times O) + (b \times O^2) + r \tag{D.3}$$

where,

V is speed
O is space occupancy (refer Equation D.1)
a and b are regression coefficients
r is the residual

Hence, V is the function of space occupancy (O). The speed can be estimated if O is known. Every edge which is not equipped with ITS infrastructure obtains the congestion level and space occupancy estimation from the mobile network every minute as data from the mobile communication is available on minute-to-minute basis. The estimated space occupancy for the current time is used as an input to the regression model of that infrastructure edge to estimate the speed. The data for space occupancy and speed can be used to assess the traffic flows, level of congestion and traffic pattern at an optimum cost without a need for expensive ITS infrastructure.

D.8 SUMMARY

The data from mobile phone, GPS (vehicle and mobile devices) and road side ITS infrastructure are assimilated and analysed using the model estimate traffic flows in the city. The model can be used for estimation of congestion, speed and pattern of traffic flow in a city. The model usage mobile network, GPS probe data from the vehicles and smart phones and minimal road side ITS infrastructure for this purpose. Since use of expensive sensors has been minimised, the model offers a cost-effective ITS solution for traffic management in a city. The model has wide scope for scaling up to cover whole city at a very low cost.

Bibliography

Calabrese, F., M. Colonna, P. Lovisolo, D. Parata, and C. Ratti, "Real-time urban monitoring using mobile phones: A case-study in Rome", *IEEE Transactions on Intelligent Transportation Systems*, Vol. 12, 2011.

Chaturvedi, M. and S. Srivastava, "Optimal ITS Infrastructure Deployment using Cellular Network and GPS Probes", PhD Thesis, DA-IICT, Gandhinagar, Gujarat, India, 2016.

Chaturvedi, M. and S. Srivastava, "Real-time vehicular traffic estimation using mobile infrastructure", *IEEE Conference on Advanced Networks and Telecommunications Systems (ANTS)*, 2013.

Varshavsky, A., M.Y. Chen, E. de Lara, J. Froehlich, D. Haehnel, J. Hightower, A. LaMarca, F. Potter, T. Sohn, K. Tang, and I. Smith, "Are GSM phones the solution for localisation?", *Proceedings of the 7th IEEE Workshop on Mobile Computing Systems and Applications (WMCSA)*, 2005.

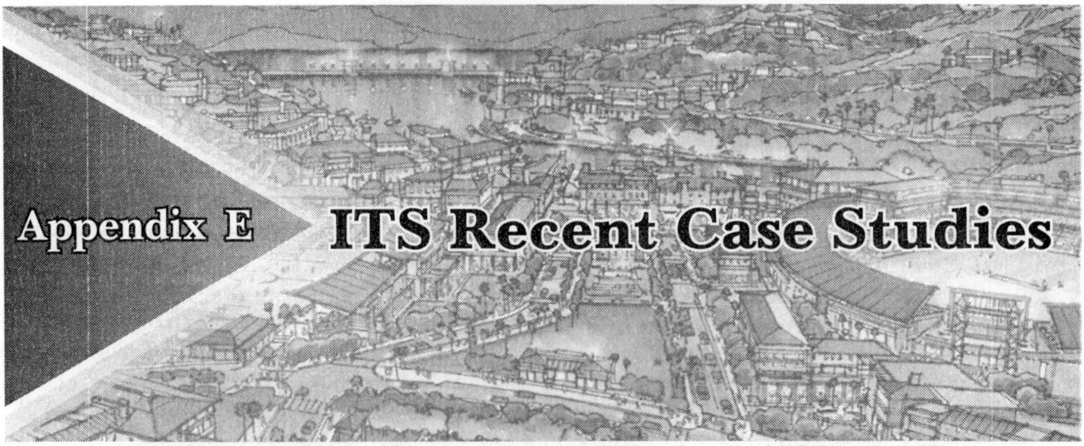

Appendix E ITS Recent Case Studies

Sl.No.	ITS project details
1.	**Intelligent Transport System in BRTS Bus Services and Rajkot City Bus Services**
	Rajkot Municipal Corporation introduced bus rapid transit system (BRTS) bus services from Gondal Chowk to Madhapar Chowk (10.70 km) and Rajkot City bus services from 10/10/2013. Daily approximately 17,500 passengers enjoy rides in BRTS and 28000 passengers avail rides in Rajkot municipal transport service (RMTS) on working days. Rajkot Municipal Corporation (RMC) has installed 112 closed-circuit television (CCTV) cameras on BRTS Pilot Blue Corridor for safety of citizens and control BRTS bus operation. RMC has set up control room, which includes CCTV camera server, server storage, work stations, live video wall, etc. Server of fare collection of Rajkot BRTS and City bus service is also taken place. Real-time update of ticketing and passengers data is displayed. RMC provides live tracking GPS system of BRTS buses (10 + 01), RMTS buses (90 + 09) and two (02) vehicles of checking squad. RMC has also installed the automatic announcement system as pilot project in two BRTS buses and offline CCTV camera in one BRTS bus as a pilot project. Virtual ticketing through PayTM mobile application is introduced on September 1, 2016 in Rajkot BRTS. Google map with multi-modal transportation has been implemented at 18 BRTS bus shelters connected with Wi–Fi which includes BRTS bus service, city bus service and pedestrian way (walking).
2.	**Intelligent Traffic Management System, Pune**
	Pune Mahanagar Parivahan Mahamandal Limited (PMPML) provides public transport service in Pune and Pimpri-Chinchwad city and its suburban region, through bus rapid transport (BRT) and non-BRT routes in these cities. It caters to about 11 lakh passengers per day. Apart from PMPML, the public transport infrastructure in the city is weak (trains run on limited route and with very sporadic frequency and metro still at a planning phase). PMPML is called lifeline for these two cities. However, PMPML was facing several issues related to fleet management, maintenance, operations and man-power management. To overcome the same, and to ensure that a reliable, efficient and sustainable public transport facility is provided to citizens, PMPML has planned and implemented an integrated test management system (ITMS) project.

	The project includes implementation of automatic vehicle location system (AVLS), automatic fare collection system, command and control centre, mobile app (PMP eConnect), incident management system, enterprise management system, grievance management system and automated rostering for scheduling and planning, etc. Recently, MI card (single mobility card/ pass) has also been launched for citizens.
3.	**Integrated Intelligent Transport System (IITS), Navi Mumbai**
	Navi Mumbai Municipal Transport (NMMT) is the transport wing of Navi Mumbai Municipal Corporation, which operates bus services in Navi Mumbai. NMMT buses serve the entire Navi Mumbai city as well as to certain parts of Mumbai, Thane, Kalyan, Dombivali, Badlapur, Taloja, Panvel and Urban. NMMT has 483 buses with 80 AC Volvo buses, 2 hybrid Volvo buses and 401 non-AC buses, which includes mini, midi and standard diesel buses, 150 CNG buses and 75 low floor Jawaharlal Nehru National Urban Renewal Mission (JNNURM) buses that run on 72 routes. These buses are operated from Turbe, Ghansoli and Asudgaon depots. NMMT has constructed state of art Ghansoli Depot under JNNURM funding. The project implemented by NMMT includes following ITS components: (a) Automatic vehicle location system (b) Passenger information system (c) Mobile application for citizen (d) Command and control centre and incident management system (e) Automatic fare collection system (f) Planning and scheduling management (g) Inventory management system (h) Financial management system (i) Business intelligence
4.	**Intelligent Traffic Management System for Surat Sitilink Limited**
	Surat, the second largest city in Gujarat and 9th largest in India is the fastest growing city in the country. It has been ranked third cleanest city of India and has been a part of first 20 smart cities to be developed. Surat covers an area of 326.51 km^2 with the population of 44.6 lakh (Census, 2011). There is large number of migrant population in the city from various parts of India due to economy generating textile and diamond industries. The city had negligible mass transit system and relied majorly on intermediate public transport (auto rickshaws) before BRTS and city buses were operational. Surat Municipal Corporation (SMC) has formed its owned company Surat Sitilink Limited that is responsible for the operations of public transport services (BRTS and city bus) in the city. Currently, Surat BRTS inaugurated first corridor of 10.3 km length in January 2014 and till date, it has expanded operation to 99 km out of planned 102 km with average daily ridership of 95,000 with 130 buses on road. Surat city bus services are operational since November 25, 2016. Surat city bus service was planned with a network of 284 km (with SMC area) with 1000 buses integrating with BRTS. Currently, the bus service is operational on 27 routes with average daily ridership of 65,000 and 216 buses on road. City bus services cover an area of 85% within SMC. A total of 46 routes including urban and sub-urban have been planned which would get operational substantially. City bus network was planned and made operational in such a way that BRT stations act as an interchange for both the system. This provides easy and convenient transfer for passengers. Single ticketing system with fare integration is developed to make travel convenient for the people of the city.

	The ITMS project constitutes component of automatic vehicle location system for all transit vehicles, emergency services and SMC vehicles; Depot management system (DMS) to digitise all the depot activity; Incident management system (IMS) to enable SMC to provide quick response during critical time of any emergency; and Planning system to prepare dynamic schedule. All the activity is being monitored from SMAC centre in real-time and any deviation is resolved as per predefined standard operating procedure (SOP). Apart from this, project is also aimed to integrate with automatic fare collection system (AFCS) and common card payment system (CCPS). ITMS project will act as parent project for any master data to these projects and information on real-time will be exchanged for dashboard of AFCS project.
5.	**Intelligent Transport System with Mobile App at Mysore City**
	Mysore City with a population of around 10 lakh is faced with problems of (a) growing number of vehicles, (b) ever increasing urban population, (c) meeting the demand in terms of civil and travel infrastructure, (d) unavailability of bus arrival and departure information in real-time, etc. In order to attract people to patronise public transport, Karnataka State Road Transport Corporation (KSRTC) has successfully implemented the intelligent transport system (ITS) with mobile app, nation's first successful project with value added features. The ITS has been able to promote the use of ITS enabled city buses of KSRTC as sustainable transport modes and enable commuters to make informed choices on travel modes by developing an integrated network in an effort to reduce passenger wait times. KSRTC aimed to bring about this modal shift in the city of Mysore by improving the perceived image of KSRTC services. The project improved the reliability of KSRTC city services through effective measures of travel demand management and meeting commuter expectations. The intelligent transport system project implemented at Mysore addresses the critical issue of road congestion by offering state-of-art technologies and attractive, convenient, comfortable, value added services to encourage the usage of bus services against individual personal vehicles. The ITS of Mysore City is aimed at offering: Real-time monitoring and tracking of buses and help in reducing road congestion and other transport issues; Dynamic passenger information system (PIS) based on geographical positioning system (GPS); advanced display and communication technologies; central control station (CCS) and intelligent display boards. The core proposition is to improve the attractiveness of the public transport, thereby gaining new users from private transport and retaining existing users. Provision of easily accessible relevant travel information to passengers before and during their journeys is seen as a major basis for increasing attractiveness of the public transport. The primary objectives to establish an intelligent system in Mysore city are: (a) improve quality and convenience of public transport system in the city and ensure the delivery of safe, fair, reliable and environment-friendly transport system; (b) promote use of sustainable transport modes and enable commuters to make choices on travel modes by developing an integrated network in an effort to reduce passenger wait times; and (c) optimise operations, improve fleet utilisation, schedules, and vehicle availability with accurate information.
6.	**Intelligent Transport System Project for BMTC, Bangalore**
	Bangalore has road length of around 11,000 km with around 6.7 million registered vehicles. Two wheelers constitute the highest percentage (nearly 70%) of the vehicle population. Modal share reveals that walking and public transport are widely used transportation modes, constituting 28% and 26% of total trips, respectively. Without mass transit, systems that are well-planned, cheap and convenient, personalised modes of transport will see a disproportionate increase, resulting in increasing vehicular congestion and an over-burdened road infrastructure.

	Bangalore Metropolitan Transport Corporation (BMTC), which was formed in the year 1997, is the sole service provider of public bus transport in the city. It operates 6,100 plus buses every day, carrying around 52 lakh passengers. It renders both AC and non-AC bus services with the man-power of 34,560. BMTC operates 74,833 trips covering 1.3 million kilometre daily with 43 depots, 10 traffic transit management centre (TTMC) and 53 bus stations. With growing demand placed on public transport system, BMTC aims to serve the citizens of Bangalore by providing safe, secure, cheap and convenient transport services within the city. BMTC's intelligent transport system, fundamentally, comprises of 4 main elements: (a) Electronic ticketing system: 10,000 plus electronic ticketing machines that are general packet radio service (GPRS) enabled and EMV compliant (b) Vehicle tracking system: 6,400 plus vehicle tracking units (c) Real-time passenger information: 35 passenger display units and BMTC mobile app (d) Establishment of a centralised 24/7 command and control centre The key aspects of the intelligent transport system as detailed above are supported through complete backend automation of processes at depots such as ticketing and cash collection process, and day end reconciliation processes.
7.	**Hyderabad Traffic Integrated Management**
	Existing standalone signals in Hyderabad are of routine type and are 20 years old, poor visibility and they do not work under power failures. Further no foot paths and signals for pedestrians. Hyderabad is most developing city in India (fourth largest) having about 39 lakh various types of vehicles and 600 vehicles are being added every day. The existing usable road is 8.32% against 14–18% in other Metros. The vehicle density is 723 per km which is again the second largest in country. Hence, it needs revamp. Scope of the project: (a) The project envisages provision of the signaling services round the clock, without any break in the service at 221 junctions. In all 221 junctions, the integration of virtual loops cameras for traffic flow management, which is managed remotely from the traffic command centre (TCC). Alfa numeric timers are installed at all junctions. (b) Provide all 221 signals with optical fibre communication (OFC) leased line connectivity and un-interrupted power with automatic fall back option with 3G GPRS dongle to send the data to the TCC. (c) Virtual loop cameras at every junction to determine traffic flows continuously and manage the traffic intelligently. (d) Established (17) variable message systems (VMS). (e) Establish TCC with video wall to monitor live of all 221 signals and synchronise the signals across the city for smooth traffic flow.
8.	**Intelligent Transport System, City Bus Service of Bhopal**
	Bhopal is best strategically located city in Madhya Pradesh. Railways as well as roads connect it to other cities in the state and country. Broad gauge railway lines to almost all the major cities like Delhi, Mumbai, Chennai and Nagpur connect the City. The total road length of the city is 1,020 km of which 66% is managed by Bhopal municipal corporation (BMC). The other agencies involved in construction and maintenance of the city roads are Bhopal development authority (BDA), public work department (PWD), Bharat heavy electrical limited (BHEL) and capital project administration (CPA). As Bhopal advances on the growth path, there are certain challenges ahead of it—the biggest of them being traffic congestion. The potential and beauty of the city is getting lost amidst increasing traffic problems. The crisis, chaos and risk of

	commuting as prevalent in other major urban centres of India, do exist in Bhopal. The situation is likely to get further grim in the coming years, if adequate steps in the improvement and planning of urban transport are not taken timely. There is urgent need to address the issue of urban transport as an important component to shape urban development and provide quality transport service to the community in the city. The road infrastructure has not expanded in tandem with the increase in the number of vehicles in the city. To cater for the intra-city travel demand, the public transport of Bhopal City comprises of buses, midi buses, mini buses, magic's, etc. To answer the needs of the growing traffic and to ensure that people move around in the city in an efficient and comfortable manner, a special purpose vehicle (SPV) was introduced by Bhopal City Link Limited (BCLL). BCLL was incorporated in 2006 to provide organised dimensions to public transportation. Comprehensive mobility plan (CMP, 2014) is guiding blueprint to address mobility. BCLL's ITS strategic plan includes: (a) automatic vehicle location system (AVLS) of buses, (b) passenger information system in buses and bus stops, (c) automatic fare collection (AFC), (d) smart cards, (e) management information system (MIS), (f) CCTV surveillance system in buses, and (g) traffic priority signals. BCLL is responsible for integration of all the above ITS projects. The components are linked with each other and the related facilities to end users that may be commuter or BCLL or bus operator. All BCLL buses are equipped with automatic vehicle localisation system (AVLS), two CCTV cameras (front and rear), passenger information system and passenger announcement system. All city bus depots are connected with Wi–Fi for transfer of CCTV surveillance videos. In first phase, AFC is started in 44 BRT bus stops out of 77. In second phase, AFC will start in balance 33 BRT bus stops and 75 signals are installed in BRT corridor for smooth functioning of traffic (at bus stop: 57, minor junctions: 16, and at major junctions: 2). These signals are equipped with bus priority signals.

Source: Compilation of Entries, Awards Committee Meeting for UMI 2017, Institute of Urban Transportation (IUT), Delhi, September 15, 2017.

Appendix F

Electronic Toll Collection (ETC) System, India

National Highway Authority of India (NHAI) has launched programme for electronic toll collection (ETC) on toll plazas on National Highways called FASTag. FASTag uses radio frequency identification (RFID) technology for making toll payments directly from the prepaid account linked to it. This device is fixed on the windscreen of the vehicle and allows the vehicle to drive through toll plazas without stopping. Each tag will have a unique code, which will help the toll plaza to identify the vehicle and automatically deduct the tax as it drives by. The toll fee is automatically deducted from the device. Once purchased, FASTag has a validity of 5 years. The user needs to recharge the FASTag as per his/her requirement. The device offers non-stop movement of vehicles through toll plazas and the convenience of cashless payment of toll fee with nation-wide interoperable electronic toll collection services. These tags, available both online and offline at toll plazas and designated banks, can be recharged once the balance drops to ₹ 0.

The two core components of the ETC system are RFID transceivers and tags, which interact for the toll transaction to take place. As the vehicle reaches the toll plaza, the passive tag on the vehicle's windshield is read by the transceiver installed at the toll plaza. This exchange of data takes places between the tag and transceiver, transceiver and toll lane controller, toll lane controller and toll plaza server, and toll plaza server and central ETC system. The class of the vehicle for charging applicable toll fee is identified from automatic vehicle identification (AVI) system. The data exchange flows are shown in Figure F.1.

The three broad functions of the ETC system are tag distribution, toll road user account registration, and recharge and toll transaction. For managing all the functions, a virtual hub called the central ETC system (CES), primarily a web application (portal) with access to all stakeholders has been created. Be it point of sale (POS) operator issuing a tag or tag user putting money in the account, be it managing toll plaza concessionaire account or serving the government through a management information system (MIS), each stakeholder interacts with CES through Web. The core functions of the CES include management of accounts of

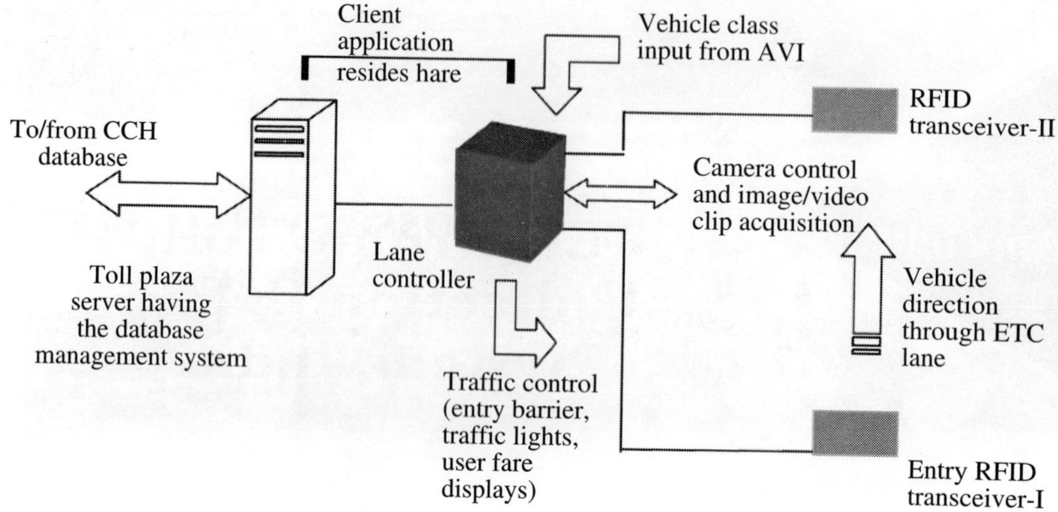

FIGURE F.1 Data exchange for ETC system.[1]

concessionaires, users, POS, distributors, manufacturers, tag issuance, recharge management, interaction with toll plaza application, toll transaction processing and settlement and public interface functions. In addition, certain basic applications at the toll plaza and POS and user services support desk are operated to serve other functionalities. The toll plaza application has two components, viz. server side, which interacts with CES, and lane controllers, which capture tag information, perform preliminary validation and push data to the server.

As per NHAI, from October 1, 2017, all lanes of all 371 NHAI toll plazas in the country will become FASTag enabled.

1 *Source:* Report by Apex Committee for ETC Implementation, Government of India, September 2011.

Glossary

Adaptive Cruise Control (ACC): It is an optional cruise control system for road vehicles that automatically adjusts the vehicle speed to maintain a safe distance from vehicles ahead.

Advanced Traffic Management System (ATMS): It is integration of various subsystems (such as CCTV, vehicle detection, communication, variable message systems, etc.) into a coherent single interface that exhibits real-time data on traffic status and predicts traffic conditions for more efficient planning and operations.

Advanced Traveller Information System (ATIS): It offers to users of transportation systems, travel-related information in advance or on real-time basis to help them in decision making on route choices, estimate travel times, and avoid congestion.

Advanced Vehicle Control and Safety (AVCS) Systems: These are ITS tools used to improve the transportation system safety by supplementing drivers' abilities for maintaining vigilance and control of the vehicle by enhancing the accident avoidance capabilities of vehicles.

Area Traffic Control (ATC): It is a traffic responsive system that uses vehicular data collected through sensors/detectors/cameras and optimises the traffic signal settings to reduce the overall vehicle delays, queue length and stops on road network.

Asset Condition Monitoring (ACM): It is a system that remotely monitors an asset's critical components to provide early warning of issues and potential failures. ACM is used as a trigger to fine-tune the timing of corrective or preventive maintenance, as well as correcting or replacing assets at risk before they fail.

Automated Vehicle Locator System (AVLS): It is a collection of hardware, software and communication system to track the vehicles to get real-time information of its locations.

Automatic Fare Collection (AFC) System: It is a contactless smartcard/token based end-to-end solution for access control, fare, collection and payment. It allows entry to the station only to the passenger having valid fare media, collects exact fare as per the prevailing fare structure from the fare media presented by the passenger and allows exit from the station.

Automatic Number Plate Recognition (ANPR) System: This system is collection of laser system, infrared system, communication system, hardware, and software clubbed as a system together to recognise the number plate of moving vehicle.

Automatic Passenger Counter (APC): It is an electronic device which can be installed in transit vehicles including buses and rail for accurately recording, boarding and alighting data. This technology can improve the accuracy and reliability of tracking transit ridership over traditional methods of manual accounting by drivers or estimation through random surveying.

Automatic Speed Detection System (ASDS): This system is collection of laser system, infrared system, communication system, hardware, and software clubbed as a system together to measure the speed of moving vehicle.

Automatic Train Operation: Under automatic train operation (ATO), the train is driven by the system, i.e., acceleration, regulation, and stopping of the train are performed by the system. The train driver is required to close the doors and push start button to start the train. The ATO system ensures automatic control of train running from station to station, while remaining within the safety envelope calculated by the ATP.

Automatic Train Protection (ATP): It ensures the safety of train operation including separation of trains, enforcement of speed restriction, opening of platform side doors only and safe operations through interlocking.

Automatic Train Supervision (ATS): It is an intelligent interface where one can see all trains running in real time in entire network from a remote central location called *operation control centre* (*OCC*). ATS subsystem monitors the system status continuously and provides the appropriate controls to direct the operation of trains in order to maintain intended traffic patterns and minimises the effect of train delays on the operating schedule.

Automatic Vehicle Classification and Counting (AVCC) System: It is a system that automatically counts and classifies the vehicles in defined categories crossing in its range.

Automatic Vehicle Location (AVL): It is a means for automatically determining and transmitting the geographic location of a vehicle. This data, from one or more vehicles, may then be collected by a vehicle tracking system for a picture of vehicle travel.

Central Traffic Control (CTC): CTC for Railways is the control centre which consolidates train routing decisions that were previously carried out at station level. The system consists of a centralised operation control centre that controls railway interlocking and traffic flows in portions of the rail system designated as CTC territory.

Close Circuit Television Camera (CCTV): It is a camera (fixed or PTZ) to capture the images in still or moving objects within its view range. The CCTV camera is used for traffic surveillance and monitoring system, incidence and stop vehicle detection system, journey time monitoring system and wrong way vehicle detection system, and pedestrian detection system.

Communication based Train Control (CBTC) System: It is a continuous and automatic train control system utilising high-resolution train location determination, independent of track circuits, continuous, high-capacity, bidirectional train-to-wayside data communications, and train borne and wayside processors capable of implementing automatic train protection (ATP) functions, as well as optional automatic train operation (ATO) and automatic train supervision (ATS) functions.

Condition Based Maintenance (CBM): It maintains assets based on whether the asset's condition still remains within required parameters. The aim of condition-based maintenance is to select the most appropriate intervention for the state of the asset at the most appropriate time.

Contactless Smart Card (CSC): It is a device that includes an embedded integrated circuit that can be either a secure microcontroller or an equivalent intelligence with internal memory or a memory chip alone. The card connects to a reader with direct physical contact or with a remote contactless radio frequency interface.

Contactless Smart Token (CST): Like CSC, the token also includes an embedded integrated circuit. However, it is used for single journey only. So it is issued from originating station on payment of specified fare and it is valid up to the destination station only.

Continuous Automatic Train Control (CATC): It is an IT based system to automate the train working in order to ensure safety along with high speed and frequency of trains. In CATC system, most of the train control functions are performed by the system depending upon the system configuration adopted by the railway operator.

Dedicated Short-Range Communications (DSRC): These are one-way or two-way short-range to medium-range wireless communication channels specifically designed for automotive use and a corresponding set of protocols and standards.

Distress Call Management System (DCMS): This is a system (consisting of hardware, software, communication and men/women ware) used by anybody at the time of any incidence/difficulty/accident for getting help. It is highly useful system used to save the lives of human being. This system also helps the police/disaster management control room to dispatch the police vehicle/fire brigade/ambulance at desired (from where complaint is received) location.

Driverless Train Operation (DTO): It is a system of train operations without a train driver where starting and stopping are automated (performed by system) but a train attendant operates the doors and drives the train in case of emergencies. The train attendant performs mostly commercial functions inside the train.

e-challaning: It is a system used for challaning the traffic rules' violators in real time and online. It is a mobile electronic (or electro-mechanical) device capable of challaning traffic rules' violators instantly and maintains the integrity of system.

Economic Rate of Return (ERR): It is the interest rate at which the cost and benefits of a project, discounted over its life, are equal taking into account the effects of economic factors such as price controls, subsidies, and tax breaks to compute the actual cost of the project to the economy and socio economic benefits such as reduction in congestion, accidents, pollution are considered to compute actual benefits of the project.

Electric Vehicles (EVs): These are the vehicles which use electric motors or traction motors for propulsion. An electric vehicle may be powered through a collector system by electricity from off-vehicle sources, or may be self-contained with a battery or generator to convert fuel to electricity.

Electronic Road Pricing (ERP) System: It is a transport demand management tool used to mitigate road congestion. Based on a pay-as-you-use principle, motorists are charged when they use priced roads during peak hours. ERP rates vary for different roads and time periods depending on local traffic conditions. This encourages motorists to change their mode of transport, travel route or time of travel.

Electronic Ticketing Machine (ETM): It is an electronic machine used for dispensing tickets, add value in smart card, reading the ticket or smart card. It can accept currency notes, coins on debit/credit cards for making the payment.

Global Positioning System (GPS): It is global navigation system that is used mainly for navigation purpose. It is a network of satellites that continuously transmits coded information, which makes it possible to precisely identify any location on the earth. It is also used for managing and monitoring of the vehicles and transmitting the related data to traffic management centre in real time.

Intelligent Signaling System (ISS): It is a type of traffic signal that optimises signal timings by sensing vehicular data and pedestrian data at an intersection. The system helps in reducing the vehicular delays and also ensuring safe and convenient movement for pedestrians and non-motorised modes of transport.

Intelligent Speed Adaptation (ISA): It is an in-vehicle system that supports drivers' compliance with the speed limit. In case of potential speeding, a human driver can be alerted, or the speed reduced automatically. ISA is a collective term for various different systems used for this purpose.

Intelligent Transport System (ITS): It is a set of interrelated systems of IT, communication, electronics and electrical that work together to deliver transportation services.

Interactive Voice Response System (IVRS): It is a technology that provides automatic interaction between human being and computerised database using telephone keypad or using speech recognition.

Near Field Communication (NFC): It is a short-range wireless technology for two-way interactions between electronic devices. NFC technology enables contactless digital transactions between NFC enabled devices when these devices are brought closer within a stipulated range (around 20 cm).

Parking Management System (PMS): It comprises the access control system, revenue management, security system, boom barrier and statistical information used for efficient management of parking lots.

Passenger Car Unit (PCU): It is a vehicle unit used for expressing highway capacity. One car is considered as a single unit. Cycle and motorcycle are considered as half car unit. Bus/truck causes a lot of inconvenience because of its large size and is considered equivalent to 3 cars or 3 PCU.

Passenger Information Display System (PIDS): It is a system that provides the different types of information (like expected arrival of bus/train, fare, route, etc.) about the transport system (or for different types of transport system) to the passengers in various forms at bus stop, metro/railway station, transit point, terminal, home, mobile phone, etc.

Pedestrian Detection System (PDS): It is an ITS tool used to detect the presence of pedestrians on road, intersections, etc. Pedestrian detection systems use LIDAR, RADAR, or camera technology to detect a pedestrian. The PDS in the vehicle warns the vehicle driver and may apply brakes automatically if an impending impact is anticipated to either avoid the collision or reduce the collision speed. The PDS equipped signals may adjust the signal as per presence of the pedestrians.

Portable/Mobile Variable Message System (VMS): It is a type of VMS that is self-sufficient (power supply, etc.) and portable/mobile system (fitted on vehicle) used for communicating messages to road users at any location on network.

Public-Private Partnership (PPP): It is a long-term contract between a private party and a government entity, for providing a public asset or service, in which the private party bears significant risk and management responsibility, and remuneration is linked to performance of the private party. The performance is measured in terms of key performance indicators as specified in the concession agreement signed between the public authority and the private party.

Public Transport (PT): It is a system of vehicles such as buses and trains that operate at regular times on fixed routes and are used by the public on payment of fare or free.

Radio Frequency Identification (RFID): It is a kind of wireless communication which uses radio waves to identify and track the objects. An RFID system has *readers* and *tags* that communicate with each through radio waves.

Real-Time Passenger Information (RTPI): It means any information available to transit providers or customers about the current status of vehicles, including approximate locations and predictive arrival times.

Red Light Violation Detection System (RLVDS): It is a collection of hardware, software and communication system to capture the evidence (still photo or video based) about the vehicles violating red light at intersection in real-time automatically.

Split Cycle Offset Optimisation Technique (SCOOT): It is a tool for managing and controlling traffic signals in urban areas. It is an adaptive system that responds automatically to fluctuations in traffic flow through the use of on-street detectors embedded on the road.

Stop line Violation Detection System (SLVDS): It is a collection of hardware, software and communication system to capture the evidence (still photo or video based) about the vehicles violating stop line at the time of red light at intersection in real-time automatically.

Supervisory Control and Data Acquisition (SCADA): It is a category of software application program for process control, the gathering of data in real-time from remote locations in order to control equipment and conditions.

Traffic Management Centre (TMC): It is the hub of transport management, where data is collected, and analysed and combined with other operational and control concepts to manage the complex transport network. It is the control centre for communicating transportation-related information to the media and the motoring public.

Traffic Signal Priority (TSP): It is a term for a set of operational improvements that use technology to reduce dwell time at traffic signals for transit vehicles by holding green lights longer or shortening red lights.

Train Integrated Management System (TIMS): It is a train information system that performs continuous monitoring of control information in the driving cab and instruments for monitoring as well as checks on train status to ease the burden on operators and conductors, to properly provide emergency information to the train crew and to streamline the checking procedure.

Transport Demand Management (TDM): It is the application of strategies and policies to reduce the travel demand or to redistribute this demand in space or in time so as to ease out the peak hour demand or demand in the congested corridors.

Unmanned Train Operation (UTO): It is a system of train operation in which train operation is fully automatic and there is no attendant in the train even for emergency duties.

Variable Message Sign (VMS): It is electro-mechanical system in which the messages, signs, movie, photo, and symbol are displayed in illuminated form.

Vehicle Miles Traveled (VMT): It is defined as a measurement of miles travelled by vehicles within a specified region for a specified time period.

Vulnerable road user (VRU): It is a term used to describe those road users who face the highest risk of being involved in a traffic accident on roads.

Wireless Application Protocol (WAP): It is a technical standard for accessing information over a mobile wireless network. A WAP browser is a Web browser for mobile devices such as mobile phones that uses the protocol.

Wrong Way Vehicle Detection System (WWVDS): It is a collection of hardware, software and communication system to capture the evidence (still photo or video based) about the vehicles coming in wrong direction (at intersection or corridor) in real-time automatically.

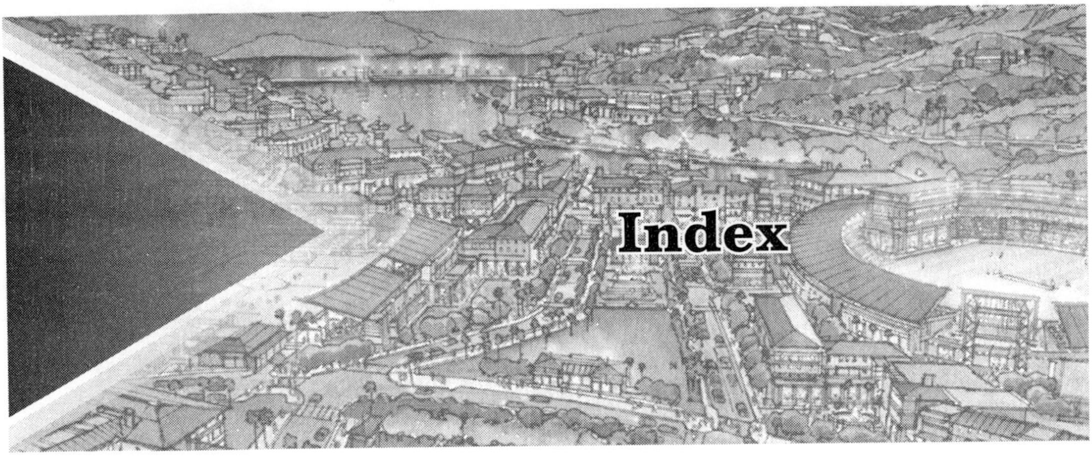

Index